Creations of Fire

Chemistry's Lively History from
Alchemy to the Atomic Age

Creations of Fire

Chemistry's Lively History from Alchemy to the Atomic Age

Cathy Cobb
and
Harold Goldwhite

PERSEUS PUBLISHING
Cambridge, Massachusetts

Many of the designations used by manufacturers and sellers to distinguish their products are claimed as trademarks. Where those designations appear in this book and Perseus Publishing was aware of a trademark claim, the designations have been printed in initial capital letters.

Copyright 1995, by Cathy Cobb and Harold Goldwhite

All rights reserved. No part of this publication may be reproduced, stored in a retrieval system, or transmitted, in any form or by any means, electronic, mechanical, photocopying, recording, or otherwise, without the prior written permission of the publisher. Printed in the United States of America.

Cataloging-in-Publication Data available from the Library of Congress

ISBN 0-7382-0594-X

Perseus Publishing is a member of the Perseus Books Group

Find us on the World Wide Web at http://www.perseuspublishing.com

Perseus Publishing books are available at special discounts for bulk purchases in the U.S. by corporations, institutions, and other organizations. For more information, please contact the Special Markets Department at the Perseus Books Group, 11 Cambridge Center, Cambridge, MA 02142, or call (800) 255-1514 or (617) 252-5298, or email j.mccrary@perseusbooks.com

First paperback printing, December 2001
1 2 3 4 5 6 7 8 9 10—05 04 03 02 01

Dedication: To Marie, as always

Acknowledgments: James Bohning and Marjorie Gapp of the Chemical Heritage Foundation; and the staff of the John F. Kennedy Library and Creative Media Services, California State University, Los Angeles, for assistance with illustrations.

H. G., 1995

Dedication: To Monty, Mathew, Benjamin, and Daniel. I couldn't have done it without you, and without you there would have been no reason to try.

Acknowledgments: The staff of the Augusta College Reese Library, Augusta, Georgia; the Augusta College Office of Public Relations; the faculty and staff of the Department of Chemistry and Physics, Augusta College; and the staff of the Library of the University of South Carolina at Aiken, Aiken, South Carolina.

Thanks: I would like to thank the people who taught me chemistry—my mentors Marshal Cronyn and Richard Martin—and the people who taught me how to write: Savage Williamson, the first editor who ever had the faith, and Barbara Jordan, who taught me the value of a good quote. I would like to thank the Plenum editors for all their hard work, excellent guidance, and patience, especially Melicca McCormick, Linda Regan, and Deirdre Marino-Alessandri.

I would like to thank my father, mother, Judi, Marshal Cronyn, and Dave Karraker (for wading through the text in various crude forms and offering invaluable suggestions and support), my friends Gary and Mara, Paula, Debbie, Alice, Barbara, Daphne, and Doris (who have always believed in my wildest schemes, though I have never known why), and the people who trusted me with their valued books, Drs. Harold Kelly, Ann Wilbrand, Kutty Pariyadath, Monty Fetterolf, and Sam Meyers.

C. C., 1995

The chemists are a strange class of mortals who seek their pleasures among soot and flame, poisons and poverty ... yet among all these evils I seem to live so sweetly ... may I die if I would change places with the Persian King.

John Joachim Becher, alchemist, ca. 1650

it). To borrow a phrase from Bernard Trevisan (alchemist, 1300s), we "looked for where most books were in agreement. . . . And in this way . . . found the truth."[1]

The approach we chose is also not that of a technical treatise. A quick scan through the pages reveals few formulas or equations and recondite matter painted with a rather broad brush. We strove to convey an intuitive understanding of concepts so that the interested reader who goes on to the details encounters them on friendly terms.

The approach we *did* choose is that of a humanized history of chemistry: one that tracks social history along with chemical history and portrays the personalities of the people creating the history as well as the events themselves.

We chose this approach because it quickly shows that chemical inspiration is not limited to any one culture or climate, but extends all over the globe. World politics dictate the rate and location of chemical discovery, and chemical discovery changes the politics of the world. The approach also shows that chemical heroes are not limited to scholars and academics, but include hedonists and hermits, dullards and daredevils, saints and charlatans, doctors and lawyers, and men and women working in garages and kitchen sinks. The creators of chemistry are capable of great inspiration but also of petty bickering, jealousy, obstinacy, chauvinism, and pride. Chemical theory is not the product of biologically different brains but the fruit of ordinary, fallible, human minds.

But did we choose this approach because it shows the great chemists are not as great as we believed them to be? No. We chose it because it shows that the potential for greatness is something within us all.

Preface

The history of chemistry is a story of human endeavor—and as erratic as human nature itself. Progress has been made in fits and starts, and it has come from all parts of the globe. Because the scope of this history is considerable (some 100,000 years), it is necessary to impose some order, and we have organized the text around three discernible—albeit gross—divisions of time: Part 1 (Chaps. 1–7) covers 100,000 BCE (Before Common Era) to the late 1700s and presents the background of the Chemical Revolution; Part 2 (Chaps. 8–14) covers the late 1700s to World War I and presents the Chemical Revolution and its consequences; Part 3 (Chaps. 15–20) covers World War I to 1950 and presents the Quantum Revolution and its consequences—and hints at revolutions to come.

There have always been two tributaries to the chemical stream: experiment and theory. But systematic experimental methods were not routinely employed until the 1600s—and quantitative theories did not evolve until the 1700s—and it can be argued that modern chemistry as a science did not begin until the Chemical Revolution in the 1700s.

We argue however that the first experiments were performed by artisans and the first theories proposed by philosophers—and that a revolution can be understood only in terms of what is being revolted against. Therefore we begin our story with the work of healers, artists, clothiers, and metal workers and show how early philosophers—explicitly or not—used the observations of artisans to develop the first chemical theories. One of these theories—the four-element (fire, water, earth, and air) theory of Aristotle—became the focus of experimental efforts for some two thousand years.

The reasons for that longevity are twofold. The theory had intuitive appeal: The four elements were supposed to be present in some admixture in all materials, and indeed, when wood burned, it was possible to see the air (smoke) rising, water (sap) dripping, earth (ash) forming, and fire (flame) leaving. The theory also held promise: If all materials were composed of an admixture of elements, then it should be possible to change the proportion in one material to create another—that is, perform a transmutation. The most attractive transmutation was from base metal to gold—and this transmutation was most diligently pursued from the first millennium BCE to almost two millennia after.

The efforts of alchemists—the testers of the theory of transmutation—constituted a divergence in the development of chemistry (the mainstream, plodding accumulation of chemical fact by the artisan would be the downfall of the theory), but it was an important divergence. Alchemists developed techniques with applications in practical chemistry, and they influenced the image of the chemical practitioner: Chemical experimentation became associated with secrecy, magic, mysticism, and fraud. Even after alchemists turned from the production of gold to more altruistic endeavors, such as the production of medicines (which we call the Chemical Reformation of the 1500s), it took a while for the image of chemistry to change. Ultimately though, this redirection—coinciding as it did with the Scientific Revolution of the 1600s—lent new respectability to chemical experimentation and inspired enough serious reconsideration to result in the Chemical Revolution.

The Chemical Revolution—led by the French chemist Lavoisier—was a revolution against imprecision in thought and experiment. The evidence of smoke vaguely supported the idea of air as an element, but the precise thinker questioned whether smoke was the pure element of air, and if so, why air sometimes took on different properties and reactivities. The lightness of ash vaguely supported the idea that something left the wood during heating, but the precise thinker asked why metals *gained* weight when heated. The Chemical Revolution rejected—firmly and finally—magic as an explanation, reliance on authority as proof, and nonverifiable speculations as chemical theory. The Chemical Revolution established—firmly and finally—the need for accurate quantitative measurements in experimental analysis, the need for clear, explicit language in analytic thought, and the need for verifiable experimental results to support chemical theory.

The result of the rejection of the four elements of Aristotle was the introduction of dozens of others. Chemists used their new tools of measurement and thought to discover new elements and new laws governing their interactions. Chemistry became a science; chemical reactions became controllable; chemical production became an industry; and the chemist became a professional.

For all the progress though, there remained one large theoretical gap: What forces held elements together? The electrochemical theories of the 1800s explained how opposite electrical charges held some species together, but they failed—as did every other theory—to explain the existence of the simplest compound, molecular hydrogen, formed from two identical atoms of hydrogen. Without an explanation, it was impossible to make further predictions except one: Another revolution was needed. This time the revolution was much quicker in coming. It arrived with the new century, the 1900s: The Quantum Revolution.

Quantum theory, the basis of the Quantum Revolution, was the result of a marriage of efforts: physicists worked with chemists to gain an understanding of radioactivity, reactivity, and the structure of the atom. Rutherford, the originator of the planetary model of the atom, was a physicist by training, but he won the Nobel Prize in chemistry. G.N. Lewis, a chemist, had the insight (born of an extensive knowledge of chemical reactivity) to suggest the two-electron bond—the foundation on which theoreticians built quantum theories of the chemical bond. The explosion of experimental information and theoretical understanding begun before the Quantum Revolution—and now quickened by it—forced a separation of subdisciplines. Biochemistry, organic chemistry, inorganic chemistry, analytical chemistry, and physical chemistry became separate fields, though the interplay among them was essential to the advancement of each.

And advancement is what chemists have achieved. We can now predict the properties of materials before they are created—and postulate what materials were present at the inception of life itself. What is over the horizon for chemistry? It is hard to say. Surely Aristotle would have been surprised by the results of Lavoisier—and Lavoisier would have been amazed by the conclusions of Lewis. So no doubt will we be astounded—and delighted—with the chemistry to come.

Contents

Part I

1. ca. 100,000–300 BCE: Prehistoric Chemist to Chemical Philosopher—The Seeds 3
2. ca. 300 BCE–600 CE: Alexandria and Alchemy 27
3. ca. 200 BCE–1000 CE: From Rome to Baghdad 51
4. ca. 1000–1200: Alchemy Translates from East to West 69
5. ca. 1300–1500: The Evolution of European Alchemy 89
6. ca. 1600: Philosophers of Fire 107
7. ca. 1700: The Search for System and Phlogiston 127

Part II

8. ca. 1700: Révolution! 151
9. ca. 1800–1848: Après Le Déluge 171

10.	ca. 1800–1848: The Professional Chemist	185
11.	ca. 1848–1914: Thermodynamics—The Heat of the Matter	213
12.	ca. 1830–1914: Organic Chemistry—Up from the Ooze	235
13.	ca. 1848–1914: Inorganic Elements and Ions—New Earths and Airs	257
14.	ca. 1848–1914: Analytical, Industrial, and Biochemistry—Creations of Coal	283

Part III

15.	ca. 1914–1950: Quantum Chemistry—The Belly of the Beast	309
16.	ca. 1914–1950: Polymers and Proteins: Links in the Chain	337
17.	ca. 1914–1950: New Materials and Methods—Organic and Inorganic Chemistry Grow	359
18.	ca. 1914–1950: Chemical Kinetics—Boom or Bust	377
19.	ca. 1914–1950: Radiochemistry—Dalton Dissected	391
20.	The Best Is Yet To Come	415

Endnotes 431
Annotated Bibliography 449
Index 453

part ONE

part
ONE

chapter ONE

ca. 100,000–300 BCE: Prehistoric Chemist to Chemical Philosopher— the Seeds

There are those who envision prehistoric peoples as grunting, stooping, quasi-humans with hair in unusual places, so there are no doubt those who would take exception to the title with which this chapter begins. Prehistoric implies before written record. If we define science as the systematic recording and interpretation of observed phenomena, then prehistory is prescience—"prehistoric chemist" is an erudite oxymoron—and we are forced to jump to the first millennium BCE for our story to begin.

But too much would be lost with such a view: It did not take pencil and papyrus for the chemical art to begin. Prehistoric peoples, tossing bits of rock into fires and contemplating the color changes that resulted, were in essence practicing chemistry. They—and their early civilized progeny—accumulated the respectable repertoire of chemical materials, techniques, and observations on which further developments were built. They used paint. They controlled fire. They fashioned clay pots. They gathered and worked metals. They practiced the healing arts. We elaborate here on some of these accomplishments because of their "chemical" nature; that is, they involve observing the properties of materials and what happens when these materials interact.

CHEMICAL TECHNOLOGY IN AN AGE OF STONE

Current evidence suggests (with some debate) that modern humans originated in Africa as the Cro-Magnon about 100,000 BCE and eventually merged with or displaced the Neanderthal, Dali, and Ngandong.[1] It has also been suggested that in anatomy, behavior, and language, the Cro-Magnon people were fully modern 35,000 years ago and a Cro-Magnon could be taught to fly an airplane and could walk by us unrecognized on a street.[2]

The Old Stone Age, or Paleolithic Age, extends to about 8000 BCE. Paleolithic humans were food-gathering nomads who learned to use language, control fire, and fashion tools of stone and bone. There is evidence that during the upper Paleolithic period people cooked food, which could easily be considered the first chemical process. Cooking uses heat to break down the chemical networks in vegetable and muscle fiber and accomplishes a sort of predigestion that takes some of the burden off teeth, stomach, and intestine.

It is also known—from the existence of skulls with healed wounds—that Paleolithic people cared for their sick and practiced a primitive form of medicine. The practice of burying various implements and food with corpses has been interpreted as having a religious significance, but it could as easily have been an early attempt at sanitation. Burying possessions of the dead (especially the last thing eaten) would be effective for disease control. Proving that Paleolithic people had collections of healing herbs, or pharmacopoeia, is difficult because herbs—unlike stone artifacts—decay and would not have survived in a form that revealed their use. However in our era isolated cultures have been found that still practice what could be classified as Stone Age technology, and these cultures generally know of and use herbs for antiseptics and analgesics. This preparation of materials for medicine is a form of chemistry that we find in every age of our story.

Cave paintings, dating from about 20,000 years ago—perhaps naturally preserved by salt deposits from seeping rainwater—show that the Paleolithic Cro-Magnon also practiced art. The art may have been a form of sympathetic magic—imitating a desired result to bring about that result[3]—because game animals and successful hunts were subjects of art, as well as prosperously plump, large-breasted women. Materials used to create paintings were charcoal and sharpened lumps of colored clay. Clay can acquire a red color from mercury sulfide (cinnabar), red and yellow from different iron oxides, and brown from manganese oxide. Dishes found near the paintings show that pigments were mixed with fat for ease of application.

About 8000 BCE, when the last ice age ended, the Middle Stone Age, or Mesolithic Age, began. People of this period are said to have tamed the dog and hollowed out logs to make crude boats. They also made the first pottery by sun baking clay, a chemical process that transforms loose, liquidy hydrated silicates into a strongly bonded network. Pottery appeared in Japan as early as 10,000 BCE and in the Americas around 5000 BCE.

The Mesolithic people also used paint, and as shown by two human skulls colored with cinnabar found in graves dating from the Mesolithic period, they apparently used paint to adorn the human body. Cinnabar has a dark red color, and it has been suggested that the red skulls were a blood symbol or had some other deep, significant, mystical meaning. But cinnabar is also an effective treatment for nits and lice—which inhabit hair—and coating the skulls with cinnabar may have been the loving gesture of some friend or relative who wished to give a departed one a bit of comfort in death.[4]

The period from about 6000 to 3000 BCE is considered the New Stone Age, or Neolithic Age. During this time, people learned to raise food and produce fire by friction—perhaps the first controlled chemical reaction. They domesticated animals and invented the plow, the wheel, and the sail. They learned how to spin and weave and make kiln-fired pots. Then somewhere between 6000 and 7000 BCE, a new material, copper, was hammered into shape.[5] This material allowed people to fashion new tools that, with the development of agricultural techniques, permitted the growth of farming communities in permanent locations. Many food-gathering nomads became workers of the land, and somewhere around 4000 BCE, civilization began.

CHEMISTRY AND CIVILIZATION TECHNOLOGY TAKES ROOT

With the advent of civilization, chemical technology matured. Techniques requiring permanent structures (such as furnaces for smelting metals) could now be developed, and chemical processes could be recorded, repeated, and refined. We can piece together a picture of the methods used from artifacts that remain. The most durable of these are the metals, and metallurgy is the first chemical technology whose history we are able to reconstruct with some surety.

METALLURGY

Metals were used by virtually all civilizations even when other seemingly elementary technologies, such as the wheel, were not. Most metals however react easily with oxygen, sulfur, and halogens—such as chlorine, fluorine, and iodine—and exist in nature as ores; that is, complex matrices of salts and silicates, such as feldspars, pyrite, and bauxite. Silver and gold on the other hand are fairly unreactive, and these may exist in nature as the pure metal (such as the nuggets of gold found in riverbeds during the U.S. gold rush of 1849). Copper, too, though most often found as an ore, can be found as deposits of native metal. These three metals then can be worked by techniques as simple as hammering and mined by techniques as simple as gathering.

Copper, Silver, and Gold

Because of its abundance relative to the others, copper was probably the first of these metals to be collected. Worked copper beads dating from about 9000 BCE, possibly used for decoration, have been found in northern Iraq.[6] Native North American burial mounds from about 2000 BCE have been found to contain copper spear heads, chisels, and bracelets. Pre-Columbian inhabitants of Ecuador worked native copper by hot hammering, and they made small copper axes, bells, and sewing needles.[7] Copper also appears in the earliest remains of Egyptian, Mesopotamian, Indian, and Chinese settlements.

Gold nuggets are relatively soft (hence the legendary bite test of gold coins), and these can be hammered together into sheets. This property of gold, as well as its color, makes it desirable as a decorative material—but fairly useless for anything more. The practical use of gold for making corrosion-resistant electrical contacts and capping teeth is a purely modern development. This makes all the more ludicrous the amount of human suffering that occurred for the sake of gold. In Egypt gold was mined and processed by slaves, and because there was an ample supply of condemned prisoners, prisoners of war, and political prisoners (sometimes whole families with children), it was cheaper to work them to death and replace them than prolong their lives by humane treatment.

Silver can be found alloyed with gold in a material called electrum. A number of methods can be used to separate silver from this alloy and recover the gold, but presumably the earliest way was by heating electrum in a crucible with common salt, sodium chloride. With time, heat, and repetitive treatments, the silver was converted into silver chloride, which passed into the slag: the layer of impurities that floats on molten metal during processing. However silver was not always the unwanted product. In Egypt between the thirteenth and fifteenth centuries BCE, pure silver was rarer and more costly than gold.[8]

By 3000 BCE the Sumerians, perhaps while heating copper to make it more malleable, had discovered that more copper could be retrieved from the fire if the metal were heated with certain types of dirt and stones—that is, certain earths. These earths were the metal ores, and the process they discovered, *smelting,* reduced metal salts to pure metal by the action of carbon in the charcoal fire. The process of changing metal salts into pure metal is known as *reduction* because the metal without the accompanying oxygen, halogen, or sulfur of the salt weighs less than the ore. Eventually metal workers learned to distinguish various metal-bearing ores by color, texture, weight, flame color, or smell when heated (such as the garlic odor of arsenic ores), and they could produce a desired material on demand.

Bronze

The Sumerians also mixed copper and tin to create a new material: bronze. They found the new material relatively easy to cast and much harder than copper alone. Bronze could be used to make more durable hoes and spades, and knives that retained a cutting edge for a longer time. Bronze was such an important discovery that an entire era of history, the Bronze Age, was once identified by its use. However this term has lost its chronological meaning because different cultures discovered the use of bronze at widely different times. Some cultures, such as those in Finland, northern Russia, Polynesia, central Africa, southern India, North America, Australia, and Japan, had no Bronze Age but went directly from stone to iron. Egyptian bronze objects however date from as early as 2500 BCE and possibly as early as 3000 BCE.

For the Egyptians to make bronze, tin ores probably had to be imported—perhaps from Persia—and artisans working in Mesopotamia during this era had to import their tin too. An arsenic-copper bronze could be made when tin was scarce, but the practice seems to have died out. (Fumes from the process would have caused arsenic poisoning, so it may have been the artisans, rather than the art, that disappeared.) With bronze, then, a picture of a growing network of Mediterranean trade begins to emerge. In fact there is some evidence that the Assyrians, an exceptionally ruthless, unisex warrior society that conquered the Mesopotamian valley ca. 1200 BCE, engaged in some creative—if unethical—entrepreneurship. They apparently duped the ancient Turks into believing that the nearest tin supply was in the Hindu Kush—and charged the Turks accordingly, though the metal actually came from a mine on the Turkish coast.[9]

Next it was noticed that copper smelting went better when certain iron-oxide-containing earths were added. The iron-oxide minerals may have accidentally been included with the copper ores because they are found in the weathered upper zones of copper sulfide deposits. Iron oxide improved copper smelting by acting as a flux: a material that aids the removal of impurities by combining with them to form a floating crust or slag. As a side product, bits of the new material—iron—were formed.

Iron

Iron was known in Egypt perhaps as early as 3000 BCE, but it was called the metal of heaven, implying that the first samples were meteoric in origin. As such iron was rare and considered a novelty rather than a commodity. However Mesopotamian and northern Syrian specimens of smelted iron (from iron ore, not meteors) may have been produced as early as 3000 BCE, and an iron foundry was operating in southern Africa as early as 2000 BCE.[10]

The first smelted iron was a spongy, pasty mass in a semi-liquid slag because the actual melting temperature of iron is 1500 degrees Celsius, whereas the temperature attainable by early charcoal furnaces was only about 1200 degrees Celsius. This product, later known as bloom, required repeated heating and hot hammering to eliminate the slag. Tempering was achieved by repeated cold forging and reheating.

Workers eventually increased the ratio of fuel to ore and employed better bellows to raise the temperature of the ovens. But even so, pure iron is softer than bronze and an inferior material for weapons and other applications requiring durability. It remained to be discovered that heating iron in the presence of carbon, which incorporates a very small of amount carbon into the iron matrix, would significantly increase the strength of the material.

This heating of iron in the presence of carbon, usually provided by a charcoal fire, is called carburizing. Hindu physicians were among the first to discover carburized iron, or steel, and use it for surgical instruments. In China carburizing was done directly by using carbon-rich iron ore. This technique was also discovered by an obscure

Indo-European tribe called the Hittites, who employed it immediately to make better weapons. With these better weapons, they were able to march across Asia Minor and in 1200 BCE arrive menacingly at Egypt's door.

CHEMICAL COMPOUNDS

In addition to pure metals and metal alloys, ancient chemical workers also had a significant store of chemical compounds that they could employ in various processes. As with the pharmacopoeia however, these materials do not come down to us intact, so we must surmise their use from archeological artifacts such as pestles, mortars, mills, strainers, stills, and crucibles. One Mesopotamian find was a double-rimmed earthenware pot, probably used for extracting plant oils or essences. Raw material could be placed between two rims, a lid placed over the vessel, and a solvent (water or oil) boiled in the bottom. Vapor from the boiling solvent would have condensed on the lid, run down over the raw material, extracted the desired ingredient, and dripped back into the bottom of the pot—basically the same principle used in coffee percolators today.[11]

We also have an idea of some other commonly used compounds because one fine third-millennium Mesopotamian day, an anonymous Sumerian physician took a four-by-six-inch clay tablet and made a list of his favorite prescriptions.

The Legacy of the Sumerian Physician

Some ingredients in the prescriptions were salts whose names appear many times in our story: sodium chloride (known now as ordinary table salt and found as natural deposits or obtained by evaporating sea water), sodium carbonate (known later to Europeans as soda ash and found as natural deposits or produced when plants rich in sodium are burned), and ammonium chloride (known later to Europeans as sal ammoniac and obtained by burning large masses of coal or from heating well-weathered camel dung). Minerals, such as alum, or potassium aluminum sulfate, and gypsum, or calcium sulfate, were gathered and ground for medicines, though gypsum was also used as a type of mortar.[12] Sodium nitrate, or potassium nitrate, both of which later Europeans called saltpeter or niter, also appears in prescriptions. (Many medicinal properties have been ascribed to niter through the ages and it has been prescribed for conditions as diverse as sexual impotence, asthma, and excess sexual desire, though most of these treatments have proved ineffective. The value of niter really skyrocketed when it was found to be an essential ingredient of gunpowder.) The colorless crystalline compounds that make up niter are produced by the action of bacteria on nitrogenous waste, such as urine and manure, and these compounds are familiar to anyone who has had contact with stables. Judging from later practices, the Sumerians probably collected niter from the sides of sewers and animal yards, and it no doubt contained a variety of contaminants. Of course none of the compounds were the clean, well-formed crystals seen on the pharmacist's shelf today. They were more often dirty mixtures: Even one of the commonest chemicals—water—was most often a murky, possibly brownish, material.

Used either directly or in powdered or pulverized form, other ingredients in the prescriptions came from animals (milk, snake skin, turtle shell), plants (myrtle, thyme), or trees (willow, pear, fir, fig, and date). Sometimes prescriptions called for the desired constituent to be extracted with boiling water, but if it were not water soluble, the constituent could be extracted with beer or wine. The wine or beer, of course, contained ethanol—a chemical as important then as it is now. Ethanol is produced when certain single-celled fungi called yeast consume sugars in a process called fermentation. Yeasts that cause fermentation are found naturally on grapes, so fermentation can occur spontaneously in high-sugar grape juice. Conditions that encourage fermentation and ethanol production may have been known and exploited for as long as 10,000 years—making fermentation one of the oldest, if not most celebrated, chemical processes. Ethanol is an excellent extractant for organic, or carbon-containing, materials, such as plant oils or essences. The organic end of the ethanol molecule, consisting of carbon and hydrogen, mixes well with plant materials, which are also organic. The alcohol end of the ethanol molecule, consisting of oxygen and hydrogen, mixes well with water, which also contains oxygen and hydrogen, in the familiar ratio H_2O. Organic materials mixed with the ethanol could be made into aqueous solutions, and the extracts so obtained could have been stored, as today, in solid, liquid, or powdered form.

The physician also used vegetable oils and animal fats as extractants and salves. Unfortunately knowledge of whatever curative value the herbal remedies may have had is lost because our physician—perhaps constrained by tablet size or the cumbersome nature of the cuneiform script—failed to record what diseases they were used to treat. Some of the therapeutic effects, however, may be surmised. For example, in the twelfth prescription, the physician advises:

> Sift and knead together—all in one—turtle shell, the sprouting [sodium containing] naga-plant, salt, and mustard. Wash [the sick spot] with quality beer [and] hot water; scrub [the sick spot] with all of it. After scrubbing, rub with vegetable oil and cover with pulverized fir.[13]

The washing prescribed in this treatment cleaned the afflicted area, and the salt and alcohol no doubt acted as antiseptics. Two of the prescriptions on the tablet called for an alkali (sodium or potassium) salt to be used together with natural fat—which produced a soap—and this would again help clean a wound or diseased area if externally applied.

Interestingly our Sumerian physician did not record incantations or magic rituals to be used with the medicines, indicating that rituals were either not used or not considered important enough to take up limited space on the tablet. However in later Babylonian society, which produced one of the first extensive written pharmacopoeias,[14] sorcerers became more important than physicians in treating disease. In fact disease was taken to be demonic possession caused by sin, and drugs were used to exorcise rather than cleanse the patient. Made deliberately disgusting to drive out the demon (the poor patient had no choice but to stay), remedies consisted of raw meat, snake flesh, wood shavings, oil, rotten food, crushed bones, fat, dirt, and animal or human excrement. We mention this here because the demonic theory surfaced again—in Europe of the 1600s.

The need for such drastic medical measures can be appreciated by reflecting on the quality of life in these ancient cultures. People died routinely from tooth decay, minor wounds, infections, and fever that today could be treated in an hour with over-the-counter drugs. People suffered from abscesses and blindness. Pain and parasites were an accepted fact of life. Women had about the same chance of surviving childbirth as they had surviving marching in war—which also happened quite often—and a child's chances of surviving infancy were not much better.

Understandably ways of escaping life often became as important as ways of preserving it, and one common escape was through art. Though this expression of art often took the form of painted objects and walls, another familiar vehicle was used: personal adornment. Evidence of the importance of this particular form of art is found in the biblical story of Joseph. In the story, the boy was thrown into a pit and left to die by his brothers. His crime? They envied his coat of many colors.

Pigment and Dye

Cloth was made from wool, cotton, and linen, but the most common was wool. Dyes were generally *organic,* the name given to the class of compounds formed from carbon and hydrogen, or *hydrocarbons,* which at the time were extracted from animals and plants; *inorganic* is the name given to all compounds other than hydrocarbons, which were mainly derived from minerals. Several inorganic salts were used as *mordants:* substances that fix the dye by chemically bonding with both the dye and the protein structure of the cloth. A favorite blue dye was indigo from the indigo plant (and we will see this dye make chemical and social history again in the early 1900s CE [Common Era]). Fermented in vats, the dye was colorless, but a cloth dipped in it and exposed to air would turn deep blue in seconds. Other plants processed for dyes included saffron, which produced a yellow dye, and madder, which produced a red dye that has been found on the cloth of Egyptian mummies. In Mesopotamia red dye came from *kermes,* pulverized dried scale insects found on Mediterranean oaks. The most costly dye was the purple dye obtained from the gland of a mussel found only in very specific locations on the Mediterranean coast. Each mussel harvested provided only a tiny amount of a creamy fluid with a garlic smell (some reports say as many as 12,000 mussels were required for an eyedropper-sized sample of dye[15]), but once processed it made an exceptionally striking purple dye.

In addition to decorative clothing, cosmetics were also used to enhance personal appearance, as were perfumes. One Babylonian text specifically mentions that women worked as perfume preparers as well as men, so women were involved in at least this area of early chemical technology, and they were most likely involved in others. As time went on, these chemical technologies became more advanced, so that the artisans were able to produce many more interesting and useful materials. One of the most interesting is a material that like gold initially had no purpose other than the beauty it provided: glass.

Glass

Silicon forms roughly twenty-five percent of the earth's crust, and it is second only to oxygen in abundance. Most of the ground we stand on is silicon-oxygen *silicates*, or what is commonly known as sand. Glass, formed from fused silicates, has been created naturally in lightning strikes, volcanoes, or where meteorites have scorched the Earth. Egyptians used glass as a glaze since Neolithic times, but purposeful production of glass as an independent material probably did not occur until 3000 to 2000 BCE in Mesopotamia.

Initially artisans made glass from sand or quartz (silicon dioxide) and crude sodium carbonate, which in Egypt is found in dry lake beds near Alexandria. Without sodium carbonate, 1700-degree-Celcius fires are required to melt the sand (adding sodium carbonate lowers the melting point by producing a sodium oxide flux). Glasses produced with sodium however are somewhat soluble in water, and this material was most often used for art, though an occasional vessel might be created (for a privileged few) to hold ointments or other materials that might be absorbed by ceramic pots.[16] Around 1300 BCE artisans found that incorporating calcium oxide, another chemical compound known to the ancients, reduced the solubility of the glass. Called *lime* or *quicklime* by the later Europeans, calcium oxide is formed by heating calcium carbonate found in shells or natural deposits of limestone or chalk. Once this technique was found, large-scale glass factories began operating in Egypt. The glass of these early civilizations was molded, not blown. It was generally cloudy and blue (colored by copper compounds and sometimes cobalt), although some other colors were achieved. The Assyrians used tin oxide and lead antimonate to color their decorative glass white and yellow, respectively. It has been speculated that the Assyrians knew how to make aqua regia (a mixture of nitric and hydrochloric acids) because they used gold salts to give a red color to their glass and aqua regia should have been necessary to dissolve the gold.

OTHER MATERIALS AND METHODS

While the preceding technologies make an impressive list, they are by no means an exhaustive one. For instance we have not gone into the Hindu practice of leather tanning nor the Meso-American metal-casting arts. One other technology we should probably quickly include, though admittedly more for the mystery that surrounds it than for any particularly high degree of technical expertise, is the famous art of Egyptian mummification. To create a mummy, Egyptians gutted a corpse, filled the cavity with wine and perfume, drew the brains out piece by piece through the nostrils with an iron hook (perhaps the most mysterious part of the process), then steeped the body in a bath of natron (sodium-aluminum-silicon-oxygen salt) for 70 days. This process killed bacteria that cause decomposition and dehydrated cells so that future bacteria would not find a pleasant home. They then wrapped the body in waxed cloth bandages, smeared it with gum, and sealed it in a sepulcher away from corrupting moisture and air. All in all the process was not much more mysterious than salting pork.

With the development of mummification, glass, and dyes, we see societies becoming successful enough to be able to dedicate time and materials to tasks not directly associated with the day-to-day business of staying alive. This trend culminated in a profession that required no manual labor and produced no product but thought: philosophy.

Civilization and chemical technology gave the philosophers time to think—and something to think about.

CA. 2000—300 BCE: THE PHILOSOPHERS

While the collection of chemical facts continued to be enlarged by the artisan, these facts were interpreted by the philosophers—who also served as mathematicians, astronomers, anatomists, and physicists, as well as theologians and political theoreticians. In fact not until the 1800s did European scientists begin to think of their work as separate from that of philosophers. Though philosophy is common to all cultures, the most influential in the development of modern chemistry were the philosophers of Greece. These thinkers derived hypotheses about the nature of matter and material interactions that helped and hindered chemical developments over the next 2000 years.

EARLY GREEK CIVILIZATION

Around 2000 BCE the Minoans appeared as the first Aegean civilization. Centered on the island of Crete just off the main peninsula of Greece, the Minoans were an affluent and comfortable society. The Cretian soil grew grape vines and olive trees and obtained metal by trading timber, grapes, olive oil, wood, and opium with the Greek mainland, Cyprus, Egypt, and the Levant coast of the Mediterranean.[17] Cretians knew about concrete and ceramics, and they used these materials to create indoor sanitary systems that were unavailable in northern Europe until around 1700 CE. The principal deity was female, and Minoan women enjoyed complete equality with men, including the professions of bull fighter and boxer.

This peaceful society was invaded and overrun by Mycenaeans from mainland Greece around 1500 BCE. As a militaristic society, the Mycenaeans built massive fortified walls around Greek cities and concentrated their efforts on manufacturing daggers, swords, helmets, and shields. To provide materials for their armaments, they extended their sphere of trade to the coast of Asia Minor, Rhodes, Syria, and the central Mediterranean. After about 1200 BCE a wave of invasions by Indo-European Hellenic tribes from the Balkans led to the fall of the Asia Minor base, Troy (the legendary Trojan War), and Greece entered a dark age that lasted about 400 years. This period ended around 800 BCE when the Indo-European groups, now settled in their new home, reestablished trade with Asia and adopted a version of the Phoenician alphabet. The increased efficiency, accessibility, and versatility of writing that the phonetic (that is, Phoenician) alphabet brought about—along with inexpensive writing materials, such as papyrus from Egypt—has been compared to the invention of movable type or in recent times to the development of the computer.[18] One of the products of this rich new medium of thought was the development of a class of teachers who taught for hire the children of the wealthy. These teachers—the philosophers—sought to derive (as all teachers do today) reasonable explanations for the observed world.

GREEK PHILOSOPHERS

Explanations for the observed world were of course not new. They were and still are the essence of religion. These new thinkers however did not necessarily invoke a deity in their explanations; that is, they attempted to keep their reasonings rational (based on reason) as opposed to mystical (based solely on intuition or faith). However just as speculation on the nature of the world was nothing new, neither was rational thought. The Babylonians, who had developed sophisticated mathematics by 1600 BCE were at least in that area engaged in rational thought. In Africa, Egyptian mathematical knowledge, as demonstrated by land division, surveying, and the construction of the pyramids, surely implied the capacity for rational thought. The ancient Indian philosopher Kapila believed that all philosophic questions could be reasoned out without resorting to mystical explanation, and the Indian Nyaya Sutra, which may have been written anywhere between the third century BCE and the first century CE, lists principles of argument and common fallacies of thought. Special attention here though needs to be paid to the Greeks. For reasons political rather than philosophical (to be explored shortly), it was the fate of the Greeks to expound the philosophy of nature, the natural philosophy, that would have the greatest impact on future chemical events.

Thales, Anaximander, Anaximenes, and Anaxagoras

In modern chemistry we credit some one hundred elements—and counting—with being the basic stuff from which all matter is formed, but there was no reason for this to be the first assumption, and it was not. The initial thrust of the philosophical effort was to discover the one basic stuff of nature—the primal substance, the material from which all else is formed. This effort began for the Greeks around 600 BCE, on the Aegean shores of Asia Minor, in the Ionian city-state of Miletos, with the first of the Ionian philosophers, Thales.

Little is known of Thales' life other than that he was born into a distinguished family. We do not know why he chose a life of teaching and philosophy, just that he did. Thales believed that the basic stuff of nature should be one material, and he taught, possibly influenced by Babylonian religious beliefs, that this basic substance was water. Thales did not invoke religious justification for his theory, but instead reasoned from observation that water could turn into air (evaporation) or become solid (freeze), and thus form all things. This set the tone for future Greek schools of thought: explanations for natural phenomena based on observational reasoning rather than transcendental inspiration.

Following Thales, his pupil Anaximander (in the mid-500s BCE), proposed that the basic material of all things was something he called apeiron. Anaximander did not have an exact description for the material, but he claimed that worlds formed and disappeared as bubbles in this apeiron. He theorized that our solar system came into being when a mass of apeiron broke away from the infinite in a rotary motion that caused the heavy materials to concentrate at the center and the edges to condense into the heavenly bodies. This compares with modern cosmological theory,[19] and it gives the impression of a curiously accurate intuition about the physical world. On the other hand Anaximander visualized Earth as a flat cylinder, thought that animals originated from inanimate matter and humans from fish—which reminds us of the speculative nature of these theories.

Anaximenes, who lived around 550 BCE, may have been a pupil of Anaximander, but the former proposed that air was the primal substance. According to Anaximenes, air rarefied was fire, and condensed air formed everything else (water, earth, and stones). In support of his theory, he pointed out that air expelled through puckered lips becomes cold, whereas air expelled through an open mouth is hot: compressed air condenses and expanded air is transformed into fire.

The last notable Ionian philosopher, Anaxagoras, lived around 400 BCE. Although born of wealthy parents, he apparently abandoned his inheritance to pursue natural philosophy. Obviously little impeded by custom, Anaxagoras also tossed aside the tradition of seeking one primal substance and argued, "How could hair come to be from what is not hair and flesh from what is not flesh?"[20] He proposed the existence of seeds: extremely small portions of everything that exists in the visible world. These seeds were never created or destroyed, but they constituted all material and formed new materials when mixed or separated. However in contrast to later atomic theories, Anaxagoras believed that all matter was infinitely divisible and the resulting smaller portions contained portions of every other substance.

It is important to pause here and note what is evolving: The identity of the primal substance or substances was debated, but all the debaters seemed comfortable with the assumption that whatever it was, this primal element (or elements) would be found in some portion in all matter. This was the assumption under which chemists would labor for the next 2000 years. One reason for its persistence was the intuitive appeal of arguments offered by philosophers, backed by everyday observation, but another reason was the social upheavals that scattered Greek philosophers throughout the Mediterranean like Anaxagoras' seeds.

This upheaval began when Persians invaded Miletos and Ionian philosophers migrated to other parts of Greece. Anaxagoras traveled to Athens where he is said to have proposed an accurate explanation for eclipses. He was eventually prosecuted on a charge of impiety when he asserted that the sun was an incandescent stone somewhat larger than the region of the Peloponnese. Despite this initial seemingly unimaginative reaction by the citizens of Athens, Athens was destined to become an important center of philosophy.

Pythagoras, Zeno, and Empedocles

Prior to Anaxagoras' arrival in Athens, other philosophic traditions had found a foothold in this fertile ground, including the school of Pythagoras. Pythagoras himself, born on an Aegean island, had been instructed in the teachings of the Ionia philosophers, but through travel he had also become acquainted with Babylonian and Egyptian mathematics. The blend of these influences prompted Pythagoras to found a communal secret society devoted to mathematical speculation and religious contemplation. The followers of Pythagoras (who were said to have included men and women on equal footing) proposed mathematical theories of matter, including five elements envisioned as geometric solids: Earths were cubes, water was an icosahedron, air was an octahedron, fire was a tetrahedron, and ether (a nebulous element that filled spaces unoccupied by the other four) was a dodecahedron.

A contrasting school of thought was that of Zeno of Elea. Zeno insisted that matter is continuous, and he rejected the idea that it could be broken down into individual particles of various elements. He argued that if two material particles are separated, then whatever separated them must also be a thing; therefore material is continuous. The continuity of matter was debated until the 1900s but without much more input from Zeno. He joined a conspiracy to overthrow a tyrant in his home city-state, Elea, and after refusing under torture to reveal the names of his co-conspirators, his ideas died with him.

A third important school of thought that influenced the philosophies of Athens was that of Empedocles. It is difficult to decide with what trumpet blast to announce this name, but let it suffice to say that Empedocles was the originator of the four-element theory of matter—another idea that dominated Arabic and European chemical thought until the close of the 1700s.

Though an aristocrat by birth, Empedocles was a disciple of the equalitarian school of Pythagoras and refused the crown offered to him by his home city-state when he assisted in overthrowing the ruling oligarchy. Instead Empedocles instituted democracy. A prodigious writer, he produced a treatise on medicine as well as the natural philosophy in which he proposed that four elements (which he called *roots*)—earth, water, air, and fire—formed the basis for all things, and two forces—love and hate—governed their intermingling. According to Empedocles all four elements are found in all matter, but different materials possessed different proportions of each.

Empedocles is also known for his proof by physical means that air is a material body. Using a water clock (a cone with holes at the bottom and top, which when placed in water sinks slowly to the bottom, serving as a rough measure of time), Empedocles showed that when he placed his finger over the opening at the top of the clock, water would not fill the cone. When he removed his finger, air rushed out. Given such a fine demonstration of logic and what we know of his democratic beliefs, it is all the more curious that in the end, Empedocles decided to declare himself a god and threw himself into a volcano to convince his followers of his divinity. Legend has it that the volcano threw back his sandals, and thus his divinity denied.

Democritus and Leucippus

The last of the stage-setting philosophies that we examine is the important Greek atomistic theory: A theory whose precepts and name reemerge continuously in our story. Though the beginnings of the Greek atomistic theory are often attributed to Democritus alone, it is likely that these notions originated with Leucippus, his teacher. Democritus however was extensively traveled and he was known as the laughing philosopher for his derisive assessment of humankind. The writings of the teacher may have erroneously been credited to the more famous pupil.

The essence of the theory was that matter is made up of an infinite number of solid atoms of elements in constant motion. These atoms were thought to form materials with characteristics determined by the atoms' shapes (for instance membranes were woven from hook-shaped atoms)—an idea that still had appeal in the chemical philosophies of Western Europe in the 1600s. The atomic nature of matter is an accepted theory today, though the atoms of modern theory are vastly different from those conceived by Democritus, as we shall see.

Socrates

Such was the rich medium that nurtured Socrates. Short and stout, with bulging eyes and a snub nose, Socrates was born of working-class parents (his father may have been a sculptor and his mother a midwife), and he served as a foot soldier in the Peloponnesian War—an experience that may have influenced his philosophy. Concerned with developing rules for a peaceful, ordered, ethical society, he used a deductive method of reasoning that starts with a supposedly irrefutable truth apparent to all rational beings, then draws conclusions from this truth using clearly defined rules of logic. This may have worked well for devising social systems—the irrefutable truth would be something that all people in that particular society agreed on—but it does not always work well in chemistry. In chemistry, nature dictates the rules, and they may not be immediately apparent to all rational beings. Socrates rejected proof by analogy (for example, the first plant was not watered, and it died; therefore the second plant will die if not watered) and inductive reasoning (for example, the first, second, and third lemming drowned on falling into the sea, therefore the fourth lemming will drown too) in favor of his deductive reasoning. But these forms of reasoning are very useful in chemistry (and would serve plants and lemmings in good stead, too). Future chemists who tried to adhere to strict Socratic logic were seriously hampered in this regard.

Socrates also rejected experimentation as a method for arriving at truths, believing that the fundamental nature of the world could be discerned by mental reflection alone. So although his followers experimented with the material world in the same sense that we all do—by walking, eating, breathing (and observing what we walk on, eat, and breathe)—they did not engage in deliberate experimental manipulation of materials. Though this did not hold for long, and chemists were soon up to their elbows in experiments, it did have a residual damaging effect. Often chemists felt more comfortable believing in what they reasoned they should be seeing rather than in what they actually saw.

Socrates was eventually tried on the charge of corrupting the youth and introducing new gods and sentenced to death. (We may note here that none of the characters in our chemical history ever endured legal persecution for their chemical views, while political views caused several to suffer such a fate—a word to the wise for anyone choosing an occupation.) Apparently friends offered to help Socrates escape, but he believed that the verdict from a legitimate court should be obeyed. He drank the hemlock poison and left his star pupil, Plato, to carry on the debate.

Plato

Plato came from a distinguished Athenian family; his father even claimed to be descended from the god Poseidon. Plato's early ambitions were probably political, but perhaps after the death of his teacher, he became convinced that Athenian politics were no place for a person of conscience. He turned to a life of philosophy and teaching instead. However in contrast to his teacher, Plato believed that natural philosophy was a worthy area for study as long as one removed the taint of atheism (and threat of retribution) by making natural laws subordinate to the authority of divine principles.

A careful observer and cataloger of thought, Plato believed everything could be understood by intellectual effort. As such his natural philosophy, based on the four elements of Empedocles, had a very clear, logical presentation. For example in the following excerpt from Plato's dialogue, Timaeus, metals are referred to as waters because they melt, and gold is described as the perfect metal, other metals being the same as gold but in impure form.

> Congealments[:] . . . that which is the densest . . . is formed out of the finest and . . . is that most precious possession called gold. . . . This is unique in kind and has both the glittering and yellow color. A shoot of gold which is so dense as to be very hard, and takes a black color, is termed adamant. There is also another kind, which has parts nearly like gold, . . . and this substance, which is one of the bright and denser kinds of waters, when solidified is called copper. There is an alloy of earth mingled with it which, when the two parts grow old and are disunited, shows itself separately, and is called rust.[21]

Plato accepted the notion that all materials were made of different proportions of the elements, but he refined this idea by adding that under the right circumstances, materials could, be changed into one another or undergo transmutation. To promulgate his ideas (and support himself), Plato founded the Athens Academy, an institute for the systematic pursuit of philosophical and scientific teaching and research. To this academy came a 17-year-old student who was to remain as a pupil and a teacher for the next 20 years: Aristotle.

Aristotle

Though orphaned at an early age, Aristotle came from a prominent Macedonian (northern Greek) family, and his family saw to his education. At the academy Aristotle accepted the idea of the four elements and transmutation, but he rejected the idea that particles were made of atoms and empty space. Air, he reasoned, would rush in and fill any void. "Nature," he is said to have stated, "abhors a vacuum."[22] Aristotle expanded a bit on the basic four elements by giving them qualities, a sensory property associated with each: Hot and dry were associated with fire; hot and moist, with air; cold and moist, with water; cold and dry, with earth.

Aristotle used observation as well as pure reasoning to arrive at his ideas. He reasoned that wood when burned produces smoke (air), pitch (water), ash (earth), and fire; therefore wood is composed of these elements. Flint when sparked produces fire; therefore fire is an element in rock. Some rocks thrown in water bubble as air trapped in crevices escapes. To someone with no other evidence to go on, this would prove that air is an element in rock. Some crystals dissolved in water turn the water cold; some crystals dissolved in water heat the water, so hot and cold would appear to be qualities inherent in these materials.

Though Aristotle did not engage in systematic experimentation, his reasoning from observation represents an application of inductive logic, which has a more intuitive appeal than purely deductive reasoning when applied to natural phenomena, and this probably helped the general acceptance of his ideas. Aristotle's attitude was also teleological—that is, he believed the universe was governed by a purpose—and he interpreted phenomena in light of how they fit into a general scheme. This teleological approach was in harmony with Hebrew theology and later with Christian and Islamic thought. For this reason organized religion later supported Aristotle's views and lent Aristotle prolonged authority. The views of Aristotle—accepted by many succeeding writers and philosophers—became embodied in European common speech and thought. Aristotle's theory of the formation of metals—a moist, vaporous exhalation combined with a dry, smoky one—was still used in Europe in the 1600s. The existence of Aristotle's fifth element—the rather vague and immaterial ether—was still argued in the 1900s.

In this chapter we have seen how primitive people acquired an impressive knowledge of usable materials to make tools, heal wounds, preserve meat, and color their lives. This acquisition of new materials continues throughout our history—with some amazing materials to come. We have also seen how the early Greek philosophers reflected on these materials and their interactions, using methods of rational logic (as opposed to mystical speculation) to arrive at the theory that matter is composed of a basic set of elements and all matter contains all elements. These early philosophers also proposed the possibility of transmutation: the theory that under the right conditions, materials could be changed into one another. The Aristotelian version of Greek philosophy was to have the greatest influence on the subsequent development of chemical thought because of its intuitive appeal, its fit with other popular modes of thought, and because Aristotle served as the tutor of the son of Philip of Macedonia. The boy, Alexander, was evidently much taken by the things he learned at Aristotle's knee. In his subsequent career as Alexander the Great, this one student spread Greek culture and Greek philosophical ideas from the borders of China to the coast of Spain.

chapter
TWO

ca. 300 BCE–600 CE: Alexandria and Alchemy

Beginning in 335 BCE, the events of 10 short years reshaped the Mediterranean world, seeded Greek culture from Egypt to the Indus, and reset the course of chemical history. The agent of this change was one man: the boyish, amiable, 23-year-old Alexander the soon-to-be-Great.

Alexander was born to the business of conquering. His father, Philip of Macedonia (the region directly north of Greece) is credited with developing several military techniques, including the 10-deep phalanx of pike carriers (an impenetrable and prickly front) and the use of catapults in siege warfare (to keep defenders away from walls while battering rams were applied). Philip's goal was to unite the Greek city-states (under his jurisdiction of course) and to this end he conquered Thrace, Thebes, and Athens. As part of his plan to assimilate Greek ways, he had his son educated by the leading Greek educator (and former Macedonian), Aristotle.

From these two influences—warrior–father Philip and philosopher–mentor Aristotle—Alexander's personality emerged: a mix of hard brutality and supple idealism. When his father was assassinated, Alexander assumed his place by driving his armies from Macedonia to Syria, crushing Greek rebellions without mercy and selling whole populations into slavery. But when Alexander captured the family of the Persian king Darius, he declined his prerogative with the women and offered them the courtesies due their royal status. On conquering the remains of the Persian ex-empire, Alexander pushed his weary troops into regions beyond that required to secure his borders, but on the death march back from the Indus, Alexander poured his water ration on the ground so they could see he would not take water when they had none. He murdered one friend in a drunken rage, but on the death of another, grieved until his own health was threatened. Alexander crushed free city-states and was resented by the Greeks as a tyrant, but he revered the gods of Egypt and was welcomed in that land as a deliverer.

Alexander's ability to shrug off contradiction allowed him to be one of the great homogenizing influences of the ancient world. He dreamed of uniting East and West into one enlightened empire. He founded some 25 cities throughout Asia and North Africa—most of them called Alexandria—and peopled them with Greek veterans. In this manner he produced the Hellenized world: Greek culture with a Persian blend.

Alexander died of fever at age 33. There was some rumor that he was poisoned. This could have easily been done by giving him a drink of bad water (in those days it was well-known which sources had tainted water and which had good). However he had been badly wounded in battle, and he could not have lived much longer anyway. As was customary for his people and his age, Alexander had much stronger emotional and physical relationships with men than with women, and this, plus campaigning, kept him from impregnating his Persian wife Roxana until just before his death. With his heir apparent in utero, Alexander is said to have whispered on his death bed only that his realm go to the most worthy. His generals fought for some 40 odd years after his death to determine the most worthy, and they ended by dividing his kingdom. When the dust and disputes were settled, Ptolemy had Egypt and the surrounding areas, and Seleucus had

Babylon, which included Asia Minor, Mesopotamia, and Persia. Though our path soon leads us to Babylon, it is with the Greeks of Egypt that our story of chemistry begins.

ALEXANDRIA OF EGYPT

Alexandria of Egypt is a port on the mouth of the Nile, opening onto the Mediterranean. The ancient city was about 4 miles in length, built in a regular pattern with streets crossing at right angles. Ptolemy, who had himself been a student of Aristotle, quickly seized power in Egypt. While the wars for the rest of Alexander's empire raged—cities looted, libraries destroyed, and scholars killed or dispersed—Ptolemy built in Alexandria a center of learning for the Mediterranean and the Near East.

In Alexandria grew the hybrid culture that Alexander had sought: a rich blend of philosophical traditions and practical knowledge. These two types of learning—the philosophical and the technical—would merge in a way unexpected even by Alexander. They created a new sort of philosophic technology: *alchemy*.

ALCHEMY

Though *alchemy* refers to an Alexandrian practice, the word actually has Arabic origins, so it is used rather prematurely here. *Al* is an article in Arabic, and *alchemy* is derived from *alchymia*, which means *the chymia*—a general word for the practice of chemistry.

The origins of the word chymia are not known. Some historians have speculated that it comes from an Egyptian word for black, meaning the black soil of the Nile, some sort of black magic, or even the black eye powder used to adorn the eyes and discourage biting flies. Zosimos, an Alexandrian alchemist, offered another possibility. He said the word was revealed to humankind by angels who fell in love with mortal women. This is an explanation that the authors, as historians, must reject but, as chemists, prefer.

Though derived from a general word for chemistry—meaning the technology of the artisan—the word alchemy came to denote an offshoot from practical chemistry, which grew in parallel to the technology of the artisan. It was an extremely important offshoot—alchemical techniques and observations fed back into practical chemistry (such as medicine, metallurgy, and art)—but beyond that alchemy is important because it was a primary hypothesis testing. The object of science is to explain and to predict, and Aristotle, with his explanations of matter, had predicted transmutation: changing one material into another. This was the hypothesis the alchemists put to the test and would continue to test for the next 1000 years.

For alchemists the goal of transmutation varied. Most often it was the production of gold, although it could be medicine or even transmuting old into young or earthly body into soul. The techniques of transmutation also varied: Some alchemists took a pragmatic approach, dissolving, melting, combining, distilling, but others used only magic incantation. Most used both. The stakes were high (gold and eternal life), and the practitioners inevitably attempted to hide their methods and results, causing the practice to become associated with secrecy, mystery, and strange happenings in the night. This caused the popular perception of alchemists to be less than good and that of other chemical workers (healers, miners, and artisans) to sometimes suffer. But alchemy served as the hope of the bewildered chemical worker, and it provided the inspiration to stay at the fires when the potters and the weavers had gone home.

Of course from our enlightened perspective, the notion of universal transmutation may seem a bit bizarre, and we are tempted to question the acumen of our forebears. Carefully considered however alchemy as a first proposition is not all that unreasonable. Alchemists believed in transmutation because they *saw* transmutation every day of their lives in cooking, dyeing, bodily functions, or producing metals from ores. They also knew of the striking transformation of mercury sulfide ore, cinnabar, into liquid mercury: Heating the ore drove something off and left a puddle of silvery metal behind. If this were not sufficient proof for transmutation, then they only had to show that heated again, the puddle reformed as a red solid. It was not the return of cinnabar, as they thought—it was an oxide of mercury that also happened to be red—but in the absence of additional information, transmutation seemed as plausible an explanation as any.

Furthermore when our first chemical forebears made a yellow or gold-colored powder, they had no reason to believe that it was not gold, so it was reported as such. When assay techniques became better, they realized that the product was not gold, but having believed for generations that the procedure produced gold, they blamed their own techniques for the failure rather than the recipe.

Lastly, the seemingly exhaustive and thankless labor of the alchemist can be understood by realizing how tantalizingly close they kept coming to their goal. Producing a yellow metal was *almost* producing gold. As far as they knew, it might even *be* gold, just not of such good quality as the stuff from the ground. And there were workers out there feeding their families by making near gold that was pretty darn good—good enough, as they said, "to deceive even the artisan."[1]

TECHNICAL TRADITION OF ALEXANDRIA: THE UNKNOWN ARTISAN

Knowledge of this near gold comes to us because an artisan decided to record the techniques for its production on papyri—and because the Swedish diplomat who came across the papyri in the 1800s had the sagacity to acquire them and ship them to Europe for study. After some 85 years of study, scholars realized that two of the scrolls contained descriptions of chemical processes.

The papyri seem to be in the same handwriting and to have been written about the end of the 200s CE. They were most likely copies of older works because slight copying errors can be found. The original was probably written sometime after 100 BCE.

The first papyrus contains mostly recipes for dyeing and mordanting (fixing dyes) and preparing imitation gems. The following recipe is for the preparation of copper acetate (*verdigris*), a green material used in making artificial emeralds. Note that the copper is specified as Cyprian copper. As we mentioned, because materials used were rarely pure, materials nominally the same sometimes had a different behavior chemically due to contaminants they contained. Therefore the same material from different sources was sometimes treated as an entirely different material.

> Clean a well-made sheet of Cyprian copper by means of pumice stone and water, dry, and smear it very lightly with a very little oil. Spread it out and tie a cord around it. Then hang it in a cask with a sharp vinegar so that it does not touch the vinegar, and carefully close the cask so that no evaporation takes place. Now if you put it [in the cask] in the morning, then scrape off the verdigris carefully in the evening ... and suspend it again until the sheet becomes used up ... The vinegar is ... unfit for [consumption].[2]

The second papyrus however seems to be concerned mostly with metals, including very matter-of-fact recipes for making false gold.

> Manufacture of Asem [the alloy of silver and gold]: Take soft tin in small pieces, purified forty times; take 4 parts of it and 3 parts of pure white copper and 1 part of asem. Melt, and after casting, clean several times and make with it whatever you wish to. It will be asem of the first quality, which will deceive even the artisans.[3]

The recipes may have been used by an artisan in a workshop. They are practical, with no reference to the mystical and no attempt to obscure the content—contrasting sharply with later alchemical writings. How they came to be preserved is a mystery. One thought is that they may have been hidden in a mummy case for safekeeping. The owner may also have wished to be buried with valued possessions, and written procedures might be an artisan's most valuable possession.

The roots of this pragmatic approach can be traced to mainland Greece, for while it is true that the center of learning shifted from Athens to Alexandria, Athens did not become a ghost town overnight. In fact Plato's Academy and Aristotle's Lyceum were active for another 700 years. Wandering Greek philosophers—doing private tutoring, making public speeches, and collecting whatever donations they could—carried Greek natural philosophy throughout the Hellenized world. And although traditionally the Greek philosopher was to reason without resorting to the common labor of experiments, the successor to Aristotle at the Lyceum, Theophrastus (ca. 315 BCE), was more than an immaculate philosophizer. Theophrastus was an astute observer of practical chemistry, as seen from his descriptions of the manufacture of lead acetate (white lead):

lead is placed in an earthen vessel over sharp vinegar, and after it has acquired some thickness of a kind of rust, which it commonly does in about ten days, they open the vessels and scrape it off.... What has been scraped off they then beat to a powder and boil (with water) for a long time and what at last settles to the bottom of the vessel is white lead.[4]

He also described the use of plaster of Paris:

> The stone from which gypsum is made by burning is like alabaster. Its toughness and heat when moistened is very wonderful. They prepare it for use by reducing it to powder and then pouring water on it and stirring and mixing well with wooden tools, for they cannot do this by hand because of the heat. They prepare it in this manner immediately before using, for in a very little while it becomes hard...[5]

This practical approach to chemistry might have continued unmodified if Alexandria were not its next port of call.

Alexandrian Mysticism

In Alexandria however this pragmatism floated in a sea of other *isms*: Aristotleism, Stoicism, Epicureanism, Neoplatonism, Gnosticism, Zoroastrianism, Mithraism, and probably others whose names have been lost. Of major importance to chemistry was Judaism and its outgrowths: Christianity and Islam.

Judaism can trace its roots to around 1800 BCE, when according to tradition, a man called Abram was commanded by his god to change his name to Abraham as a sign of his new status as the ancestor of a people. Indeed Jews, Christians, and Muslims—three of the major religious influences in the modern world—all regard themselves as children of Abraham. The first resulting civilization, the Hebrew, was in many ways identical to that of its neighbors. The Hebrews were equally fierce, territorial, and brutal. Their command of chemical technology was also comparable to that of surrounding civilizations. They knew about gold, silver, copper, iron, lead, tin, and the means of purifying and working them. The Hebrews did however have one peculiarity.

According to tradition the deity of Abraham insisted on being the *sole* divinity—introducing the then novel concept of monotheism.

Monotheism helped guarantee the longevity of the culture: A monotheistic religion could not be diluted by the assimilation of other gods or customs. Equally importantly Abraham was from a nomadic people and his god was a nomadic god, not associated with particular shrine. This made the religion transportable and allowed the Hebrew civilization to survive as a culture even after the Hebrew people were driven from their land and dispersed.

This dispersal (*Diaspora*) occurred around 590 BCE when Babylonians destroyed Jerusalem and deported most of its inhabitants to Babylon. During the Diaspora, as a minority population, the Hebrews found cultural tolerance to be necessary. A humanitarianism developed that became a cornerstone in the Jewish and offspring religions. In addition the Jews of the Diaspora found that to preserve their religion, each had to become responsible for a personal covenant with their god and all individuals had to learn and understand the Torah—the collection of sacred writings. Without a temple the book and learning became sacrosanct. This tradition of the sanctity of writings was jointly held by several of philosophies, and was an important influence on the early intellectual development of Alexandria. When Ptolemy founded the Museum in Alexandria, a research and teaching institution modeled on the Lyceum at Athens, it was said to have employed a hundred professors (paid by the state), housed a zoo, a botanical garden, an astronomical observation room, dissecting rooms, and a library of a-half-million scrolls.

ALCHEMY AS PRACTICED

From remnants of these scrolls we learn how the rich philosophic tradition combined with practical technology to give us Alexandrian alchemy, and at least in the beginning, the practical approach was dominant. A common tactic for instance was to try to make a material appear more goldlike, on the theory that if it assumed enough of the qualities of gold, it would eventually be gold—not an unreasonable approach. When it was found that metal surfaces could be dyed white

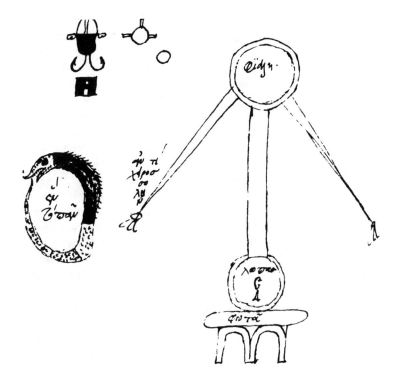

Figure 2.1. The alchemical sign of the serpent Ourobouros symbolized the eternal cycle of changes: life and death, seasons, transmutation. (Courtesy of the John F. Kennedy Library, California State University, Los Angeles.)

with arsenic sulfides and yellow with polysulfides, the search was on for just the right dye to produce actual gold. Another theory involved treating metals as living organisms that could be brought to gold perfection by isolating the soul of gold and transferring it to base metal. As was common to all the approaches, the methods used included distillation, *sublimation* (transforming a heated solid into a gas with no liquid phase), *fusion* (melting), *solvation* (dissolving), filtration, crystallization, and *calcination* (heating to a high temperature without melting; this usually results in *oxidation*—incorporating oxygen from the air). Because of their ability to effect change in materials, certain reagents also became common in alchemical practice. Two important ones were mercury, the liquid metal, and sulfur, the stone that burns.

In a process that is fascinating to watch, mercury poured on many clean metals unites with the metal to form mercury-containing alloys, or amalgams. This is in fact a well-known method for refining gold: Gold is pulled from ore or another matrix into the mercury. The mercury can then be driven off by heat, leaving behind the pure metal. Surface amalgams formed this way were used by the alchemists to create a silvery appearance. When treated with sulfur, the amalgams assumed a yellow color, or were "transformed" into gold.

Also of interest were such corrosive substances as acids, hydroxides, and ammonium chloride, which on heating turns into ammonia and hydrochloric acid and readily attack metals. Because arsenic sublimes like mercury, forms soft alloys like mercury, and has a sulfide that looks like the sulfide of mercury, alchemists regarded arsenic as a kind of mercury, and much use was made of arsenic's property of coloring metals. *Vitriols* (copper and iron sulfates) were also of interest to the alchemist, probably because of their colors. Copper sulfate forms beautiful blue crystals and solutions, and iron sulfate is green.

We can confidently write about the preceding materials and techniques because of the efforts of a few early alchemical practitioners to preserve their knowledge in writing. Three are considered here: pseudo-Democritus, Mary the Jew, and Zosimos.

Pseudo-Democritus

Around 100 CE pseudo-Democritus (the original Democritus was the Greek philosopher of ca. 450 BCE) wrote a treatise called *Physica et mystica* (*Natural and Mystical Things*), a recipe book for dyeing, coloring, and making gold and silver.[6] Though it is impossible to identify all the ingredients (the names were imprecise and the chemicals were impure), he generally used whitish powders to color amalgamated surfaces to look like silver and used yellow colorings to give the appearance of gold.

Although these recipes are still basically practical, we see mysticism beginning to creep in. The recipes call on Greek and astrological theories and contain references to shadowy magical practices. Each recipe ends with "nature triumphs over nature," "one nature rejoices in another nature," or some other equally enigmatic phrase.

Mary the Jew

This progression from the practical to the obscure continued with the work of the Alexandrian alchemist know as Miriam, Maria, or Mary the Jew. Although no complete works by her have been found, enough fragments exist to establish her as an historical fact. She was also mentioned by Zosimos and Pliny, a Roman historian. There is reason to believe female alchemists were not unusual; frequent references to a Cleopatra (the alchemist, not the queen) can be found, though only the existence of Mary is known with any certainty.

Mary approached alchemy as a fusion of the rational, the mystical, and the practical, and she is remembered for the practical. Mary introduced several types of apparatus, including a three-armed still, a hot-ash bath, a dung bed, and a water bath. The water bath, a double-boiler, is still known in France as the *bain-marie*.

In her theory Mary frequently refers to a medicine by which metals are transformed. She also attributes gender to the metals, stating that the secret is to "Unite the male with the female, and you will find what you seek."[7] She writes that silver does this readily (perhaps with reference to the action of mercury), but copper couples "as the horse with the ass, and the dog with the wolf."[8]

Zosimos

We do not know why some alchemists chose to record their work and others did not—it may have been a matter of literacy or lack thereof—but occasionally we find some clues about personal motives. Zosimos, who probably lived about 300 CE, addresses each of his 28 books to his sister, Theosebeia, who was probably an alchemist herself. In at least one letter he refers to her work by describing a failed procedure tried by another alchemist: "Paphnutia was much laughed at, and you will be laughed at too, if you do the same."[9] What is not known however is whether *sister* was meant figuratively or literally. Zosimos' writings may have been the loving efforts of a brother to aid a sibling in her chosen trade or correspondence with a female colleague.

Zosimos relays practical information (he says it is good to hold the nose when working with compounds of sulfur); he describes such operations as fusion, calcination, solution, filtration, crystallization, and sublimation; and he systematized his materials into bodies (metals), spirits, vapors, and smokes. But with Zosimos the obscurity of alchemical writing deepens: Alchemical procedures were revealed to him in dreams. In the following account of one of Zosimos' dream, it has been proposed that the dome refers to the glass condenser usually placed over a distilling apparatus (which shared the same name as *dome* in Greek) and the 15 steps refer to the operations of distilling, fusing, and so forth.

> While saying these things, I fell asleep and I saw standing before me at an altar shaped like a dome, a priest sacrificing. There were fifteen steps to mount to this altar. The priest stood there, and I heard a voice from above saying—"I have accomplished the act of descending the fifteen steps walking toward the light. It is the sacrifice that renews me eliminating the dense nature of the body. Thus by necessity consecrated, I become a spirit." Having heard the voice of him who stood at the dome-shaped altar, I asked him who he was. In a shrill voice he answered in these words, "I am Ion, priest of the sanctuaries, and I undergo intolerable violence . . ."
>
> After this vision, I awoke again and said, "What is the meaning of this vision?" . . . In the dome-shaped altar all things are blended, all are dissociated, all things unite, all things combine, all things are mixed and all are separated . . . Indeed for each it is by method, by measure, by exact weight of the four elements that the mixing and the separation of all things takes place . . .
>
> Thou wilt find what thou seekest. The priest, this man of copper . . . he has changed the color of his nature and has become a man of silver. If thou wishest, thou wilt soon have him a man of gold.[10, 11]

Thrice Great Hermes

Another rather ethereal alchemical author of this period is Hermes Trismegistus, or Thrice Great Hermes. Texts attributed to him were Greek translations of Egyptian sacred texts attributed to Thoth, the god of wisdom. Fragments found on temple walls and in papyri show that some 42 books covered various subjects, including alchemy. Though the true author or authors will probably never be known, the books found an audience when they were revived by medieval European alchemists, and these are of interest to us now because they are the origin of the term *hermetic art* in reference to the practice of alchemy. The term *hermetically sealed* also comes from this source, and it refers to a method for storing certain alchemical potions, probably in jars with wax-sealed lids.

ALEXANDRIAN DECLINE

By the late 300s CE Alexandrian alchemy had moved almost completely from the material to the mystical. In Alexandria this may have been due to the influence of Eastern philosophies, but it may have also been for another, simpler reason: Material methods did not work. They did not produce gold. Without a systematic experimental method with which to proceed, alchemists moved to the only other thing available to them: magic. In the process the objective also shifted from solid gold to a golden soul: Spiritual transmutation of the alchemist became the new goal. Materials in the procedures slowly became less important until secret ritual and incantations were the only ingredients left. Now an alchemist could practice without ever breathing fumes or lighting a fire. This is not to say that material techniques might not have seen a revival, but unfortunately about the same time, Alexandria as a whole was undergoing an intellectual decline.

There was a gradual philosophical shift. In the 100s CE Gnosticism became prominent and with it the belief that direct revelation from the deity was the only source of knowledge. In the 200s Neoplatonism became important and with it a contempt for reason and science. In addition there were two other events around 0 CE that on their own were not disastrous but taken together resulted in stagnation and decline in Alexandria.

One was the defeat of Cleopatra, the last of the Ptolemys, which put the city under Roman rule. The other was that Jesus of Nazareth began to teach in the Roman province of Palestine. The teachings of

Jesus were adopted by a small Jewish sect, and—primarily through the work of Paul of Tarsus—these eventually evolved into the separate religion of Christianity.

Starting about 100 CE these two events worked together in a way that eventually ended the Hellenistic Age. Around 100 CE a Jewish revolt against Rome resulted in the annihilation of the Jewish population, the destruction of a large portion of Alexandria, and the loss of scholars and scholarship. In the 200s the political climate became more repressive, culminating in the emperor Dioclectian's order that all alchemical writings be destroyed—he was afraid that gold and silver producers could create enough money to fund a rebellion.

Into this climate of repression philosophies spreading from Persia and Babylon continued to bring uncompromising dualistic ideas of good and evil, boasting extreme asceticism and mystical beliefs. Christianity assimilated these ideas, and when Constantine (ca. 330 CE) made Christianity the official cult of the Roman empire, suppressive measures came with it. The Christians sought to destroy pagan philosophy and alchemy. In Egypt pagan temples and libraries were sacked. Around 400 CE in Alexandria, the last pagan philosopher, the Lady Hypatia, was lynched by a mob of monks, and shortly thereafter the Museum at Alexandria closed and its library was scattered. Around 500 CE the Academy in Athens was closed, and the remaining philosophers fled to Persia. The Hellenistic Age was at an end.

There was however a savior for the wealth of Hellenistic scholarship. In the 400s the Christian church was violently divided over the issue of the exact nature of the divinity of Christ, and Nestorius, a monk, was excommunicated. He fled to Syria with his followers. These people took with them what they could of Hellenistic learning, and in the refugee monasteries of these dissidents, manuscripts were accumulated and stored.

Around 500 the Nestorians were banished from Syria. They moved to Persia where they were protected by the Shah. There they founded schools in the style of Alexandria and translated the works of Plato, Aristotle, and some of the early alchemical writers into Syrian. Astronomy, astrology, medicine, and alchemy were among the subjects taught at their schools—and from these schools we will see the science of Alexandria resurface.

In these schools alchemy was accepted as a serious subject of study along with medicine and astronomy because, as we have seen, alchemy appeared to be a reasonably logical theory. It found its inspiration in natural events and its confirmation in the appearance of near success. So the question may then arise: If the theory was as defensible as it sounds, then why did it not occur to other people in other places?

The answer of course is that it did.

OTHER ALCHEMIES OF THE ALEXANDRIAN PERIOD: INDIA AND CHINA

The Chinese and Indian civilizations that flourished around the same time as Alexandria each developed an alchemical system of their own. There certainly was some information exchange—India, on the well-worn route of conquest, is thought to have had wandering bands of artisans (similar to modern Gypsies) who traveled from the Indus to Thrace and even Gaul[12] before 0 CE, and the Jews of the Diaspora established communities in China and India as early as the 100s CE and maintained links with each other, as well as with Rome and Greece. But these systems were sufficiently different from the alchemy of Alexandria to show that similar people contemplating similar grains of sand can come up with similar ideas on their own. In fact it is instructive to ask why all societies did not develop an alchemy. The answer—different for each society—may have depended on climate or simply the relative value of gold.

The absence of alchemy in extreme northern regions might be understood by considering climate: Daily survival no doubt took precedence over alchemical speculation. Southern Africa however was certainly warm enough—and had no lack of technology or societal organization: Beginning around 200 BCE, the kingdom of Kush spread iron-making through the central south. But as we have mentioned at this time gold was useless except as decoration or a medium of exchange, and the people of southern Africa did not really need trade. Their growing season is essentially year-long, and southern Africans have always practiced hunting and gathering along with land cultivation. Ancient cultures in southern Africa were said to have prized iron over gold because it was more useful in hunting and farming. In addition if they had really wanted gold, the people of southern Africa had it readily available in and on the ground—which may provide the answer to the next question: Who didn't try to manufacture gold? People who could go out and pick it up.

In Meso-America this may be the answer that applies. In 1534 Pedro Sancho reported a plenitude of Inca gold: "Amongst other things there were sheepe of fine gold very great, and ten or twelve statues of women in their just bignesse and proportion, artificially composed of fine Gold . . ."[13] In 1586 Lopez Vaz reported "[Panama] is the most richest Land of Gold then [sic] all the rest of the Indies."[14] Not much motivation for bending over gold-brewing caldrons here. This argument may also apply to Japan, which is rich in mineral and surface gold, but not rich in mercury: the perennial ingredient in alchemical recipes. The observed transformations and interactions of this liquid metal were among the more persuasive arguments for transmutation, and mercury may have been as necessary for the development of alchemy as the need for gold. There were however two regions that met our conditions (a plenitude of mercury, a scarcity of gold, and motivation): India and China.

In some ways the alchemies of India and China represent separate microcosms of chemical history, but in other ways they are inseparable from the general history. The development of alchemy in these two regions roughly parallels developments in the West—sometimes leading and sometimes lagging. Therefore we introduce these two developing sciences here, along with their technical and philosophical bases, and follow them through Alexandrian times. We will then return in subsequent chapters to see how they contribute to the whole.

CHINESE TECHNICAL TRADITION

Medicine is always an inspiration for developments in chemistry, and the pharmacopoeia of China is legendary. Some herbal and mineral remedies that were taken over by European medicine include iron (for anemia), castor oil, *kaolin* (a clay used to treat diarrhea); *aconite* (a tuber used as a sedative); camphor, cannabis, and *chaulmoogra oil* (used for leprosy). The active agent of *rauwolfia*, used in China to induce purgative vomiting, has been isolated, and it is currently used to treat high blood pressure and some mental conditions. The bronze age did not arrive in China until around 1500 BCE, and iron appeared only about 500 BCE, but by the beginning of their alchemical age, around 100 CE, the Chinese had knowledge of zinc and brass (a copper–zinc alloy), mercury, sulfur, and several of the common salts, such as niter. In the 400s BCE an administrator made a list of known materials that included lead oxide, lead carbonate, sulfur, calcium carbonate, hydrated calcium sulfate, ferric oxide, niter, talc or soapstone, hydrated magnesium silicate, *potash alum* (the double sulfate of potassium and aluminum), *malachite* (basic copper carbonate), azurite, and others. The Chinese knew of *corrosive sublimate* (mercuric chloride) and *calomel* (mercurous chloride), and they could distinguish between them. They used mercury to extract gold and silver by amalgamation, and toward the end of the Alexandrian period, they used a tin–silver amalgam for dental purposes. This material repertoire of China, combined with the philosophic tradition that we examine next, provided fertile ground for alchemy's growth.

CHINESE PHILOSOPHICAL TRADITION

Around 525 BCE a philosopher called Confucius (the latinized form of K'ung-fu-tzu, or Master K'ung) set down a system of values by which people were assigned roles in a fixed society. The resultant civil bureaucracy, Confucianism, held China together through changing dynasties, empires, and social upheaval, and it kept China intact as a political entity for 2000 years. The Confucian philosophy concerned itself with social systems, and not natural philosophy—at least not to any great extent. There were however other philosophies that arose in reaction to the rigidity of Confucianism, and these were concerned with the nature of the world.

One such group of philosophers, aptly named the Naturalists, proposed a world built of five elements—metal, wood, earth, water, and fire—where all material substances arose from combinations of these five. They also employed the dualism of yin and yang. Yin, the female principle, is the moon—negative, heavy, earthy—with the characteristics of dryness, cold, darkness, and death. Yang, the male principle, is the sun—positive, active, fiery—with the characteristics of wetness, warmth, light, and life. Taoism (pronounced *dowism*) may actually have been pre-Confucian, but it increased in popularity in reaction to Confucianism. This philosophic group also embraced the concept of five elements and yin and yang, believing for instance that all minerals or metals are essentially the same; these just differed in the amount of yin and yang. The Taoists were also anti-intellectual, rejecting any search for knowledge and advocating, at least at first, a return to a simpler way of life. They believed that life, regulated on the principles of Taoism, could be prolonged, and they searched for peace and longevity in a rustic life spent in quiet contemplation.

Corruption of the ideal however eventually ensued. The search for longevity turned into a search for immortality, and Taoism turned from quiet contemplation to the practice of magic. The secret, they believed, must be in the nature of gold: Gold is incorruptible and eternal; therefore those who manage to incorporate gold into their bodies achieve an immortal state. The Taoists began to search for ways of making and ingesting this potable gold. They began the search for elixirs.

CHINESE ALCHEMY

At first these efforts received support from high places. The First Emperor in the 200s CE, was apparently obsessed with the idea of achieving physical immortality, and he was often under the influence of one Taoist magician or another. The current court favorite was showered with money and honors until the emperor became disillusioned and issued the ultimate dismissal. Finally a magician from the Shantung coast persuaded the emperor to consume elixirs made from mercury transmuted into gold, and significantly the emperor died during a trip to the Shantung coast. At least two other emperors died as a result of elixir poisoning or debility caused by it.

This brings up the same type of question we asked about Alexandrian alchemy: Why did the alchemists continue in the face of such obvious evidence of the futility of their course (the dead emperors)? The answer, according to the scholar Joseph Needham,[15] again may be found in their seeming partial success. The bodies of those who die from mercury and other heavy-metal poisoning tend to have delayed decomposition, probably due to the poisoning of bodily bacteria as well. Therefore those who die after taking elixirs may have appeared to have cheated death—at least a bit—so the deaths inspired further research instead of withdrawal of the theory.

All this deadly experimentation however may not have been for naught. Chinese alchemists did find some mercury compounds that were excellent for getting rid of fleas and lice (soap was still a luxury in those times), and other scholars credit the richness of Chinese cuisine to the Taoist habit of experimentally eating all sorts of organic and inorganic substances. Another outcome may have been the discovery of anesthetics and other items of their tremendous pharmacopoeia. According to Edwin Reischauer and John Fairbank:

> Most Chinese scientific inquiry, for that matter, seems to have grown out of the activities of the curious, experimenting Taoist alchemists and magicians, and perhaps one reason why later Chinese thinkers turned their backs so emphatically on scientific experimentation was its association in their minds with Taoism.[16]

One is tempted to say that it left a bad taste in their mouth.

Around 150 BCE an imperial edict demanded public execution for those who counterfeited gold. This time it may not have been inspired by a fear of fortunes amassed for rebellion but simply by the death of too many prominent people. Alchemy however continued to be practiced. By 140 CE the atmosphere had again become tolerant enough that a Chinese book on alchemy appeared.

Wei Po-Yang

Very little is known from direct sources about the life of this first author on alchemy. Wei Po-Yang may have been from a *shaman* (magician–technician) family, though in Po-Yang's first work, *Tshan Thung Chhi* (*Kinship of Three*), he says of himself only that he kept away from government service.

Although the *Kinship of Three* is extremely obscure, one of the more decipherable descriptions is that of making a mercury–lead amalgam. In the following, yellow sprout refers to the yellow lead oxide film that forms on the surface of molten lead.

> From the very beginning of Yin and Yang, lead ore encloses the "yellow sprout." . . . The "flowing pearls" [mercury] . . . has a tendency to escape . . . Eventually when [they] get the "golden flower" [they] turn and react with it, melting into a white paste or solidifying into a mass. It is the "golden flower" that first undergoes change [for] in a few moments it melts into a [viscous] liquid. [The two substances now fuse together and] assume a disorderly appearance like coral or horse-teeth. The [essence of] Yang then comes forth to join it, and the nature of things is now working in harmony. Within a brief interval of time [the two substances] will be confined within a single gate.[17]

Po-Yang says that eating gold promotes longevity, and he discusses making potable gold (gold-colored powders or liquids). Though he is well aware of the nature of true gold, he and other alchemists thought of potable gold as a different kind of gold.

Ko Hung

The other important writer on alchemy and maker of potable gold, Ko Hung, lived ca. 300 CE, and we know a few more details of his life. Ko Hung started out as a officer in the military, where he fought in the suppression of some rebellions. However he shunned a military career, and with the support of his father who had risen to governor of Shao-ping, Ko Hung managed to study alchemy and medicine with the scholars. He traveled and studied plants and mineral substances, especially in the south. He later used his bureaucratic connections to be appointed magistrate in a region close to some cinnabar deposits that he needed for his work. On the way to assume the post however, he passed through some mountains and decided to stop for awhile. He ended up living in the mountains, writing his book, and doing alchemical experiments until he died. Somewhere in this sequence of events, Ko Hung married. His wife, Pao Ku, was also an alchemist, and this was apparently not exceptional. A number of women appear in drawings and written records of alchemy, and descriptions of several female adepts can be found in the works of Ko Hung.[18] From his writing it can be seen that Ko Hung is also well aware of the nature of true gold. When describing a colleague putting a pinch of something into boiling lead and tin to make silver, Ko Hung adds:

> Of course these are counterfeit things. For example, when iron is rubbed with stratified malachite [a copper carbonate], its color changes to red like copper. Silver can be transformed by the white of an egg so that it looks yellow like gold. However, both have undergone changes outside, but not inside.[19]

But in his monumental *Pao Phu Tzu* (*Book of the Preservation of the Solemn Seeming Philosopher*), Ko Hung gives recipes for making gold out of mercury, lead, and other ingredients, many of which were really just yellow-colored mercury-containing precipitates. He ascribes to these mixtures a number of beneficial properties, and along with recipes, he describes some physiological effects that include visions and symptoms of mercury poisoning.

INDIAN TECHNICAL TRADITION

We know about early Indian metallurgical technology from such artifacts as steel found in graves from the 500s–600s BCE and a wrought-iron pillar near Delhi some 1500 years old. Information about other chemical technologies comes almost exclusively from medical writings: the *Charaka*, ca. 100 CE; the *Susruta*, ca. 200 CE; and the *Vagbhata*, ca. 600 CE. These writings show that India had an extensive pharmacopoeia (intoxicating plants, laxatives, diuretics), metallurgy (gold, silver, tin, iron, lead, copper, steel, bronze, brass), and a repertoire of other practical chemicals (alcohol, caustic alkalis, chlorides, and sulfates of iron and copper).

Indians used alcohol as an analgesic, stopped bleeding with hot oils and tars, removed tumors, drained abscesses, repaired anal fistulas, stitched wounds, performed amputations and cesarean sections, and they knew how to splint fractures. Their fumigation of wounds is one of the earliest efforts at antiseptic surgery. They were however forbidden to cut the deceased, so anatomical studies had to be performed on bodies sunk in a river for several days, then pulled apart. This did give them a fairly good understanding of bone structure and muscles but a poor understanding of the soft organs, more subject to decomposition.

In the *Susruta* physicians are advised to treat skin lesions with caustic alkalis, wait the space of time it takes to say one hundred words, and then neutralize with acid.[20] The procedure would be agonizing, no doubt (lye, a caustic alkali, is used as a drain cleaner), but it would effectively sterilize a wound. Caustic alkali is also recommended for making incisions and to remove growths; it was regarded as superior to cutting instruments because it cauterizes the wound as it is made.[21]

INDIAN PHILOSOPHICAL TRADITION

Early Indian natural philosophies had much in common with China, Alexandria, and Greece. As early as 1000 BCE the *Vedas* (a collection of Hindu sacred writings) identified five elements—earth, water, air, ether, and light—and suggested that animated atoms of these elements combined to make all things. Tantrism, an important philosophic, social, and religious system, in some ways similar to Taoism, may have developed quietly from 100–300 CE, but it began to be influential ca. 400 CE. The essence of Tantrism is the search for spiritual power and the ultimate release from earthly ties. Techniques include sacred syllables and phrases (mantras), symbolic drawings, *yoga* (a concentration technique), and secret rites. The Tantric cult provided India with an ingredient for alchemy as essential as the belief in transmutation, lust for gold, or the availability of mercury: a touch of the mystic.

INDIAN ALCHEMY

The *Vedas* hint at an association between gold and long life, and the idea of transmuting base metals to gold appears in Buddhist texts from 100–400 CE—roughly the same time as in Mediterranean texts. The practice of Indian alchemy is verified in the 600s by accounts of Chinese travelers and Indian alchemists with particular skills (such as preparing strong mineral acids to transform metals by dissolving them) were welcomed at the Chinese imperial court. Indians, like Chinese alchemists, were more concerned with making gold elixirs than money, but they wanted to use the elixirs for medicines, not immortality. Tantrism provided other routes to immortality. There is evidence that the adepts knew how to color metal and make "gold," but they placed little importance on this skill. Whether or not all Indian alchemists adhered to such noble goals however must remain conjecture because detailed records were not kept until the 600s CE. Early Indian chemical history is disappointingly poor in information on individual personalities. Considering the number of women who held positions of power and prominence—there were monarchies headed by women; women worked as accountants, judges, bailiffs, and guards; some early Indian tribes were named after women; and women fought in armies (not as an occasional Joan of Arc but as part of the regular troops)[22]—there may very well have been female alchemists paralleling those in China and Alexandria, but this is currently uncertain. Nāgārjuna (which means *a name to conjure with*) is the only early alchemist of whom we have any concrete knowledge.

Nāgārjuna

The name *Nāgārjuna* was used by at least two other authors from 100–800 CE, so it is not clear exactly which Nāgārjuna was the alchemist. A translation of a Sanskrit text into Chinese in the 400s CE gives the name of the author as Nāgārjuna, the Buddhist philosopher who was the founder of the dialectical Mādhyamika logic and associated with the beginnings of Tantrism. This particular Nāgārjuna is quoted as saying:

> By drugs and incantations one can change bronze into gold . . . By the skillful use of chemical substances, silver can be changed into gold and gold into silver . . . By spiritual power [an adept] can change even pottery or stone into gold . . . [and] One measure of a [certain] liquid [prepared] from minerals can change a thousand measures of bronze into gold.[23]

Despite the difficulty in dating events in early Indian history, these quotes confirm the practice of alchemy by at least this time.

Modern scholars were not the only ones to have difficulty with dating. The Chinese traveler Hsüan-Chuang, writing in the 600s, reports that "Nāgārjuna was deeply versed in the techniques of pharmacy, and by eating certain preparations he had attained a longevity of several hundreds of years, without any decay either in mind or body."[24]

To add to the dating difficulties, there is the obscurity of the Tantric writing (there is a story of a Tantrist poet who chanted a poem to a thousand people—but he was understood by only one). Tantric texts could be read on several different levels: the yogic, the liturgical, the sexual, or the alchemical. For instance the same word is used for diamond, thunderbolt, penis, emptiness, or vacuity; and it is difficult to tell erotic imagery from alchemical technique.

Tantric mysticism, as with the mysticism of Alexandria and China, had its dark side too. A Chinese traveler ca. 600 CE described ascetics who engaged in necrophilia and necrophagy, drank from skulls, and ate feces and other filth. Sex was associated with sadism, and there were ritual cruelties, including human sacrifice and mutilations. These practitioners were however most decidedly on the fringe. In fact they rejected the Hindu social system in general, including the caste system. But they had their influence: The art of alchemy continued to develop in India, but it veered from its practical pursuit of medicines into this realm of deviant magic.

So we now have some new threads to follow: the chemistry of China and India. These chemical traditions were in many ways separate, but the cultures were connected by war, and they had some alchemical elements in common. In subsequent chapters, we see interchange increase world wide due to improvements in travel, maps, trade, and major amalgamating forces known as empires—not in the Alexandrian sense of whirlwind conquest but rather stable political systems that successfully held and administered such vast regions as the empires of Rome and Islam.

chapter
THREE

ca. 200 BCE–1000 CE: From Rome to Baghdad

In the period 200 BCE–1000 CE ideas and traditions of the Greeks spread from Rome to Baghdad, and along the way the practical store of chemical knowledge grew and the theory changed. But at the end the theory arrived in recognizable form at Europe's door.

ROME CA. 200 BCE–600 CE

The original site of Rome consists of a group of seven hills: high ground that was ideal for defense from other humans and malaria-bearing mosquitoes—these mosquitoes will also play an influential role in the later European development of chemistry. The communities scattered throughout the hills consolidated, and they were ruled by monarchies until about 500 BCE, when this system was replaced by a republican form of government. Around 400 BCE the

Romans set off to conquer Italy, and by the 200s they had control of the peninsula.

The Romans then engaged Carthage for control of the Mediterranean. In this war Hannibal crossed the Alps into Italy—in 15 days with elephants, foot soldiers, cavalry, and battle gear. In this passage he reportedly used vinegar to break up large stones blocking his path (meaning, we assume, stones larger than an elephant could handle). This however could not have been vinegar as we know it today—a five percent solution of acetic acid—because this material will not dent rocks unless one has a great deal of time (millennia) and a great deal of vinegar (oceans). Hannibal may have used some sort of explosive, or he may have used one of the strong salt solutions sometimes referred to as vinegar in those days. These strong salt solutions do not freeze at normal water-freezing temperatures (which is why salt is poured on icy roads to form a salt solution that freezes only at low temperatures). The salt solution could have been poured into the rocks during the relatively warm Alpine days, then when the temperature fell to the nighttime extremes, the salt solutions would have finally frozen, expanded, and cracked the rock. If this were the rock-breaking mechanism, the success of these methods certainly depended on luck and time, and it is doubtful that Hannibal used it routinely on the 15-day march. But if it were used once and worked, that may have been enough to make it into the history books. (But then Hannibal is also said to have catapulted vessels of poisonous snakes onto the decks of his enemy's ships, which seems a bit too dramatic and difficult to be real: If he had vinegar strong enough to crack rocks, why did he not catapult *this* onto the enemy ships?)

As it was however Hannibal managed to harry Rome for quite a while until Rome employed his tactics and attacked him on his home front, Africa. Forced to return to Africa, Hannibal was eventually defeated, leaving the Mediterranean in Roman control. Greek cities that had sided with Hannibal were made Roman colonies, and the Romans, like the Egyptians, assimilated Greek culture, so that educated Romans became bilingual in Latin and Greek. Seemingly unstoppable now, Rome went on to conquer areas in southern Europe, Macedonia, Greece, Gaul, Britain, Egypt, Asia Minor, and Persia.

IMPERIAL ROME

Administering such a large area with a mixed society of slaves, citizens, plebeians, and provincials—and with a powerful standing army—proved to be too much of a strain on the republican form of government. An effort to stabilize and centralize resulted in Julius Caesar's dictatorship ca. 40 BCE. From there the succession of Roman emperors is a fascinating story of heroism, depravity, genius, idiocy, and the full range of brilliance and breakdown in human behavior. Through it all however the emperors had one thing in common: Their principal job was administrative.

Each emperor handled piles on piles of papers every day. Every citizen had a right to petition, and they routinely did. Laws, decrees, and appointments had to be read and acted on. Armies had to be positioned, garrisoned, and fed. All of this administrative nightmare (accomplished without Fax machines or conference calls) inspired a passion for record keeping, and enlightening records on technology were also kept.

A pragmatic people, the Romans applied their energies to political and military systems and spent little time contemplating the secrets of nature. They imported their chemical knowledge from Greece and conquered territories, and they did not add much that was their own. They did however do an excellent job of cataloging this knowledge: The encyclopedias and compendiums they assembled were still recognized as authorities in Europe in the 1600s.

The authors of these volumes had the advantage of a large territorial range and varied populations from which to draw. For example Dioscorides (actually a native Greek) traveled as a surgeon with the armies of Nero and wrote a pharmacopoeia of about a thousand simple drugs, including opium and mandragora, used as surgical anesthetics, and such inorganic preparations as mercury, lead acetate, calcium hydroxide, and copper oxide. Equally representative of this genre is the 37-volume *Natural History*, which the author Pliny claimed to contain 20,000 important facts, extracted from 2000 volumes by a hundred authors. Pliny is credited with the origin of the word encyclopedia because of his stated attempt to bring together the scattered material belonging to the encyclical population or *enkyklios paideia*.

Pliny's efforts made him the de facto authority on scientific matters up to the time of the European Middle Ages.

Pliny

Pliny lived in the first century CE, around the time of the reign of Nero. Born in Gaul, Pliny served in the army on campaign in Germany, then studied law, and then devoted himself to scholarly study and writing. On the accession of Emperor Vespasian, with whom Pliny had served in Germany, Pliny went to Rome and assumed various official positions.

Pliny wrote many books; however his *Natural History* is the only one that has been preserved. In this work, Pliny recorded whatever information he could find and not having a mechanism to verify each fact, he recorded most of it unverified. A well-known example is his description of the unicorn and the phoenix, written just as sincerely as his description of lions and eagles. Some have seen this as a fault on his part—almost a laziness—but we think the circumstances of his death refute that: He succumbed to vapors caused by the Vesuvius eruption that destroyed Pompeii because he went ashore for firsthand observation.

Pliny was probably just guilty of a forgivable credulousness; he certainly saw enough in his life to allow him to believe anything was possible. Also given his dull encyclopedic task, he should be forgiven for preferring more colorful explanations. Pliny was after all a product of his time. Seneca, the Stoic philosopher who killed himself at Nero's command, also compiled an encyclopedia no more critical than Pliny's. Celsus (a name we will hear again) wrote a comprehensive treatise on medicine compiled from hearsay if not directly translated from the Greeks. And when we read in modern histories of how Hannibal used vinegar to crack rocks, this lessens our criticism of Pliny.

Pliny's reports sometimes reflect his travels. He talks about fumes from a silver mine in Spain that were dangerous, especially to dogs. (Carbon dioxide may have been the culprit. Although not usually thought of as poisonous, its density would cause it to accumulate near the ground where dogs would be trying to breathe, and it could kill by asphyxiation.) Pliny also reports that Gauls—indigenous northern

Europeans—dyed their hair red with soap. (The soap may have just taken dirt off a naturally red-headed people). And Pliny did strive to be comprehensive. He recorded processes involving metals, salts, sulfur, glass, mortar, soot, ash, and a large variety of chalks, earths, and stones. He describes the manufacture of charcoal; the enrichment of the soil with lime, ashes, and manure; the production of wines and vinegar; varieties of mineral waters; plants of medical or chemical interest; and types of marble, gems and precious stones. He discusses some simple chemical reactions, such as the preparation of lead and copper sulfate, the use of salt to form silver chloride, and a crude indicator paper in the form of papyrus strips soaked in an extract of oak galls that changed color when dipped in solutions of blue vitriol (copper sulfate) contaminated with iron.

Pliny's reports also show that almost all of the elements for alchemy could be found in the technology of Rome. They were acquainted with mercury and mercury–gold amalgams (used to recover gold from the ashes of clothing decorated with gold-thread), and they knew how to make fake gold by using bronze dyed with ox gall for stage crowns. But Pliny discusses tests to differentiate real from artificial gold, and there is no mysticism nor claims of manufactured authentic gold. Pliny's writings also show that the Romans had and used a variety of organic and inorganic dyes, but whereas the Alexandrians had taken the existence of dyes as evidence of transmutation, Pliny just reports them as a matter of fact and draws no further inference. He mentions the use of indigo, purple, white, orange, green, red, black, and various shades of these, and he says the colors were used in murals, on statues, ships, and the funeral pyres of gladiators.

When reporting medicines, however, Pliny's lack of discrimination becomes an impediment. Not only is every possible remedy reported (a headache is cured by touching an elephant's trunk to the head, and the cure is more effective if the animal sneezes), but every possible preparation seems to be recommended for the same disease, and each cure seems to have universal restorative powers. For instance *bitumen* (crude natural tar) is said to stop bleeding, heal wounds, drive away snakes, treat cataracts, leucoma, leprous spots, lichens, prurigo, gout, fever, and "straighten . . . out eyelashes which inconvenience the eyes."[1] Rubbed on with soda, bitumen soothes aching teeth; taken in wine it calms a chronic cough, relieves shortness of breath, and checks diarrhea. Bitumen and vinegar are said to dissolve accumulations of clotted blood and relieve lumbago and rheumatism. A poultice of bitumen with barley flour is used to draw together severed muscles. Burning bitumen detects epileptics, and fumigation with bitumen checks prolapse.

Some of these claims of course may have had some basis, such as the use of bitumen to stop bleeding or seal wounds. If taken with wine, the wine itself may have been the active agent. Bitumen said to "dispel . . . congestion of the womb . . . [and] hasten . . . menstruation"[2] may have referred to its use as an abortificant (effective because of general toxicity to the system). But in most medical matters Pliny was out of his league. An informed compilation of the medical arts was assembled only slightly later and then by a Greek: Galen.

Galen

Galen of Pergamos, ca. 150 CE, was the son of an architect, but he seems to have had no choice but to go into the medical trade. The city in which he was born had a shrine to a healing god, and many distinguished personalities from Rome visited the shrine for cures. The high priest of the shrine kept a troop of gladiators, which gave Galen ample opportunity to examine wounds and judge the effects of medical treatment. Galen's father financed his studies in Asia Minor, Corinth, and Alexandria, and on his return he became chief physician for the gladiators, which again increased his practical knowledge.

Shortly thereafter, like other ambitious Greeks, Galen traveled to Rome. He soon achieved a reputation for devising cures when others could not, and he was not modest about his success. He left when the plague came to Rome, though he claimed to be escaping the constant harangue of his envious enemies rather than the disease. After the plague Galen returned, and he was appointed physician to Commodus, the heir to the throne. This light duty gave him time to write—and write he did.

Galen wrote on his anatomical studies, which were based on primates because dissection of the human body was illegal at the time. The anatomy was close to human but with enough important exceptions to cause confusion for later generations of followers. Galen also wrote on his medical philosophy; he believed that good health required a balance between four humors: phlegm, black bile, yellow bile, and the blood. He studied the function of arteries and veins, and he came close to elucidating a theory of circulation. Galen had a powerful influence on medicine for the next 1400 years, and at the end of that period, we will hear of him again.

Galen also believed a physician should be a philosopher, and in his writings he commented on, and criticized, the impact of Judaism and Christianity on Roman life. His concern was understandable. The impact by his time was starting to be considerable.

RISING CHRISTIANITY

In the beginning followers of Jesus were just one Jewish sect among others and a small one at that. It was primarily through the work of Paul, a Hellenized Jew of the Diaspora, that this sect became Christianity, with its broad and elaborate theology. It was Paul who took Christianity to the Roman Empire.

By the 100s CE the Christian church had a hierarchy of officers—bishops, presbyters, and deacons—and in general these early church administrators were fairly hostile to pagan science and philosophy. Part of this was just Roman anti-intellectualism, but they also felt the need to defend their fledgling religion against the questioning methods associated with the doctrines of Plato and Aristotle. One such early Christian, Irenaeus (ca. 180 CE), commented that the heretics "strive to transfer to ... matters of faith that hairsplitting and subtle mode of handling questions which is, in fact, a copying of Aristotle."[3] Another, Tertullian (late 100s to early 200s), said, "Philosophers ... indulge in stupid curiosity on natural objects"[4] and asked the famous rhetorical question, "What has Athens to do with Jerusalem?"[5] We see the affect of these attitudes again in medieval Europe, where they serve to slow, though not stop, scientific inquiry. Eventually this new, more rigid religion gained momentum in the Roman Empire: The austerity of its structure offered stability to a culture attempting to shore its own crumbling form.

DECLINING ROME

Starting around 200 the Roman Empire began to disintegrate—a process referred to by historians as the Decline of Rome, and these same historians have been guessing the cause ever since. Some historians have credited the Decline to a general moral decay, but while there were excesses—institutionalized torture; chronic matri-, fratri-, and homicide; bloody games that passed for public entertainment; and sexual practices that can only be described as extremely creative—most societies of that time were far from demure. Other historians have pointed to imperialism, with its accompanying class strife and dependence on the military. As evidence of this they give the succession of 12 emperors who nearly all came to violent deaths—and almost all at the hands of the soldiers who put them on the throne. From our perspective of chemistry though one other theory is especially interesting. This is the theory that the emperors of the Decline suffered from lead poisoning.[6]

It is true that one of the symptoms of lead poisoning is irritability, and from the number and forms of capital punishments, it can readily be substantiated that these emperors were easily annoyed. It is also true that the Romans used lead in early water systems and drinking vessels, and they added lead to wine to suppress souring. (Souring of wine is caused by the buildup of vinegar—acetic acid. Lead forms a salt with the acetate ion.) But before accepting the theory that lead poisoning caused the Decline, some caution must be applied. At least some Romans recognized the hazards of lead and probably limited their intake. For instance the architect Vitruvius noticed the poor health and deep pallor of lead workers and recommended earthenware pipes for conducting water. So the Decline is probably still best explained in more prosaic terms: pressures for change from within and without and leaders inadequate to meet the challenge of change.

Whatever the cause, the empire split in two in the late 300s: The eastern half, the Byzantine Empire, survived more-or-less intact for the next thousand years, but the western empire dissolved into a confusion of loosely adhering wartime federations. In the 400s these were overrun by Attila the Hun, and Rome itself was sacked by Goths.

In the east Emperor Justinian in the 500s strove to protect what was left of the old Roman Empire by throwing up intellectual walls.

He closed the remaining learning centers in Athens, and he had all pagan icons destroyed. He demoted Jews in status, and he restricted their freedom to practice religion. He alienated the *Nestorians* and *Monophysites* (heretics who refused to accept the relationship of God the father to God the son described by the Justinian church), and he drove them from the empire. In this last act however, he actually promoted the preservation of some Hellenized culture. When these heretics fled and found refuge in Persia, they took their Greek learning with them. This is the trail that we take up next.

THE ARABS CA. 600–1000

In the biblical story of Abraham, an Egyptian slave, Hagar, bore Abraham a son. But when Sarah, his wife, had her own son, she demanded that Abraham send Hagar away. Hagar and the child wandered in the desert, but they managed to survive. According to the Islamic account, the son, Ishmael, founded a line of his own: the Arabs. Out of this line, ca. 570 CE, came Mohammad ibn Abdullah: Mohammad, the founder of Islam.

The practice of making raids on neighboring communities for food and booty was well established in Arabic culture (having arisen naturally in this region of meager resources), and followers of Islam excelled at raiding. Eventually Mohammad conquered and united the tribes in the area, and he converted all the pagan shrines to Islam. When Mohammad died his successors, the caliphs, continued this tradition, and for the followers of Islam—the Muslims—conquest became a way of life. The force was so astoundingly successful that between 640 and 720, Egypt, Persia, Syria, North Africa, and Spain all fell to Muslim invaders. Within a hundred years of Mohammad's death, the Islamic empire stretched from the Himalayas to the Pyrenees.

There are several myths associated with Islamic empire building—one being that it was entirely hostile; however Alexandria, for instance, surrendered almost without a fight, perhaps preferring the Muslims to their Byzantine rulers. Another misconception is that the war was waged to propagate a religious faith—not so, it was a war to secure booty. Islamic invaders did not actively promote conversion to

Figure 3.1. The influence of the ancient Islamic Empire is seen in many places in the modern world. The lamp of learning shown on the shield of Augusta College, Augusta, Georgia, U.S.A. has its orgins in Arabic culture.

Islam (although they did not discourage it) primarily because followers of Islam were not taxed and full-scale conversion reduced the tax base. Within the Islamic empire Jews and Christians were allowed full religious liberty, as were Zoroastrians, Buddhists, and Hindus.

The third and probably most prevalent misunderstanding is that Arabs had designs on Europe. But there is no evidence that Arabs wanted to incorporate Europe. The only Arab army venturing beyond Spain was there by invitation from a European ruler trying to settle a personal score. Arabs disliked the northern European climate, which was so much cooler than the Mediterranean, and they saw no particular revenue to be gained from a people struggling to feed themselves. They thought of Europeans as backward, and Arabs were rather put off by their hygienic practices—or lack thereof. So Spain was where the Arab conquest stopped. This toehold however turned out to be of tremendous importance to our chemical history: It opened Europe to the learning of the Arabs, the Greeks, and the East.

ca. 200 BCE–1000 CE: From Rome to Baghdad

ISLAMIC INTELLECTUALISM

These Muslim imperialists were a vibrant, fresh, and intellectually eager people. The Islamic religion requires individuals to understand the Koran for themselves; therefore the literacy rate was high. (In contrast the Roman Christian church relied on interpretation of the Bible by church leaders, which meant that only the clergy had to read.) This emphasis on literacy translated into an interest in all intellectual pursuits, including alchemy and chemical technology, and learning was acquired from native scholars, schools, and libraries. Some of the learning however was acquired by less comfortable means.

Greek Fire

Arabs first became aware of the incendiary weapon called *Greek fire* when it was used against them by the Byzantines. In fact it may have been the decisive factor that prevented the fall of Constantinople and kept the eastern Roman Empire alive. Sprayed from a pumplike device onto attacking ships, Greek fire was a viscous liquid that ignited on contact with water and burned fiercely. Perhaps invented by a Jewish architect, Callinicus of Heliopolis, the ingredients for Greek fire were kept a state secret, known only by the Byzantine emperor and the Callinicus family.

The precise composition is still unknown, but from its reported properties, some inferences can be made. It was probably some self-igniting mixture, such as quicklime in a petroleum base. Quicklime, a crude form of calcium oxide obtained by heating limestone or shells, generates a good deal of heat when combined with water. If a mixture of quicklime and petroleum is exposed to water, the heat can ignite the petroleum.

Supposedly sand, urine, and vinegar were the only effective means of extinguishing Greek fire. "Vinegar," again probably meant a strong salt solution that formed a crust after evaporating, extinguishing the flames by excluding oxygen. "Urine" probably also meant some sort of concentrated solution, such as old, evaporated collections of urine, which would contain considerable sediment. Fresh urine would not have worked because fresh urine is mostly water—and probably not many volunteers could be found for the application.

Such chemical weapons were certainly not new to the world. Besieged towns had thrown pots of burning sulfur, asphalt, and pitch on soldiers since at least 200 CE. Liquid petroleum or naphtha from oil wells, together with burning pitch and sulfur were used by Assyrians and Peloponesian Greeks. The ignition mechanism was not particularly new either. Pliny reported that quicklime mixed with petroleum or sulfur burst into flame spontaneously when wet with water. Such a mixture may have been used "magically" to light lamps in shrines from the 200s CE. The innovation of the architect of Constantinople may have been the siphonlike mechanism used to deliver Greek fire. As such it was a triumph of chemical engineering as much as of chemistry.

In any case Muslims quickly learned the trick. During the Crusades Europeans fighting Muslims in Syria and Egypt encountered Greek fire. In the end the invention may have been turned against its inventors and perhaps used in the sack of Constantinople in the 1200s. Greek fire disappeared from use in this particular form after the fall of Constantinople in 1453, but chemical weapons, incendiary and otherwise, are still with us, and they surface in this history again.

ARAB ALCHEMY

Despite their trial by Greek fire, the Arabs prevailed and began the work of assimilating the accumulated knowledge of their subject states. Baghdad became the leading intellectual center of Europe, Asia, and Africa, and learned people from all over were invited to teach in Arab courts. Among these were Hindu scholars, physicians, and scribes, and because India had some exchange with China (the Tantrists venerated 18 magician–alchemists, at least two of whom were Chinese), access to Indian knowledge meant some access to Chinese knowledge too. One particular piece of information transmitted to the Arabs and ultimately, as we will see, to the West was a formula for an explosive mixture that came to be known as gunpowder. This mixture of potassium nitrate (niter), sulfur, and carbon explodes because the solids react when ignited to form gases (carbon monoxide, nitrogen, and sulfur dioxide), which take up a lot more room than the beginning solids, and the expansion takes place very quickly.

Muslim rulers also patronized Alexandrian refugee scholars, and they had the works of Plato, Aristotle, Galen, pseudo-Democritus, Zosimos, and others translated into Arabic. In this way Arabs came into contact with the practice of alchemy and quickly made it their own. The main contribution of the Arabs to alchemy was to tone down the mystical and to take an approach more akin to the practical approach of the early Alexandrian alchemists. Perhaps the Arabs felt less compelled to invoke magic to attain results because they were as interested in the process as in the goal. Whatever the reason, the alchemy eventually inherited by Europe used methods that had come back down to earth.

Probably based on the Aristotelian qualities for the elements, Arabic alchemists proposed that all materials had natures—such as heat, coldness, and dryness—and the task of the alchemist was to prepare the pure natures, determine the proportion in which they entered into substances, and then recombine them in proper amounts to give the desired products. For instance certain organic materials when heated produce gases, inflammable materials, liquids, and ash. These were taken to correspond to air, fire, water, and earth—elements that must have comprised the original material. Each of these separate components was then distilled to isolate the pure nature of the element: heat, coldness, wetness, or dryness. "Coldness" was painstakingly isolated by evaporating water to dryness with tens to hundreds of distillations. The result, "a white and pure substance [that] when . . . touched by the smallest degree of moisture, dissolves and is again transformed into water,"[7] was undoubtedly residual salts that had been dissolved in the original impure water. (Such dissolved salts are responsible for water spots that form on dishes left to dry in air.) But these were taken by the alchemist as evidence of the validity of the theory.

Characteristics of Arabic alchemy are evident in the work of the following personages identified with it. Although the actual existence of some of these personalities as individuals is questionable, they were, as composites or otherwise, historically influential: The fairly clear written records of their work were the base on which European alchemy was to be built.

Jabir ibn Hayyan

Reminiscent of Aristotle, Jabir proposed that there were two exhalations: "earthy smoke" (small particles of earth on the way to becoming fire) and "watery vapor" (small particles of water on the way to becoming air). These, he believed, mingled to become the metals. But Jabir modified the Aristotelian approach by proposing that exhalations underwent intermediate transformations into sulfur and mercury before becoming metal. The reason for the existence of different kinds of metals, he believed, was that the sulfur and mercury were not always pure. He proposed that if the right proportions of sulfur and mercury with the right purity could be found, then gold would result.

As an Arab alchemist Jabir believed in the value of experimentation, but he could not completely avoid the mystical influences prevalent in his day. The *Jabir Corpus*, originally credited to Jabir, is written in a heavily mystical style, but the standard methods of crystallization, calcination, solution, sublimation, and reduction are clearly discussed, as well as such diverse processes as the preparation of steel and hair dye. Although Jabir probably produced some of the writings, all the works ascribed to him could not have done by one person, and these were probably really the collected works of a secret society called the Faithful Brethren or Brethren of Purity. In addition the work appears to have been completed around 1000, while the person identified as Jabir died in the 800s; different parts of the work are also written in different styles, as would occur with different authors. Some later Latin books were credited to Geber, a latinized form of Jabir, but they do not have an Arabic counterpart, and these were probably written after 1100.

Abu Bakr Mohammad ibn Zakariyya al-Razi (aka Rhazes)

A more corporeal personage, Abu Bakr Mohammad ibn Zakariyya al-Razi must have been of a wealthy family because he was able to study music, literature, philosophy, and magic as well as alchemy. He also studied medicine under a Jewish convert to Islam, and he wrote extensively on medicine, natural science, mathematics, astronomy, philosophy, logic, theology, and alchemy. He was said to be a man with a large square head, and when he taught he sat his own pupils in front, sat their pupils behind them, and other pupils behind them. If someone came with a question, it was directed to the back row first, then if unanswered, it was passed up the rows until it reached al-Razi.

Because of the extent of the Islamic Empire, Arabs knew and used many more naturally occurring chemical materials than did Alexandrian chemists. In his important alchemical work, the *Secret of Secrets,* al-Razi sets down this knowledge, classifying chemicals by origin—animal, vegetable, mineral, or derived from other chemicals—and dividing minerals into six classes (an extension of an earlier classification by Zosimos). There are *bodies*—metals (fusible substances that can be hammered); *spirits*—sulfur, arsenic, mercury, and ammonium chloride (substances that volatilize in fire); *stones*—marcasite, magnesia (substances that shatter on hammering); *vitriols* —sulfates (soluble compounds of metal with sulfur and oxygen); *boraces*—borax (a naturally occurring sodium–boron salt), natron (naturally occurring sodium carbonate), and plant ash; and *salts*—common salt (sodium chloride, table salt), potash (potassium carbonate from wood ash), and niter (potassium and sodium nitrate).

Notably systematic, al-Razi relied on observed and verifiable facts, and he almost entirely avoided mysticism. For instance the following recipe for *sharp waters,* a strong caustic solution, is very clear, and it could easily be followed in any general chemistry laboratory today.

> Take equal parts of calcined al-Qili [sodium carbonate] and unslaked lime [calcium oxide] and pour over them 4 times their amount of water and leave it for 3 days. Filter the mixture, and again add al-Qili and lime to the extent of one-fourth of the filtered solution. Do this 7 times. Pour it into half (the volume) of dissolved sal ammoniac [ammonium chloride]. Then keep it; for verily it is the strongest sharp water. It will dissolve Talq (mica) immediately.[8]

Of interest for his religious philosophy as well as his alchemical pursuits, al-Razi chose a rational source for morality, and he rejected divine intervention entirely. He found no value in traditional religious beliefs; he said that these beliefs were the sole cause of war. That he could hold this conviction and still function freely in Islam society shows the tolerance of the culture at that time.

Ali al-Husayn ibn Sina, aka Avicenna

Avicenna, to use the Westernized version of his name, was one of the most prolific and influential of the Arabic authors of his time. A Persian physician living around 1000 CE, Avicenna prepared his own medicines and investigated alchemical gold making too. The reasons for his range of accomplishments were threefold: his own natural energy, his early training, and his broad experience—though unhappily, the last was forced on him.

As a child he was said to have been precocious, memorizing long, involved works of literature. Exposed to philosophy and learning by his scholarly father, Avicenna studied Islamic law, medicine, and metaphysics. As a young man he earned a reputation as an able physician, and he gained access to the royal library of the ruling dynasty by successfully treating an ailing prince.

This ruling dynasty, though, was eventually deposed, and Avicenna became an exile. He finally found another court to employ him as physician and even twice to appoint him vizier. But this last proved unfortunate because the position subjected him to political intrigues. Avicenna occasionally had to go into hiding, and he was once imprisoned. Despite political turmoil around him however, Avicenna managed to continue his work.

He is said to have carried out his duties as physician and administrator during the day, then hold boisterous discussions with students through the night. Even in prison Avicenna continued to write. When his patron was deposed, Avicenna again suffered imprisonment, and he was finally forced to flee. He and a small group of followers eventually found another court where he completed his works. Unfortunately, as part of his duties at this new court Avicenna had to accompany his new patron onto the battlefield. Avicenna fell sick on one of these trips and died despite his own attempts at a cure.

During his lifetime Avicenna produced an impressive body of work. In some 200 medical treatises he wrote on the contagious nature of tuberculosis, described pleurisy and several varieties of nervous ailments, and he pointed out that disease can be spread through contamination of water and soil. As a chemist, Avicenna classified minerals into stones, fusible substances, sulfurs, and salts. As an alchemist he rejected the theory that metal could be treated with elixirs and made into gold. He believed if transmutation were possible at all, metal would have to be broken down into its constituents and reassembled.

His major work, the *Canon of Medicine,* is a systematic encyclopedia based on his reading of Greek physicians of the Roman imperial age, other Arabic works, and to a lesser extent his own clinical knowledge (his notes had been lost during his wanderings). The *Canon of Medicine* served as an authority for Arab and then European medicine for the next 500 years. When his authority, like Galen's, finally fell from glory, it was in Europe of the 1500s, and the event made chemical history again.

ISLAMIC DECLINE

Arabs in Islam found unity and inspiration, and on its strength they built an empire. The Islamic Empire accumulated Greek, Latin, Indian, and Chinese knowledge, imported alchemy and chemical technology from their tributary territories, and with their passion for all knowledge and science, Arabs broadened the base of these fields and made alchemy pragmatic again. Eventually though the empire began to succumb to forces reminiscent of those responsible for the Roman decline. There was a gradual breakup: In the 700s an Arab prince declared Spain independent of the caliphate, and later Egyptian Arabs broke away too. The empire was also harried by outsiders—Mongol and Hun—along with the vaguely annoying nip of another gnat: the Crusades. Though the Arab people continued in the next centuries to contribute to chemistry (in fact until modern times, for example with the 1990 Arab–American winner of the Nobel Prize for chemistry), the pennant for now was passed.

chapter
FOUR

ca. 1000–1200: Alchemy Translates from East to West

Though the Arabs started their march out of Mecca in the 600s, it took until the 1100s for Arab learning, including the alchemical tradition, to find its way to Europe. The primary problem was there was no structure to receive it. When a structure was found, it came from a reluctant benefactor: The Christian church, which eventually emerged as a stabilizing force in Europe and the monasteries became centers of learning. But church leaders encouraged only religious learning and distrusted secular education and philosophy, believing that such inquiries would erode their religion of unquestioning faith.

With an upsurge of urban development in the 1100s, the centers of learning shifted from monasteries to towns with newly founded universities. Though still treading lightly in the face of church opposition,

European academics at these universities began to study newly translated Arabic and Greek texts on mathematics, philosophy, astronomy, medicine—and alchemy. Through the work of these academics the information was eventually assimilated: Encyclopedists compiled the information in encyclopedias, Scholastics interpreted the knowledge in the context of revealed religion, and Empiricists tried the new knowledge out.

WESTERN EUROPE CA. 1000–1100 CE

The decline of the Roman Empire was followed by a period of disorder, fragmentation, and invasion from Celts, Goths, Visigoths, and Vandals. During this time the Christian church took over many of the functions of the old empire, including the administration of justice, record keeping, and most importantly the preservation of knowledge. At first the organization of the Christian church was simple, but under the influence of pagan religions, rituals began to grow, and a hierarchy of professional priests, patriarchs, and bishops became necessary to administer them. The bishop at Rome became especially important—the holder of this position eventually become pope of the Roman Catholic Church—though this authority was not recognized by the Byzantine Empire and initially not even by everyone in Western Europe.

Two events strengthened and established the Christian church as the organizing power in Europe: Charlemagne in the 800s and 200 years later, the Crusades. Charlemagne, a Frankish warrior, managed for a moment to subjugate the vast portion of Europe that now approximately includes Germany, France, Belgium, Netherlands, Switzerland, Austria, and northern Italy, and forcibly to convert the pagans within his domains to Christianity. In 800 the Pope crowned him emperor of the Romans—creating what was later known as the Holy Roman Empire—and set a precedent for papal involvement in secular affairs. But within a hundred years Charlemagne's empire had disintegrated into scattered groups that fought fairly continuously among themselves. Royal authority fell off as feudal lords grew in strength and number, and government became increasingly fragmented. By the 1000s Western European people were locked in a philosophically unproductive state. Feeling very much under siege they spent their time trying to survive. They worked from sunrise to sunset, and only the richest had other than earth floors and straw beds. Everyone, even the wealthy, had only coarse food—black or brown bread, salted meat, and fish, apt as not to be putrid; and they daily faced starvation and disease. Two centuries of battle had created a class of warring knights who now turned on each other. Compete chaos threatened until a solution was found. The knights were sent on Crusades.

THE CRUSADES

The Crusades brought to Europe something new: unbridled, institutionalized intolerance. Crusaders, heading to the Holy Land, slaughtered Jewish communities in France and Germany along their way. The Christians staying at home wanted to do their part, so they held pogroms in the crusaders' wake. By the 1100s Crusades were being launched against Europeans—against the heathen Slavs and Wends, and then north, forcing Prussian and eastern Baltic peoples to become Christian.

The new intolerance established anti-Semitism and anti-Paganism as a permanent part of European culture, radicalized an otherwise benign Islam, and had many repercussions throughout the history of chemistry. The Crusades however did add strength and unity to the Christian church, and when the revival in chemical learning came, the church was the first to stir.

MONASTICISM

Monasticism, originating in Egypt, was the basis for this revival. As an institution Christian monasticism underwent several reforms that rendered it independent of secular leaders and made it able to pursue an agenda of its own. To remain independent however, monasteries had to be more-or-less self-sufficient, so in addition to the traditional medieval monk bent over a manuscript, these monks were farmers, physicians, and artisans. As literate artisans they were able to record what had been an oral tradition of practical chemical technology inherited from the Roman Empire and introduced by invasions. One such monk was Theophilus, a Benedictine.

Theophilus

Theophilus was probably a pseudonym of Roger of Helmarshausen, a German metal worker who made a gilded and engraved portable altar that can still be found in the Franciscan monastery of Paderborn, Germany. In addition to being a practicing artisan, the scholar Theophilus produced a compendium of the known crafts of the early 1100s called *On Divers Arts*. In this three-book work he describes with notable clarity and detail oil painting, wall painting, dyeing, gilding, manuscript illumination, ivory carving, and glass and metal working, including the art of making stained glass. His recipe for making red mercury sulfide pigment is

> Take sulfur . . . break it up on a stone, and add to it two equal parts of mercury, weighed out on a scale. When you have mixed them carefully put them into a glass jar. Cover it all over with clay, block up the mouth so that no fumes can escape, and put it near the fire to dry. Then bury it in blazing coals and as soon as it begins to get hot you will hear a crashing inside, as the mercury unites with the blazing sulfur. When the noise stops immediately remove the jar, open it, and take out the pigment.[1]

Once recorded however these technologies were little modified or improved. The practical chemistry of Theophilus turned out to be the practical chemistry of the European Middle Ages. Chemical workers made dyes, soaps, and metals, but the methods were kept secret and handed down in unaltered, unimproved form. What would evolve in medieval Europe was alchemy, and in the 1100s this alchemy was just in the process of being discovered.

RECONQUISTA

The Arabs in Spain welcomed the new monastic Christian scholars in their libraries and schools, and contact with the Arabs fed a European revival. By the end of the 1000s some European nations felt strong enough to start pushing the Arabs back. They were by then aware of the treasures of texts available in the Arabic language, so when they reconquered, they were careful not to destroy these works. Arab Sicily was taken by the Normans in 1091, but Muslim physicians and other scientists stayed at the Norman court. In Spain after the reconquest of Toledo in 1085, a translation center was established, employing bilingual and trilingual Arabs, Christians, and Jews.

THE TRANSLATORS

Arab works on medicine, mathematics, astronomy, philosophy, and alchemy—with influences from India and China mingled in—were translated and preserved. Of the people who translated works pertinent to chemistry, the names of most non-European Arabs and Jews have been lost. Some names of European workers are, however, still known. For instance one Gerald of Cremona made translations of al-Razi, Aristotle, Euclid, and Galen and the *Canon* of Avicenna ca. 1150. Another two on whom we have some biographical information are Robert of Chester and Adelard of Bath.

Robert of Chester and Adelard of Bath

Probably one of the first translators, Robert of Chester (ca. 1150), was English and associated with the Christian church, perhaps as a cleric. He and his friend Hermann the Dalmatian were living in Spain and studying astrology when Peter the Venerable (a French abbot who argued for peaceful missionary Crusades) found them and asked them to translate the Koran. After finishing this work Robert translated *The Book of the Composition of Alchemy*.

Western Europe still thought of itself as the remains of the Roman Empire, and the language spoken there was called Latin, though it was evolving and merging into French, Italian, English, German. In his preface to the translation Robert states, "Since what Alchymia is ... your Latin world does not yet know, I will explain in the present book."[2]

Robert also translated the algebra of mathematician Al-Khwarizmi and introduced this part of mathematics to Europe. Part of the reason for Europe's eventual acceptance of Aristotle as an authority—and transmutation as a possibility—was the fact that it reached Europe at the same time as the mathematics of Greece and the Islamic Empire. Though the concepts of zero and negative numbers were probably Indian in origin,[3] Arabic mathematicians incorporated theses ideas into their mathematics along with the mathematical system of the Greeks, Egyptians, and Babylonians. To European scholars of the 1000s–1100s, algebra must have appeared as an oasis of reason, offering refreshing precision and clarity in a world of mysticism and confused collections of technologies.

Adelard of Bath was also English, and he also translated mathematical works, including the Arab version of the *Elements* of Euclid. He is said to have obtained a copy for translation in Spain while he traveled disguised as a Muslim student. If this story is true, then he was familiar enough with the Arabic language to pass himself off as a native speaker, which attests to his ability as translator. He must also have been an accomplished mathematician because he composed an abridged version of the *Elements* as well as an edition with a commentary. Adelard traveled in France, Italy, Syria, Palestine, and Spain before returning to England to become the tutor of the future Henry II. A prominent writer on scientific subjects, Adelard held that the new secular learning was not always compatible with traditional Christian thought, and he showed in this way that he retained some of the free-thinking spirit of his adopted Arab personae.

This influx of information whet the appetites of Europeans for more. Merchants, royalty, and popes all sent agents to Spain to learn Arabic and bring home manuscripts. The monastic storehouses began to fill—and just in time, for the Islamic world soon ceased to provide more.

DECLINE OF ISLAMIC INFLUENCE

In the 1200s the free spirit of inquiry characteristic of Muslim scholars suffered a fatal blow at the hands of a young Mongol warlord known as Genghis Khan. Bursting out of Asia, by 1227 he had become the greatest conqueror the world had ever known. His tactic was terror. Those who surrendered had only to pay tribute; those who did not were sacked, murdered, and destroyed. After he died his sons raided

Europe and Russia, and the third Great Khan turned on Islam. After destroying the Assassins en route, he moved on Baghdad. The city was stormed and sacked, and the last caliph was rolled up in a carpet and trampled to death by horses (there was a superstition against spilling his blood).

The Mongol invasion devastated major Arab cities and destroyed libraries, manuscripts, and schools. Arabs turned to the task of saving what they could rather than developing anything new. In 1260 with the defeat of a Mongol general, Mongol invincibility was shattered, and the reign of terror was brought to an end; however by this time the closing in of Arab intellectualism was complete. The Arabs, like societies before them and societies since, put up walls—mental as much as real. Their knowledge however had been preserved and by this time passed to the West.

CA. 1200: A SCHOLARLY AGE

Invasions of Europe continued into the 1200s, but by this time the former backwater of the Roman Empire had the strength to resist. Populations were still thin and feudalism still flourished, but free towns were growing and with them, centers for learning. Universities were founded at Naples, Paris, Oxford, Cambridge, Seville, and Siena.

Universities at this time generally followed one of two patterns: Italian, Spanish, and southern French universities were owned and operated by students themselves. They hired (and fired) teachers and decided on the size of their salaries. On the other hand, universities in northern Europe consisted of teaching guilds, with each faculty (arts, theology, law, medicine) headed by an elected dean. Originally a "college" was subsidized housing for poor students, but when it was realized that better discipline could be maintained when all students lived in colleges, colleges became centers of instruction as well as residences.

With the exception of the age of matriculation—which in the 1200s was 12–15 years old—students then were as they are now. There were reports of drunken confrontations between students and local toughs, and students in Paris were caught playing dice on the altar of Notre

Dame. The local populace often denounced universities as hot beds of heresy, paganism, and worldliness, and it was said that students "seek theology at Paris, law at Bologna, and medicine at Montpellier, but nowhere a life that is pleasing to God."[4]

Among all the celebration however, some teaching did occur. The teaching method was lecture, and students took notes on wax tablets for discussion later. The curriculum in the first universities included grammar, rhetoric, logic, or dialectic for a bachelor's degree and arithmetic, geometry, astronomy, and music for a master's degree—but not much history nor natural science. However monks of mendicant orders (such as Franciscans and Dominicans) taught at universities for a living, and through their teachings Arabic translations moved from monasteries to the medieval world.

At first the Christian church condemned the newly translated works of Aristotle. There were some specific points of contention, such as Aristotle's concept of the world as eternal and the idea that there were no rewards or punishments after death. But in addition the church was hostile to *rationalism*—the use of human reason rather than faith, to seek answers. Church leaders were particularity hostile to Aristotle's works on natural science, going so far as to forbid their teaching. But these works continued to be studied (the ban itself may have inspired their illicit study), and through the efforts of such apologists as Thomas Aquinas—who pointed out that reason and faith should not contradict each other if they come from the same divine source—natural science as a legitimate study was restored. By the mid-1200s it became a requirement for a Master of Arts degree.

Information also continued to come from contact with Arabia and Asia. The Muslim Empire was under assault from the Mongols, but the Mongols, like the Arabs before them, were more interested in the economic advantages of conquest than absolute intellectual rule. They tolerated local religions and customs—as long as they did not interfere with tax collection—and land trade between China and Europe became easier (it was even encouraged) during the Mongol era.

Venetian traders (Marco Polo was one) and Jewish and Islamic merchants (by 1163 there was a synagogue in China) carried out this trade. Technology, such as the Chinese use of gunpowder in bombs and rockets, found its way along the trade routes. India, active alchemically and commercially, produced the text *Sukraniti* in this period, which gives several of its own recipes for gunpowder. The Islamic alchemists were not as prolific as they had been, but contributions from them continued to trickle, through trade and translators, to the West.

Information exchanges however were not always congenial. In the opening years of the 1200s, crusaders, with mixed motives of financial opportunity, cultural jealousy, and religious zeal, attacked and sacked Constantinople, burning, looting, killing, raping, and trampling treasures and books underfoot. The Venetians, who had lately joined the crusaders, understood the value of books and managed to salvage some, but much was lost—along with the conduit of learning that Constantinople had provided between East and West.

But no matter how the new information found its way, new academics in the new institutions had plenty of it, so they did the historically logical thing: They began assembling *all* their information, new and old, into a compact and convenient form. They began writing encyclopedias.

THE ENCYCLOPEDISTS

These were not however encyclopedias as we think of them today: These encyclopedias rarely had distinctive headings or titles, making it difficult at times to distinguish one article from the next. There were no running heads, cross references, nor an alphabetical arrangement of subjects, and because they were assembled before the invention of printing, entries were handwritten. The encyclopedists themselves were church people and mostly men, though some encyclopedists were nuns, and sometimes the encyclopedias, painstakingly assembled, represented an entire life's work.

Bartholomew the Englishman and Vincent of Beauvais

One example of this special breed is Bartholomew the Englishman. Bartholomew appears to have been a Franciscan who taught at the University of Paris. His encyclopedia, *Liber de proprietatibus rerum*

(*Book of the Properties of Things*), consists of 19 volumes and used Greek, Jewish, and Arabic sources. Bartholomew reported on the Aristotelian theory of the elements along with the sulfur–mercury theory of the Arabs. He also reported that transmutation was possible— just very hard to achieve.

Vincent, a French Dominican priest and tutor to the two sons of Louis IX, labored 30 years to produce his encyclopedia, *Speculus majus* (*Great Mirror*). By *mirror*, he meant that the encyclopedia would show the world what it was and what it should become; by *great*, he must have meant the 10,000 chapters, in 80 volumes. His work covered history from creation to the time of Louis IX, summarizing all science and natural history known to the West at that time, quoting over 300 authors, and covering literature, law, politics, and economics. The *Speculus majus* contained chemical and alchemical information, but in a manner similar to Pliny, the information was unverified and unexplained. It remained for another group of scholars (collectively known as the Scholastics) to interpret and reconcile the new knowledge with religions of divine revelation, such as Judaism or Christianity. There were both Jewish and Christian Scholastics, though the Christian Scholastics were by far the more influential.

THE SCHOLASTICS

These commentators were guided by *Scholasticism*, a philosophy that taught that the best argument was backed by accepted authority. In Europe of the 1200s the highest authorities were the Bible, the Christian church leaders, and, after exposition by the Scholastics, Aristotle.

This passionate reverence for authority in the Scholastic approach encouraged a credulousness that influenced thinking for the next several centuries. Scholastics believed in astrology, magic, enchantment, necromancy, and whatever potency of animals, plants, gems, or stones they had read about in the works of the Arabs. Their readers, believing them, perpetuated these beliefs. Thus Scholasticism did much to disseminate information, but it may have impeded the development of systematic scientific inquiry. Of these new Scholastics perhaps the most influential for natural science was Albert the Great.

Figure 4.1. Albertus Magnus, the chemist's patron saint. The authors of this text hope Saint Albert smiles now. (Courtesy of the John F. Kennedy Library, California State University, Los Angeles.)

Albert the Great

Albertus Magnus, or Albert the Great, was the oldest son of a wealthy, noble German family and reportedly a very short man. Considering that the average height in this era was only around 5 feet, by modern standards he must have been very short indeed. In his twenties, he joined the Dominican order and was sent to a Dominican monastery at the University of Paris. There he came in contact with the newly translated Greek and Arabic works and became an adherent of Scholasticism. After a ban by Christian leaders on the teaching of Aristotle's work on natural philosophy, Dominicans realized there was something important afoot, and they asked Albert to explain in Latin the principle doctrines of Aristotle, so that they could read them with understanding. So inspired Albert began his work "to make intelligible to the Latins"[5] all branches of natural science, logic, rhetoric, mathematics, astronomy, ethics, economics, politics, and metaphysics.

He wrote commentaries on all the known works of Aristotle (both genuine and pseudo-Aristotle), paraphrasing the originals and adding digressions, observations, speculations, and "experiments" (by which Albert meant a process of observing, describing, and classifying). But in deference to the opinions of the Christian leaders, he said, "I expound, I do not endorse, Aristotle."[6]

Albert also reported, in the manner of the encyclopedists, practical as well as philosophical information, mentioning that mercury is a kind of poison that "kills lice and nits and other things that are produced from the filth in the pores."[7] (In the cooler climates of Europe the Eastern habit of regular bathing, though finding some popularity with relocated crusaders, had not yet taken hold completely.) Albert's *De Mirabilibus Mundi* (*On the Marvelous Things in the World*) reports superstitious hearsay: "if the wax and dirt from a dog's ears are rubbed on wicks of new cotton, and these . . . lighted, the heads of persons present will appear completely bald."[8] But it also reports technical information, such as a recipe for phosphorescent ink made from "bile of tortoise and luminous worms"[9] (the bile probably used to make an emulsion of the unfortunate glow worms). Though believing in the strength of authority, Albert was at least partially critical. When quoting dubious reports (such as those of ostriches eating iron), he was careful to add that he had not actually witnessed the event or "I was there and saw it happen."[10]

Albert believed, as some did not, that there was more to science than the science of Aristotle. He was interested in alchemy but skeptical about reported transmutations: "I myself have tested alchemical gold and found that after six or seven ignitions it was converted into powder."[11] In the end however he bowed to authority and reported alchemical gold and iron as real materials, but he said that they lacked some of the properties of the natural species (for example alchemical iron is not magnetic). Albert did however believe that the best information on materials came from alchemists, as opposed to that given by mathematicians or astrologers.

While working on his various commentaries, which took 20 years to complete, Albert became quite respected and enjoyed a reputation as an authority in his own right. He may have been quoted as often as the original Arabian philosophers and even Aristotle. Association with Albert was likewise a step toward respect and authority: Among Albert's disciples was the famous Thomas Aquinas.

Albert preferred to devote himself to study, teaching, and writing, but as so often happens in the academic world, then and now, it was assumed that because he excelled at research, he would be equally proficient at memo writing. Thus Albert was transferred to administration. This was the usual route for the educated in his time, and in Albert's case the assignments came from as high as the pope, so he complied. When he went on his official trips through the parts of Germany under his supervision, Albert, a sincerely religious person, went barefoot as a symbol of the humility of his order. Toward the end of his life however, Albert attained a position that allowed him more freedom in choosing his assignments, so he did return to teaching. He still traveled though on behest of the pope and preached to the nobility throughout Europe, urging them to support the Crusades.

There is a story that as a young monk, Albert was not especially bright, but the Virgin appeared to him and told him she would help him advance. She asked him to choose between theology and philosophy, and he chose philosophy. The Virgin, disappointed, granted his wish, but told him he would return to his former feeblemindedness before he died. He was reportedly senile from about 1278, and he died around 1280.[12]

THE EMPIRICISTS

Despite the almost complete reliance on authority however, the age also enjoyed a brief burst of experimentation. But the experiments—though groundbreaking for their day—were tinged with prior expectation and (speaking kindly) imperfect technique. These limitations can be see in the work of the empiricist Roger Bacon and in writings attributed to Ramon Lull.

Roger Bacon

Roger Bacon, also known as Doctor Mirabilis (Wonderful Teacher), was born around 1214 to a wealthy family, trained in the classics, geometry, arithmetic, music, and astronomy, then became an Oxford Franciscan. At Oxford, Bacon concentrated his research on mathematics, optics, and alchemy as well as Greek, Hebrew, and Arabic. In his day however a cleric was better off studying theology than science. Bonaventura, the general of the Franciscan order, expressed the sentiment succinctly by saying, "The tree of science cheats many of the tree of life, or exposes them to the severest pains of purgatory."[13] Bacon however continued on his course, and when he began teaching in Paris, he also began a life-long series of run-ins with the established church.

He was abruptly shunted back to Oxford, but he continued to spend time, energy, and money (supposedly his family's—a Franciscan should have no personal funds) on books, assistants, instruments, and the friendship of scholars. None of this was part of his job in the faculty of arts (when the sciences *were* studied, rational discussions were favored over experimentation), and it did not help relations with his order. But Bacon believed in the adage that no one can know that fire burns until a hand is put into the flame.

A few words of qualification are in order here. Bacon, like others of his age, believed some concepts to be self-evident, and not requiring examination. When he said that "nothing can be certainly known but by experience,"[14] he also meant the experience of faith, spiritual intuition, and divine inspiration. Bacon did however classify natural science into perspective (optics), astronomy, alchemy, agriculture, medicine, and experimental science (scientia experimentalis), being one of the first to consider experimentation as a distinctly separate pursuit. Bacon made systematic observations with lenses and mirrors, seriously studied the problem of flying with a machine with flapping wings, and conducted limited experiments with alchemy. He might have done more, had it not been for the scrutiny of his superiors.

In the mid-1200s as a result of his long-standing conflicts with Franciscan authorities, Bacon was exiled from Oxford, and as a result he was unable to continue his work. He felt (as he wrote) buried. He wrote vehement letters to the pope, trying to convince him of the place of science (and alchemy) in a university curriculum. Bacon professed the purpose of alchemy to be "to make things better . . . by art than by nature."[15] Anticipating the *iatrochemists* (medical alchemists) of the 1500s, he stated that alchemy "not only provides money and infinite other things for the State, but teaches . . . how to prolong human life as far as nature allows it to be prolonged."[16] To support his arguments, Bacon proposed to put together for the pope a grand compendium of studies in natural science, mathematics, language, perspective, and astrology. The pope however had the mistaken impression that the work was an accomplished fact and ordered Bacon to send him a copy.

For reasons not quite clear, the pope also asked Bacon to do this in secret (the pope may have thought some valuable knowledge would be included), posing a dilemma for Bacon: He was under papal orders to produce a work on the sciences—which meant going against the wishes of his superiors—but he was ordered to keep the work a secret, so he could not even tell them for whom he was doing it. Impressively, considering the impediments, Bacon managed to complete an *Opus majus* (Great Work), an *Opus minus* (Lesser Work), and an *Opus tertium* (Third Work). Ironically after such a heroic effort, the pope died before reading any of Bacon's work. For us however these works provide valuable insight into the chemical knowledge of the day. For instance in his *Opus majus,* Bacon describes gunpowder:

> that boyish trick which is performed in many parts of the world . . . [which] by the force of that salt called sal petrae [niter], such a horrible noise is produced in the rupture of . . . a little parchment . . . is felt to surpass the noise of violent thunder, and its light surpasses the greatest flashes of lightning. . . . But take 7 parts of saltpeter, 5 of young hazelwood [charcoal] and 5 of sulfur . . . and this mixture will explode if you know the trick.[17, 18]

Bacon is sometimes credited with having introduced gunpowder to the West, but if it was already used for children's firecrackers, it was already fairly well-known. Albert the Great also mentions it in his writings, and Albert probably obtained his information from the *Liber ignium ad comburendos hostes (Book of Fires for Burning Enemies)*, ascribed to Marcus Graecus. The *Liber ignium* was probably first compiled in the 700s, but it did not reach Europe until about Bacon's time. It describes among other things incendiary substances, phosphorescent substances, Greek fires, and other explosives containing niter (potassium nitrate). This work was probably written by a Jew or Spaniard (or Spanish Jew) in the 1100s or 1200s.

Although Bacon does not tell us the "trick" of gunpowder, other recipes, such as the purification of potassium nitrate with charcoal, are very clear and could easily be followed, contrasting sharply with alchemical writing in the next century.

> Carefully wash the natural saltpeter and remove all [visible] impurities. Dissolve it in water over a gentle fire, and boil it until the scum ceases to rise, and it is purified and clarified. Do this repeatedly until the solution is clear and bright. Let this water deposit the . . . [saltpeter] in pyramids, and dry them in a warm place. Take this stone and powder it and immerse in . . . water. . . . Dissolve over a gentle fire. . . . pour the hot solution upon the charcoal and our object will be achieved. If the solution is good, pour it out, stir with a pestle, collect all the crystals you can, and draw off the water.[19]

Having completed the three books, Bacon, still bravely at odds with his superiors, in 1271 wrote his *Compendium studii philosophiae* in which he lambasted the Church of Rome for corruption, pride, luxury, and avarice. But while Bacon was critical of authority, authority was critical of him. Not only did it look askance at his interest in science, some officials went as far as believing he dealt with evil spirits.

In the late 1200s Bacon was tried by the Franciscans and convicted of "suspect innovations" (novitates suspectas). It was not clear that he was actually imprisoned (which considering conditions in prisons of the time would have been very harsh treatment) or just put under house arrest. In any case nothing is reported of him again until almost the close of the 1200s, when his *Compendium studiae theologiae* appears. The exact date of Bacon's death is not known, but it was soon thereafter. He may have in the end been reconciled with the Franciscans, because he is buried at Greyfriars, the Franciscan church at Oxford.

The chemical historian J. R. Partington adequately summarized Bacon's personality when he stated that "Bacon had too good an opinion of his own undoubted genius . . . not clearly appreciating that . . . the 'Domini canes' were not chosen as the instrument of the Inquisition without foundation, and it fell to a Dominican to judge Joan of Arc."[20] But then Partington goes on to assert that this assurance was not misplaced:

> Both Albert and Roger were courageous men. Many things on which they wrote were highly suspect and regarded with much disfavor in the Church. Albert, more circumspect and calmer, overcame much of this prejudice; Bacon, rash and often violent, merely accentuated. it. After them no Churchman could or did neglect the new knowledge which they had revealed.[21]

Writings Attributed to Ramon Lull

Ramon Lull, the philosopher, was also a courageous and interesting personality, and though he probably did not try alchemy himself, his name was ascribed to several alchemical texts after his death. Born around 1230 in Catalonia (now northern Spain), Lull was the product of a romantic age. Reared in the royal court of Majorca, he was a poet, scholar, and writer in Latin, Catalan, and Arabic. He married and at first seemed content with his courtly life. Around the age of 30 however, inspired by visions of the crucifixion of Jesus, Lull turned his energies to mysticism and theology.

Lull developed a brand of theological philosophy in which he strove to relate all forms of knowledge and thereby demonstrate the godhead in the universe. He did this with complicated tables of theological propositions that he sought to interrelate. He also believed that the conversion of the Muslims could be achieved by logically refuting Islam, preferring informed missionary work over military force as a method for recovering of the Holy Land. Able to speak Arabic, Lull attempted to organize a school of Oriental languages so that missionaries could preach his ideas to the Muslims in their own tongue, but these visionary efforts met with limited success. His own proselytizing efforts met with limited success too: In Algiers, Lull was stoned to death by an unreceptive Islamic mob.

It was probably the mysticism of Lull (culminating in his famous mystical work, *The Book of the Lover and the Beloved*, still studied by philosophers today) and his complicated, cryptic logic that attracted alchemists to his writings and led them to adopt his name as their own (thus beginning the multiplication of his personalities). Alchemical works began to appear that were attributed to him but dated some years after his death. It is possible that these writings were published posthumously (alchemical works would have been an embarrassment to a practicing theologian), but the reported posthumous personal appearances are doubtful.

Whoever the author is, the works are listed here with the works of the empiricists because they contained systematic accounts of the theory and practice of alchemy, and these works are notably devoid of allegory or deliberate obscurity. Anachronistically systematic, the author(s) used the letters of the alphabet to symbolize alchemical principles, materials, and operations and arranged them in tables. Recipes were then given as combinations of these letters—including recipes for some interesting new reagents: mineral acids and an alcohol that would burn.

ALCOHOL

As we have mentioned, people had fermented alcoholic solutions in the form of beer and wine since prehistoric times. While in dilute solutions of beer and wine, alcohol will not burn, but ethanol, the type of alcohol found in beer and wine, is flammable in its pure state. The isolation of ethanol in a form pure enough to burn had to wait for improvements in the art of distillation, such as better glassware, the use of cooled coils for improved separation, and the discovery that some salts added to the distillation vessel pulled water from the alcohol–water mixture. The first people to try adding salts may have been following an Arabic alchemical notion that a dry essence should combine with a wet essence, and in this case, the alchemy worked.

Some salts absorbed enough water so that the resulting alcohol inflamed when lit.

The earliest European account of the preparation of alcohol is in a manuscript of the 1100s: *De Commixtione puri et fortissimi xkok cum III qbsuf tbmkt cocta in ejus negoii vasis fit aqua guae accensa flammam incumbustam servat meteriam.* The chemical historian Berthelot showed that by substituting the letters in the three words in cipher—*xkok, qbsuf,* and *tbmkt*—with the letters that proceed them in the alphabet,[22] the words could be decoded to *vini, parte,* and *salis,* so that the resulting passage read, "On mixing a pure and very strong wine with a third of a part of salt, and heating it in vessels suitable for the purpose, an inflammable water is obtained that burns away without consuming the material [on which it is poured]."[23] The cipher must have worked until the 1200s, because it took until then before the procedure was commonly know.

Once it became widely known alcohol was used as a medicine, appearing to be a virtually universal panacea. Externally applied it helped to heal wounds, dry sores, and remove dirt. Internally applied it alleviated pain and served as a mood elevator. By the next century a Catalonian monk, John of Rupescissa, referred to alcohol as aqua vitae, the water of life, and prescribed it as an elixir for sick metal—to make it gold—as well as for human health. Other medical practitioners used alcohol—an excellent solvent for organic materials—to extract oils from plants, and they investigated medicinal properties of extracts.

MINERAL ACIDS

The European discovery (or rediscovery as far as other parts of the world were concerned) of mineral acids was as exciting to the alchemist as the purification of alcohol to the physician. The common organic acids had been available—acetic, citric—but these are weak acids with limited dissolving powers. The mineral acids (sulfuric, nitric, and hydrochloric)—formed by heating certain salts and condensing the gas-phase product (in the presence of water vapor, which the alchemist had no way of excluding, had they known it existed)—were strong acids with much greater dissolving powers. Aqua regia, a mixture of nitric and hydrochloric acids, dissolved—lo and behold—even gold. This gave ideas to some Europeans who had been reading about a theory of transmutation and how, with the right ingredients and the right combination, one could cook up a fortune in a backroom vat. The stage was set for the European alchemist.

So although initially resisted by the established church, chemical information from Arabs and Greeks philosophy eventually found its way to Europe. Europeans learned about gunpowder, mineral acids, and probably alcohol (though they may have reinvented it for themselves based on information from Arabic texts). The information was then gathered into encyclopedias, and scholars contemplated and digested its meaning and interpreted it within the framework of their own schools of thought. There were also a few brave souls who experimented with some of the information: It was the Europeans' turn to give alchemy a try.

chapter FIVE

ca. 1300–1500: The Evolution of European Alchemy

Societies rarely develop in an orderly way. They grow by leaps and lags: boom, then famine; peace, then war; revolution, then reaction. By the 1300s Europeans had new material—alcohol and acids—and a new direction—alchemy—but European chemistry did not move ahead. Society, responding to the seemingly uncontrollable disasters of their age, cowered intellectually. From the promising beginnings of logical reasoning and exploration of the chemical arts, they backslid into superstition and fear.

Toward the end of the 1400s however, three inventions introduced from the East—gunpowder, block printing from movable type, and the compass—broke this lethargy and got things on the move again. Gunpowder made rubble out of the fortifications of feudalism;

movable type secularized learning; and the compass opened up new worlds. This revival climaxed in the first half of the 1500s with the religious Reformation in which Protestantism was established as a rival to the authority of the Catholic Church. The Reformation was such a radical break with tradition that it precipitated a period of mayhem, which was followed by a period of appreciation for law and order manifested in the growth of royal absolutism and sovereign states. Interestingly chemistry in the 1500s tracked this same sinusoidal curve: In the early part of the century, practical medicine repudiated the ancient authority of Galen and Avicenna in favor of new, radical alchemical cures; this was followed in the latter half of the century by a period of reorganization and reorder that constituted, for chemistry, a reformation in its own right.

EUROPE CA. 1300

By the 1300s European feudalism had declined; the number of towns had increased; and the newly centralized Catholic Church maintained an intellectual life of its own. Translated and compiled texts filled the monasteries; ideas from China and India filtered in through travel and trade. Everything indicated a Europe poised for progress, but for chemistry this was not to happen. For the next 200 years, European wheels spun. There was a flurry of activity, but the paths were in circles.

This was not the case outside of Europe: By 1300 Chinese and Indian alchemists were actively engaged in *iatrochemistry*—the application of alchemy to medicine—but iatrochemistry did not fully evolve in Europe until 200 years later. Chinese and Indian alchemical writers also devoted much thought to the proper design of a laboratory, another concept that did not appear in European alchemical literature until about 1500. For example an Indian treatise of this period, the *Rasaratnasamuchchaya*, contains the following description:

> The Laboratory is to be erected in a region, which abounds in medicinal herbs and wells . . . it is to be furnished with the various apparatus. The phallus of mercury [emblem of Siva, the creative principle] is to be placed in the east, furnaces to be arranged in the south-east, instruments in the south-west, washing operations in the

west, drying in the north-west. The koshti apparatus for the extraction of essences, the water vessels, a pair of bellows and various other instruments are also to be collected as also [are] the threshing and pounding mortars, the pestles, sieves of various degrees of fineness, earth for the crucibles, charcoal, dried cow-dung cakes, retorts made of glass, earth, iron and conch-shells, iron-pans, etc.[1]

A laboratory so equipped was clearly designed by someone with a knowledge of many chemical processes, such as reduction, distillation, extraction, and digestion (dissolving in acid or caustic solutions) and capable of performing some fairly intricate chemical investigations. Other sections of this work describe ideal students in the lab (though few would meet such stringent requirements today):

> The pupil[s] should be full of reverence for [their] teacher, well-behaved, truthful, hard-working, obedient, free from pride and conceit and strong in faith . . . used to life upon proper diet and regimen . . . well-versed in the knowledge of the drugs and plants and in the languages of many countries . . . (ibid.)

Requirements likewise were placed on the professor: "The instructor must be wise, experienced, well-versed in chemical processes, devoted to the Siva and his consort Parvati, sober and patient" (ibid).

It is impossible to tell if Eastern ideas on iatrochemistry and systematically designed laboratories gradually made their way to Europe or later arose spontaneously in European minds, but it is clear that these ideas appeared in India and China first and were not immediately assimilated by the Europeans. The question then arises, Why not? One answer might be found in the series of disasters—natural, political, and consequentially intellectual—that assailed Europe at this time.

DISASTERS OF THE 1300S

Starting in the 1300s a series of local famines took a heavy toll, but the real calamity struck around the 1340s: the bubonic plague, the Black Death. Within a few decades almost half the population of Europe died. Towns, especially vulnerable to the communicable disease, fell into ruin. The new centers of freedom and learning closed their doors. Progress ground to a halt.

Responding to madness with madness, political powers of the period engaged in almost incessant war. In a series of battles between England and France called the Hundred Years' War, bands of English soldiers ravaged the French countryside until the illiterate peasant girl, Joan of Arc, led a French revival. Though she was caught and burned, this just freed the English to turn to their own troubles, and in the late 1400s, suffer the internal War of the Roses.

Responding to what was believed to be the devil running amok, Europeans in this deeply religious and superstitious age became more religious and superstitious. Extreme forms of penitence, such as flagellation, increased in practice. The Spanish Inquisition was established to ferret out heretics and questionable Christians. The Jews of Spain were offered expulsion or conversion (thereby depleting Spain of many of the scholars who had been instrumental in its intellectual revival), but converts were suspect, too, and many died with the Inquisition's some 13,000 victims. Christian Europe consigned itself to an all-out war against witches, pagans, sorcerers, Jews, and anyone doing anything outside the established norm. Those accused of being witches had a particularly virulent holocaust unleashed on them: They were hunted, tortured, and executed—sometimes in mass campaigns. Many times these unfortunate women were midwifes and healers—repositories of medicinal and chemical knowledge—but it was believed that those with powers to heal could have learned these secrets only from the devil; therefore they must also have the powers to harm.[2] Alchemists had to take care that their powers were not interpreted as derived from the same source.

EUROPEAN ALCHEMY

Consequently European alchemists at this time were a cautious lot, and not given to great leaps of imagination. A number of alchemical manuscripts were produced, but they mainly repeated what had been said earlier, a number of experiments were tried, but they mainly repeated what had been done earlier. Alchemists cloaked their work in symbolism and mystery—presumably to hide "the great secret"—but the abstruseness may have served another purpose: It is difficult to accuse one of unorthodoxy if no one can discern what has actually been said.

Figure 5.1. A drawing of witches brewing by Botticelli. (Courtesy of Alinari/Art Resource, New York.)

Mercury for example was variously referred to as doorkeeper, our balm, our honey (*our* referring to the clique of alchemists), oil, urine, May-dew, mother egg, secret furnace, oven, true fire, venomous dragon, Theriac, ardent mine, green lion, bird of Hermes, and two-edged sword that guards the tree of life.[3] Birds flying to heaven and back symbolized sublimation and distillation; a devouring lion represented corrosive acid. The serpent or dragon represented matter in its imperfect state (traceable to the Alexandrian alchemy of Zosimos), and marriage or sexual union symbolized the alchemical process itself (traceable to the Alexandrian alchemy of Mary the Jew). Symbols from the pervasive Christian religion were also used to describe the alchemical process. The alchemical death and rebirth of metals paralleled the death and resurrection of Jesus; the trinity of God paralleled the trinity of salt, sulfur, and mercury, thought by some to be part of all metals.

A question that must come to mind at this point is this: If they had so valuable a secret that it had to be written in code, then why did alchemists write books—in code or otherwise—and risk someone deciphering the code and learning the secret? One possible answer may be the income gained from selling the books to other alchemists. Most alchemists were clerics (the group most likely to be literate) and impoverished (either by circumstance or vow to their church). And while they might claim to pursue alchemy for the glory of the Christian church, the Christian church did not support this rather questionable means to its ends. Equipment had to be purchased, and glassware was crude and easily broken. Many times artisans had to be paid extra to work in secret because alchemists did not have approval of superiors or the trust of their neighbors or because alchemists feared that someone might discover what they were working on and attempt to force from them a secret they did not have.

Whatever the motivation, the mystic aura of writings and the secrecy of practitioners promoted a general mistrust of the alchemical art. Another class of alchemist who did little to allay the perception were the inventive souls who shunned reliance on authority for defiance of authority—alchemists who rose above the mire of mysticism to see the true route to making gold—the alchemical swindler.

ca. 1300–1500: The Evolution of European Alchemy

Figure 5.2. An alchemist of the European Middle Ages tending his fire. (Courtesy of Alinari/Art Resource, New York, New York.

The Alchemical Swindler

In the 1300s and 1400s continuous wars and power plays created a great demand for gold, and the swindler indulged in a lively game: Find powerful nobles in need of funds; put on a good show and convince them to invest; collect all you can; then run for your life.

To gain the confidence of their patrons, these swindlers, with great pomp and mystery, mixed together strange and foul mixtures and produced gold by using caldrons with false bottoms, hollow stirring rods plugged with black wax, or chunks of minerals or charcoal containing small amounts of gold. In another trick they took a nail, half iron and half plated gold, and painted it black. When dipped in the appropriate liquid, the paint washed away, and the nail appeared to turn to gold. Coins made from an alloy of silver and gold (which was white in appearance) could be dipped in nitric acid, which would dissolve the surface silver, so that the coin appeared to turn into gold. Some of these coins can be seen today in museums. We have to admire the skill of these swindlers and ask who was the better chemist: the scholar or the con?

The game however was not without risks. An impatient patron might decide to imprison a swindler to extract the "secret" more directly—or more likely, having lost faith after a time, seize the culprits and hang them for fraud. Death was certainly the fate of many, including one female alchemist, Marie Ziglerin, burned at the stake in 1575. A certain German noble is said to have kept a special gallows, painted in gold, to be used only for alchemical swindlers. Given the distrust engendered by these swindlers—and the horror associated with the black magic practiced by some—secular and spiritual leaders issued decrees against alchemists and even the possession of alchemical paraphernalia. Dominican friars threatened to excommunicate alchemical clerics; Dante consigned his alchemical character to hell. If there had been any impetus for chemical experimentation before this time, it was certainly discouraged now.

USHERING IN A NEW AGE

Toward the end of the 1400s however, the spell began to break. Modern printing came into use around 1450, and in 1464 the Earl of Warwick used gunpowder and cannon to knock down Bamborough castle, changing warfare forever. Political change accompanied technical advance. Constantinople fell to the Ottoman Turks in 1453, and its scholars were sown throughout Europe. In the watershed year of 1492 Ferdinand and Isabella concluded the reconquest of Spain and financed the voyages of Columbus.

Philosophical changes accompanied the technical and political. In the 1300s a movement in opposition to Scholasticism, commonly known as the European Renaissance, began to stir, and by the 1400s it had wrought its change. For alchemy however this did not necessarily bode well. Petrarch, one of the motive forces for the Renaissance movement, said, "The[y] are fools who seek to understand the secrets of nature."[4] The alchemical attempt to make gold had degenerated into mysticism and magic; it had become riddled with artifice and fraud, and it was doomed to wither on the vine. However the alchemical study of material interactions still found an application once it was slightly reformed.

CA. 1500: CHEMISTRY REFORMS

The basis for reformation in chemistry like the religious Reformation undertaken by Martin Luther at the same point in history was the rejection of established authority. Such an action may not seem all that daring by today's standards, but we must remember that in the 1500s, authorities thus challenged had a millennia of precedent to authorize them—and dissidents in any arena were often subjected to a heavy hand (the Spanish Inquisition with all its associated horrors was in full force). It took courageous people to lead such rebellions and high ideals to drive them. The rejection of religious authority was so revolutionary that it led to civil uprisings, oppressions, and bloody religious wars. The reformation in chemistry was accompanied by much less bloodshed, but it, too, was a radical departure from the way things had been. The driving idea for the Chemical Reformation was that alchemy could be used for something other than the manufacture of gold; that is, it could be used to make medicines as well. This may seem mild and certainly not revolutionary until we recall that to propose this meant to reject the teachings of the time-honored medical authorities of the day: Galen, Hippocrates, and Avicenna. It was in fact such a radical stance that it required its own firebrand as forceful as Luther to promote it. This new application of alchemy to medicine, *iatrochemistry*, found its champion in Philippus Theophrastus Aureolus Bombastus von Hohenheim—aka Paracelsus.

Paracelsus

Paracelsus—a sobriquet coined by its bearer to denote his superiority to the ancient Greek medical authority Celsus—was born about 1490 (perhaps 7 years the junior of Luther), probably near Einsiedeln, in the

Figure 5.3. Paracelsus. (Courtesy of the John F. Kennedy Library, California State University, Los Angeles.)

canton Schwyz, a lead-mining region of Germany. At the age of 16 he entered the University of Basel, but he quit to study alchemy under Hans Trithemius, the abbot of Sponheim. Paracelsus did not find this study satisfying either, and he abandoned it for work in the mines in Tirol.

In the mines he learned the physical properties of minerals, ores, metals, and mineral waters, and he observed the accidents and diseases that were part of the lives of the miners. For the next 10 years, he roamed through Europe, studying in nearly every famous university. He may have traveled as far as Constantinople, Egypt, and Tartary, talking with—and learning from—Gypsies, conjurers, charlatans, sorcerers, midwives, bandits, convicts, and thieves.

ca. 1300–1500: The Evolution of European Alchemy

By the end of this time, Paracelsus had gathered quite a bit of information, much of which had to do with remedies and cures. Although it is debatable whether he actually earned any kind of a degree in medicine, he pronounced himself a physician and began to prescribe. The medicines he prescribed however were not the traditional herbal remedies described by Galen but a potpourri of his acquired folk remedies and new medicines that he himself invented, applying his alchemical talents.

Alchemy has always had transmutation for a goal, and for European alchemists this usually meant transformation from base metal into gold, although some alchemists, such as John of Rupescissa, adopted the Eastern-type goal of transmuting diseased flesh into healthy flesh with an alchemical elixir. Paracelsus on the other hand extended the definition of alchemy to any process in which naturally occurring substances were made into something new: "For the baker is an alchemist when he bakes bread, the vine-grower when he makes wine, the weaver when he makes cloth."[5] He even went so far as to give the name Archaeus to what he imagined to be an alchemist in the body that directed digestion. However the most important use of alchemy, he believed, was to prepare medicines to restore the chemical balance of a body disturbed by disease.

Accordingly Paracelsus went to work concocting chemical remedies. In what may have been the first generalized series of chemical reactions, he subjected a large number of metals to a standardized set of procedures and obtained a series of salts (solutions of which he called *oils*) for use as medicines. Paracelsus may also have been the first European to use tincture of opium (an alcohol extract of opium), which he named laudanum, to treat disease. The action of the opium may have been more an analgesic than an actual cure, and how much self-prescription (if any) he indulged in, we do not know, but we do know that his writings were not of enduring importance, partially because of a strange, rambling, and confused style. Paracelsus also used the relatively new (since Galen) medicine, distilled ethanol, and he is said to have been the first to use the word alcohol to describe this distilled essence of wine. Originally a name for an Eastern eye makeup, *al-kuhl*, or *al-kohol*, had come to mean any very finely divided powder, then "the best or finest part" of a substance.[6] That Paracelsus thought alcohol was the finest part of wine is evident by the consistent reports that he drank a great deal and often retired for the evening on a tavern floor.

But whatever cures, real or apparent, effected from salts and opium, Paracelsus' greatest triumph was the use of mercury to treat syphilis, the new disease of the day.[7] While syphilis today is characterized as a slowly developing disease causing genital sores and eventually leading to more serious symptoms if untreated, in Europe of 1495 it was described as causing pustules to cover the body from head to foot, skin to peel from faces, and to result in death within a few months. By the mid-1500s syphilis appears to have evolved into a disease closer to that known today. (It has been suggested that the microbe developed a less virulent form so that the victim would stay alive longer, ensuring the spread of microbes). But it was still a dreaded disease, and orthodox medicine, coming from Hippocrates, Avicenna, and Galen and relying on herbal remedies, had no effect on it. Topical applications of mercury however did. In fact until the 1900s a better treatment for syphilis could not be found.

Paracelsus may have heard of the treatment in his travels (Bhava Mista at this same time prescribed mercury for the syphilis brought into India by the Portuguese),[8] or the discovery may have been serendipitous, based on Paracelsus' adoption of the extension of the mercury–sulfur theory of the Islamic alchemists to a *tria prima* consisting of mercury (soul), sulfur (spirit), and salt (body). But while Paracelsus was on this one occasion very successful, there is no record of the number of people he adversely affected while experimenting with potions that were *not* effective, and it may have been considerable. He did however have a talent for observation; for instance he described the relationship between cretinism in children and the existence of goiters in their parents. His greatest contribution to medicine may have been the idea that doctors should act on what they observe rather than blindly following accepted authority.

His successes gave him the confidence (which in truth he never lacked) to criticize physicians of the time and to point out their ignorance and greed. However of the iatrochemists (that is, *spagyric* physicians), Paracelsus said:

> I praise the . . . spagyric physicians, for they do not consort with loafers or go about gorgeous in satins, silks and velvets . . . but they tend their work at the fire patiently day and night. They do not go promenading, but seek their recreation in the laboratory, wear plain leathern dress and aprons of hide upon which to wipe their hands, thrust their fingers amongst the coals, into dirt and rubbish and not into golden rings. They are sooty and dirty like the smiths and charcoal-burners, and hence make little show . . . do not highly praise their own remedies, for they well know that the work must praise the master, not the master [the] work. . . . Therefore they let such things alone and busy themselves with working with their fires and learning the steps of alchemy.[9]

Then around 1525 when Paracelsus was in his late thirties, he had the good fortune to be called to Basel to consult on a serious leg infection being suffered by Johannes Froben, a famous and influential humanist and publisher. Although amputation was being considered, Paracelsus advised less drastic treatment, and Froben survived, crediting Paracelsus with the cure. This and the medical advice he gave to another famous and influential humanist, Erasmus, who happened to be staying at the Froben house at the time, won Paracelsus the position of town physician in Basel, and shortly afterward he began giving lectures on medicine at the local university.

The lectures broke completely from tradition: They were in German, not Latin, (we note that Luther also took the radical step of translating the Latin Bible into German), and the lectures contained more practical information than theory. Not only did Paracelsus dismiss the works of Galen and Avicenna, he is said to have tossed a copy of Avicenna's *Canon* into a student bonfire and expressed the hope that the author was in like circumstances. (Again we note the similarity to Luther. Luther, threatened with excommunication by an edict from the pope—a papal bull—solemnly and publicly burned the bull.)

Paracelsus' reputation for cures grew as did his practice but unfortunately so did the ranks of his enemies. These enemies included physicians who maintained that he had no degree and therefore no qualifications, and pharmacists who felt—because he preferred to mix his own drugs—that they were being deprived of rightful income. When his patron Froben died 2 years after Paracelsus' cure (it is not known if Froben's leg was again involved), his enemies began to gain strength.

This growing hostility came to a head when Canon Cornelius von Lichtentels apparently promised Paracelsus the exorbitant fee he demanded when the canon experienced agonizing abdominal pain but when cured by a few of Paracelsus' opium pellets, refused to pay. The courts sided with the canon, and Paracelsus, never shy, voiced his opinion of the courts. Afterward his friends convinced him it was prudent to leave Basel at once.

Cast out from Basel, Paracelsus began new wanderings, which ended when he was in his fifties and the Archbishop Ernst invited him to settle in Salzburg under the archbishop's protection. A few months later however, Paracelsus died, and like many of the details of his life, the cause of his death is uncertain. Some say he was thrown from a height by his enemies, others say he died in a drunken debauch. Some say he died quietly in an almshouse, which may be the most difficult report to believe.

Though there were other chemical authorities who, like Paracelsus, believed that there was a need to combine practical chemical knowledge with theoretical research and that the direction of research should be pragmatic, they did not rush to be associated with him, primarily because of his rash outspokenness and the excesses proposed by his followers. These included using mercuric sulfide to treat epilepsy; zinc sulfate to treat nearsightedness; lead sulfide for diseases of the spleen; iron sulfide to cure diabetes; and mercuric oxide to treat all manner of malaise, even though the effects of mercury poisoning—loosening of teeth, palsy, nervous disorders, and death—were well known at the time. These other chemical authorities took a more judicious tack: They compiled books. This effort was encouraged by an increase in support for education coupled with the 1450 invention of movable type. As a result chemical knowledge could be standardized, it was no longer subject to copyist errors, and it was available in much greater quantities and at a much lower price.

THE NEW CHEMICAL AUTHORITIES
Biringuccio and Agricola

For example Vannoccio Biringuccio, an Italian, wrote *De la pirotechnia* (*Concerning Pyrotechnics*), published in 1540, which details assaying and smelting major metal ores—gold, silver, copper, tin, iron, lead, and mercury—and alloys, casting, bell making, explosives, fireworks, and some alchemy. Often cited as the first printed book on metallurgy and metallurgical chemistry, this work became a standard reference for metalworkers in the 1600s.

Georgius Agricola wrote on the methods of geology and mineralogy. Though he was acquainted with Biringuccio's work and copied some from it, Agricola emphasized individual experimentation and observation. Used as a reference for well over a century, his works included clear instructions and descriptions that were of considerable assistance to the infant chemical industries of the times. *De re metallica* (*On Metals*), written when Agricola was 61, consists of 12 books (a book then was about the size of what would now be considered a chapter) on mining, metallurgy, and geology, and it is superbly illustrated with woodcuts. Agricola discussed the geology of ore bodies, surveying, mine construction, pumping, ventilation, and water power. He described assaying, enriching ores before smelting, and procedures for smelting and refining. He also discussed the production of glass and of a variety of chemicals used in smelting operations. The work served as a textbook and guide for miners and metallurgists for the next 200 years, and it retained such an interest that in 1912, mining engineer H. C. Hoover and L. H. Hoover (later better known as President and Mrs. Herbert Hoover) provided an English translation.

Agricola included some sections on chemical theory, but they are pragmatic and skeptical of transmutation and showed an unfavorable reaction to Paracelsus. An even more vehement critic of Paracelsus' was Andreas Libau, or Libavius in the Latin form.

Libavius

Libavius was born in Germany around 1560. The son of a weaver, he attended school at the Hale Gymnasium, then at 18, he attended the University of Wittenberg (Hamlet's university). This achievement attests to his tenacity and talents because, at this time, sons of the working class rarely attended universities. (There were some universities for women, but these were rarer still.) Libavius was also riding on the post-Reformation crest: Now that religion had become more a debatable matter than a fact of life, there was a need for serious, knowledgeable, and effective religious leaders in both Protestant and Catholic countries. This gave education a boost that occasionally found its way down to the working class. The main body of Libavius' opus, *Alchymia*—considered the first chemistry textbook—is a clear and highly systematic survey of contemporary descriptive chemistry. The main work and supplements of *Alchymia* total over 2000 pages, and the book contains 200 illustrations. The main work is in four parts: "Eacheria" (which is concerned with techniques and equipment, including furnaces, sublimatories, distillation apparatus, crucibles, mortars, and vials); "Chymia" (which contains details of chemical preparations); "Ars Probandi" (which contains methods of chemical analysis); and a section on the theory of transmutation.

In "Chymia," Libavius gave clear directions for preparing aqua regia, sulfuric acid, and what may have been the first directions for making hydrochloric acid by heating saltwater in the presence of clay. Libavius may have also been the first to show that sulfuric acid can be made by burning sulfur with niter (potassium nitrate), and he proved that the acid so obtained was identical to that prepared by distilling green vitriol (hydrated ferric sulfate) or alum (hydrated potassium aluminum sulfate). He also described the synthesis of stannic chloride (which he prepared by heating tin with mercuric chloride) and the blue color given by ammonia with copper salts—both possibly for the first time. "Ars Probandi" was divided into two parts: *scevasia* and *ergastia*. The *scevasia* included information on using balances, on alchemical symbols, and methods for preparing crucibles, fluxes, and acids. The *ergastia* gave information on assaying techniques for metals, minerals, and mineral waters.

Libavius also included in *Alchymia* a design for a chemical laboratory. In addition to the main laboratory, his ideal "chemical house" contained a chemical storeroom, a preparation room, an assistant's room, crystallization and freezing room, a sand and water bath room, a fuel room, and a wine cellar. He did not however include a balance room: Precise measurements were not yet used, and chemistry was still not a quantitative science.

So we see that in the 1300s and 1400s, the chemical arts advanced little (with the possible exception of the art of the alchemical swindler). The technical advances of chemistry in the 1500s also seem pedestrian, but this period was anything but a pause. During this Age of Reformation chemistry underwent its own reform. The goal of chemistry was realigned and redefined. Mining, medicine, and alchemy were intertwined. The ancient authorities were tried and found wanting, and the task of establishing new authorities was taken on. A new emphasis was placed on chemical preparations for medicine, which had a stimulating effect on chemical research.

In many ways the Chemical Reformation set the stage for what we will soon refer to as the Chemical Revolution. But with the reformation also came reaction, and there were some hesitant steps backward—then forward—as we see in the 1600s.

chapter
SIX

ca. 1600: Philosophers of Fire

Europe in the 1600s was in an ambiguous age. The first part of the century saw the English Civil War in which Charles I was beheaded, but the latter part of the century saw the English Bloodless Revolution in which James II was politely dismissed and replaced with the monarch of parliament's choice. On the continent the Thirty Years' War marked the last major European war of religion, but the first of the pan-European nationalistic struggles for power. Feudalism for the most part had ended, but the fencing in of feudal estates caused more homelessness than liberation. World exploration and colonization resulted in an influx of new products and information, but these also reestablished the archaic practice of slavery. Europeans adopted sugar, tobacco, and coffee habits, but modern hygiene habits were still a long way off: Samuel Pepys gives an implicit picture of the odors of seventeenth-century London when he reports that his neighbor's sewage was dumping into his basement, but he did not know it until he stepped in it.

For the macroscopic sciences of physics and astronomy, it was the age of Scientific Revolution; but for chemists dealing with the contortions of unseen molecules, fundamental principles remained obscure. There was a feeling that the underlying theory for their investigations—that matter was composed of some number of elements and that these elements were present in all matter—was not providing answers, but chemists did not have the information needed for a new definition. By the end of the 1600s however, they would be closing in.

THE SCIENTIFIC REVOLUTION

The premier spokesperson for the philosophy that spurred the Scientific Revolution—René Descartes—was a soldier of fortune. French-born René Descartes joined the army in his early twenties and fought in the early stages of the Thirty Years' War. In the midst of the realities of this war, he developed his relentlessly rigorous philosophy that allowed only one basic premise—cogito ergo sum: I think therefore I am. Descartes and the English philosopher Francis Bacon proposed that truth could only be arrived at by careful, stepwise analysis that included a review at each step for oversight and accepted nothing as true unless clearly proven to be such. We find the origin of our modern scientific methods in these two philosophers. As elegantly stated by Bacon, "We are not to imagine or suppose, but to discover, what Nature does or may be made to do."[1]

To our modern way of thinking, this approach appears to be nothing more than sound experimental method, but in an age when divine revelation could be taken as proof, it was revolutionary. Descartes, Bacon, and their contemporaries—Spinoza (an outcast Dutch Jew and expert lens maker), Hobbes (a timorous English tutor), and Leibnitz (a dexterous German mathematician, philosopher, historian, and scientist)—completed the overthrow of Scholasticism by rejecting all reliance on authority, no matter how venerable. Buoyed by discoveries of Newton and Galileo, they viewed the universe as a machine governed by fixed, fathomable laws rather than divine discretion.

But Newton's laws of motion apply to macroscopic objects—cannon balls and moons—and chemists could not yet use these laws to describe attractions between atoms (if chemists had even known for certain that atoms existed). When laws governing these attractions were finally understood in the 1900s, the solution would have surprised even Newton. However the Scientific Revolution provided chemists with encouragement—it showed the possibilities.

In this wavering age however, the Scientific Revolution was not without opposition. The Catholic Church had finally incorporated the views of Aristotle into Christian dogma, and now any statement *counter* to Aristotle could be considered heretical. Galileo[2] observed the heavens through the newly discovered telescope and obtained evidence for the sun-centered planetary system proposed by Copernicus and Kepler. (Kepler used observations by Tycho Brahe and his sister Sophia—a student of chemistry as well as astronomy). However hearing the Inquisition's objections—and shown the Inquisition's instruments of torture—Galileo recanted. He probably reasoned—and rightly so—that suffering on his part would not alter the orbits of the planets and that he would be vindicated in the end. And in the end he was: The Vatican vindicated him in the early 1990s.

The eon straddling of the age was shared by chemistry. Side by side, the Rosicrucian society of mystical alchemists thrived with the French Academy of Sciences and the English Royal Society, both formed for discussion and dissemination of scientific discovery. The modern word *chemistry* came into use in the very early 1600s, but as a collective term for alchemy and iatrochemistry, and only much later in the century did it describe an independent field of study. If chemistry were practiced more often to make medicine than gold, this was only because gold and silver from the Americas made gold making less worth the effort. There were though those who would still try it: in 1603 the Scottish alchemist Alexander Seton was imprisoned and tortured in an attempt to extract the secret of his well-publicized transmutations. At the other end of the century, no less than Charles II of England undertook transmutations. Mercury fumes from his various researches may have contributed to his final illness. Even Newton, whose concept of gravity was a triumph of the mechanical view of the universe, spent considerable effort attempting to decode alchemical books.

But also during this century phosphorus was discovered by Hennig Brand; iatrochemicals were manufactured on an industrial scale by Johann Glauber; and Nicholas Lemery wrote *Cours de chymie*, a clear, concise chemistry textbook, and he made a living from chemical lectures and sales of his text. We choose Johannes van Helmont as our 1600s chemical exemplar because he also exemplifies the age: He is at times impressively progressive, at other times depressingly archaic.

Johannes van Helmont

Johannes (Joan) Baptista van Helmont, born in Brussels in the late 1500s to a family of landed gentry, had difficulty finding an intellectual home. Initially he studied arts, but believing academic degrees a vanity, he took no degree in this course of study. He studied under mystics as well as Jesuits, and he studied the classics as well as contemporary authors. He studied medicine, but after a time gave away his books to other students, saying later he should have burned them.

Figure 6.1. Johannes van Helmont and son, Franciscus Mercurius. It is a devoted father and chemist who names his son for his science. (Courtesy of the John F. Kennedy Library, California State University, Los Angeles.)

In the manner of Paracelsus, Helmont then turned to travel as a means of acquiring medical knowledge.

By the turn of the century his idealism must have given way to some practicality, because he had acquired a medical degree and begun practicing medicine. He had some success as a physician, and he was apparently able to offer some relief during a 1605 epidemic of the plague. He must have found medicine in the trenches disturbing because he soon declared "[I] refuse to live on the misery of my fellow men" or "to accumulate riches and endanger my soul,"[3] and he turned instead to private research. By Helmont's time the word chemist had come to mean someone who prepared medicines, extracts, and salts, and the word alchemist, now almost synonymous with swindle, was starting to fall out of use. Helmont, lacking a name to describe a person doing chemical research, called himself a *philosophus per ignem*, a philosopher of fire.[4]

He was affluent enough to have retired when he became (through marriage) a manorial lord. "God has given me a pious and noble wife. I retired with her . . . and for seven years I dedicated myself to pyrotechny [chemistry] and to the relief of the poor."[5] He and his wife had an unspecified number of daughters and one son, Franciscus Mercurius, a name that celebrated his father's devotion to the chemical arts.

When considering Helmont's medical theory, we must remember that in the 1600s medicine was in as much a state of flux as was chemistry. Medicine had long been a matter of tradition and superstition. And though Paracelsus had demonstrated that new remedies might occasionally be found in chemical preparations (at least mercury worked against syphilis), medicine was far from using strict empirical methods. So while Helmont progressively questioned the therapeutic effects of sweating and bleeding, he used the specific gravity of urine as a diagnostic tool, and came close to identifying stomach acid as hydrochloric acid, he regressively put faith in such foul remedies as worms from the eyes of toads.

Helmont was living in Belgium, a Spanish possession, at the time of the Spanish Inquisition, and when an article written by Helmont was published (probably without his permission) defending a rather bizarre cure for wounds, the Inquisition inquired. In the treatment the wound itself was only cleaned and bound, while the weapon that caused it was taken away and treated with medicinal ointments and salves. Ironically the technique may have had a higher cure rate than conventional treatments that applied noxious chemicals or dirty herbal preparations to the wound. The Inquisitors however were not amused by such magical methods when practiced by the laity.

Helmont was condemned for heresy, arrogance, and association with Lutheran and Calvinist groups. Helmont prudently acknowledged his error and revoked his "scandalous pronouncements,"[6] but he was still arrested, and after several interrogations, he was placed under house arrest. The house arrest was lifted after 2 years, but church proceedings against him were not formally ended until 8 years later, 2 years before his death.

Helmont's chemical theory was also a curious blend of the archaic and the advanced. Many of his day accepted the dissolution of metals in acid as evidence of transmutation, and transmutation was also used to explain the reaction (called *displacement* in modern terminology) wherein an iron horseshoe left in a stream naturally rich in copper salts eventually becomes coated with a copper layer. Helmont was able to discern the difference and denied that either of these reactions was transmutation, but this did not stop him from believing in the possibility. He gave an account of what he sincerely believed to be a transmutation of 8 ounces of mercury into gold achieved by a quarter of a grain of a yellow powder given to him by a stranger.

It is hard to know what actually happened. Mercury is well known for its combining powers, and it very well may have incorporated some yellow colored material and become a yellow-colored solid (this was how false gold was made by artisans), but it is difficult to conceive of a material that could do so in such small amounts. It may have been that the scale Helmont used was wrong, or maybe the person who gave him the powder also gave him a vessel to use with it, and something in the vessel entered into the reaction. The charlatan was still active at this time, and many were capable of pulling off such a deception.

So Helmont continued to vacillate from revolutionary to reactionary throughout his life. He rejected the Aristotelian four elements (at great personal risk, it should be noted), but then replaced them with water and air, pointing to the biblical story of the creation of the heavens and water on the first and second days. On the other hand he made extensive use of the balance, and as a result of his careful observation, he was convinced that nothing is created or destroyed in a chemical reaction.

In his famous willow tree experiment (actually suggested 200 years before), he weighed a willow seedling, then planted it in a tub with 200 pounds of soil. After watering it for 5 years, he removed the tree and reweighed it and the soil. He found that the tree had increased in weight but the soil weighed the same. He therefore concluded that the tree had converted the water into wood. The experiment of course is sorely lacking in controls: The water should have also been weighed, but that would have been difficult because of spillage, leakage, and evaporation. Also the plant's intake of carbon dioxide and output of oxygen were ignored because Helmont had no idea this was occurring. But the basic idea was there: Matter had to be accounted for in the balance. Whatever was put into a reaction should come out in one form or another.

Such ideas were not unique to Helmont. His German contemporary, Angelus Sala, dissolved a weighed amount of copper in sulfuric acid, chemically recovered the metallic copper, and found that it weighed the same as the copper he began with. Sala was also able to show that he could artificially produce a hydrated copper sulfate identical to a naturally occurring substance, a revolutionary thought for the time. But Sala did not have the prestige and influence of Helmont, so his work—though remarkable for the time—did not have the impact that Helmont's had.

Not all of Helmont's impact however had to do with his prestige. He made important contributions, such as his description of a whole new class of substances: gas. Helmont coined the word *gas*, probably based on *chaos*, and though most of the gas he produced was carbon dioxide (obtained by burning charcoal, fermenting grapes, or the action of acids on carbonate salts), he also obtained impure samples of nitrogen oxides (from the action of nitric acid on metals); sulfur dioxide (from burning sulfur); chlorine and nitrosyl chloride (from the reaction of nitric acid with ammonium chloride); and a mixture of hydrogen, methane, and carbon monoxide (from the dry distillation of organic material and from intestinal gas, which he also knew to be flammable). He called this gas phase *spiritus sylvestris* (wild spirit) because he believed it could not be restrained to a vessel or reduced to a visible state. When chemists began to tame this wild spirit—gas—

and apply their results to an unexplained problem—combustion—much needed new theories slowly emerged.

CA. 1600: CHAOS, COMBUSTION, AND THE SKEPTICAL CHEMIST

Though recognized as a subject worthy of study, gases could not immediately be attacked experimentally because of technical problems. The first of these was containment. Helmont did not always appreciate the volume of gas that would be released in his reactions, so he routinely burst the crude and delicate glassware of the day. In fact he believed gases *could not* be contained, and he left the matter there. However a Benedictine monk, Dom Perignon, showed that effervescence in his newly invented beverage, champagne, could be trapped in glass bottles with bits of the bark of a special oak tree. The resultant cork was a triumph for celebrants and chemists alike.

Another worker, Jean Bernoulli, used a burning glass (a lens used to focus the sun—soon to be standard equipment in the chemist's repertoire) to ignite gunpowder in a flask. To avoid repeating the shattering experience of Helmont, Bernoulli did his work in an open, rather than a sealed, system, running a tube from the ignition flask to a vat of water. He was able to show in this manner that gases from the reaction occupied a much larger volume that the gunpowder (and became wet in the process).

Otto von Guericke designed a practical air pump in the mid-1600s, and armed with this and new techniques for containment—corks and Bernoulli's vat—a group of young scientists took on the task of determining the qualities of Helmont's gases. Working primarily in Oxford, England, they were called, appropriately enough, the Oxford Chemists: Boyle, Hooke, and Mayow.

Robert Boyle

Robert Boyle was born to a noble family of vacillating fortunes. His father had gone to colonize southwestern Ireland in the 1500s and had become rich and influential. He lost his property in a rebellion

Figure 6.2. Robert Boyle, the skeptical chemist. (Courtesy of the John F. Kennedy Library, California State University, Los Angeles.)

however and returned to England only to be thrown into prison for misusing his position in Ireland and funds entrusted to him. Eventually acquitted (and financially recovered), the elder Boyle bought Sir Walter Raleigh's estates in Ireland. He again accumulated wealth through influence and industry, and he was made earl of Cork in 1620. Boyle was his fourteenth child.

Boyle's family background is important because his inherited wealth allowed him the freedom and finances to pursue his chemical investigations. He had as fine an education as could be acquired at the time. He was taught at home by private tutors, then at Eton, a distinguished public school. Under the auspices of an older brother, Boyle traveled in Europe and was tutored in the liberal arts and practical mathematics. In the course of his travels, Boyle made a trip to Italy to be introduced firsthand to the work of the recently deceased Galileo. He became an enthusiastic proponent of the new science. With the outbreak of civil war in England, Boyle's father, a Royalist, again suffered a setback in fortune and soon died. The young Boyle returned home to live simply and modestly in the family home. He did not take a strong interest in politics (probably wise considering the family history and the current atmosphere), and after 10 years (around the mid-1600s), apparently longing for a richer intellectual atmosphere, Boyle moved to Oxford. There he performed experiments for a time in lodgings next to University College, and he took part in a discussion group called the Invisible College.

A lifelong bachelor, Boyle was of rather frail health, though we wonder how much of his health would have been spared if he had not had an interest in chemistry. Boyle, like many other early chemists, reported *taste* among the properties of many reagents, and he was fond of dosing himself and friends with various preparations. Partially because of his bad health, he eventually moved to London to live with his sister. Boyle set up a modest laboratory at the back of her house, which became a meeting place for science-oriented intellectuals and a center for research.

Boyle was a prodigious worker, and he performed experiments on many chemical systems, but his experiments on the nature of Helmont's gases are of crucial interest here. When Boyle learned of Guericke's air pump, he immediately began investigation. The income left by his father was sufficient to have instruments made and to employ assistants. One of these, Robert Hooke—an important investigator in his own right—built an air pump. Boyle attached it to the perennial chemist's glass bulb and began experimenting. Within a matter of a few years, he had enough results to compile a book, *New Experiments Physico-Mechanicall, Touching the Spring of the Air and Its Effects*.

This was Boyle's first scientific publication and the one that established his fame. In it he reported that sound did not propagate in a vacuum and that air was truly necessary for life and flame. In the appendix of the second edition (issued only a few years later), he elaborated on his observation of the elasticity of air, noting that the volume occupied by air was inversely proportional to the pressure applied

ca. 1600: Philosophers of Fire

(squeezing on a balloon makes it smaller). This generalization became known as Boyle's law, although in France, it is referred to as Mariotte's law because Edmé Mariotte also described it, albeit later than Boyle. Mariotte gave no credit to Boyle, but he claimed no originality either, treating it instead as one of several well-known laws of air.

Boyle explained the elasticity of air by assuming that air consisted of corpuscles defined in the manner of René Descartes and that "each corpuscle endeavors to beat off all others from coming within the little sphere requisite to its motion about its own center."[7] Though lacking the mathematical treatment necessary for its full elucidation, this notion anticipated the currently accepted theory of the behavior of gases—the kinetic theory of gases—which is based on the motion of their constituent particles.

Boyle and his contemporaries considered air to be one substance, attributing differences in reactivity to differences in purity. For instance they saw that air generated by dropping steel filings into acid ignited when lit with a candle, whereas room air did not. Actually they were lucky that this was all they saw: The gas they were generating was hydrogen, and hydrogen mixed with oxygen can ignite with an explosion.

Boyle also investigated the common observation that metals, when *calcined* (heated strongly in air), increase in weight. We know now that metals heated this way gain weight because they combine with the oxygen in the air. If this process is carried out in a sealed container, the weight of the system as a whole—the metal, the air, and the container—should remain the same. Boyle heated metals in a sealed retort, but when he weighed it, he failed to account for the inrush of air that occurred when he opened the heated vessel. Boyle reported an increase in the weight of the whole system and therefore concluded that the increase was due to fire passing through the pores of the glass. When it was suggested to Boyle that he should have weighed the retort before breaking the seal, Boyle reported that he had but obtained the same final weight! It was probably a faulty vacuum seal rather than a preconception on Boyle's part that caused him to report this false result. Boyle, more than many thinkers of his age (and indeed more than many thinkers of our own age) had an independent, logical, and open mind. He understood the importance of strict experimental procedure. Rationalists, such as Leibniz, doubted the value of investigation by experiment because they believed they could arrive at truth by logical reasoning and that experiment was good only for confirmation. But Boyle conducted long disputes with rationalists, defending experiment as a means of proof.

The systematic Boyle also found rankling the imprecise language of chemical theorists of his day. To call the attention of his colleagues to these inconsistencies, he wrote in 1661 the classic didactic dialog *The Sceptical Chymist*. In it Boyle used the familiar device of a conversation among a group of friends—Themistus (defender of the Aristotelian theory of four elements), Philoponus (defender of the Paracelsian theory of three), Eleutherius (an uncommitted participant), the recorder of the conversations (whose name is not disclosed), and Carneades (the skeptical chemist)—to attack both the Aristotelian and Paracelsian systems.

In the 1600s the concept of an element was essentially the same as the Aristotelean concept: An element was a fundamental component *of all matter*. Thus if sulfur were considered an element, then sulfur must be found in everything from gold to grape juice. Boyle was not happy with this concept of an element, but he did not really offer anything in its place, though the following quotation from *The Sceptical Chymist* is sometimes given as evidence that he did:

> And to prevent mistakes, I must advertize you, that I now mean by Elements . . . certain Primitive and Simple, or perfectly unmingled bodies . . . of which all those called perfectly mixed Bodies are immediately compounded, and into which they are ultimately resolved.[8]

But this definition is compatible with what chemists of the 1600s viewed as elements as well as the modern view, so it is doubtful that Boyle meant anything new. He likewise offered no new list of what should be considered an element, but this may be to his credit. Boyle knew the definitions were faulty, but he did not feel on firm enough experimental ground to offer an alternative. He did however hypothesize that there could be more than four elements and even perhaps more than five. He would have been interested to know that today we list 109, and we are looking for more.

Boyle's contributions to chemistry, beyond his work on gases, are substantial. They include systematic studies of reactions between acids and bases; some of the earliest application of plant-derived acid and base indicators; and duplicating the process of isolating phosphorus. In *The Skeptical Chymist* Boyle seems to question transmutation, but he never really abandoned a belief in alchemy. When he died, he left Newton a sample of a red earth that he believed would turn mercury into gold. However this incongruous end was in an incongruous age. One group guarded new discoveries as alchemical secrets, while another group quarreled over who had announced new discoveries first. One of the foremost members of this latter group was the former assistant to Boyle: Robert Hooke.

Robert Hooke

Robert Hooke was by all accounts ill-natured, but he lived with such constant discomfort, real and imagined, that he could not have been otherwise. The son of an English minister, he was born with a skeletal defect that gave him a twisted frame and such poor health that he was not expected to survive infancy. Survive he did, but he continually suffered from headaches and other maladies, perhaps compounding his condition with hypochondria.

His family attempted to educate him for the ministry, but he was not diligent in this study. On seeing a clock dismantled however, he built his own out of wood, then proceeded to construct his own mechanical toys. When his father died, the family, believing that he had some talent in art, gave him 100 pounds and sent him to London to buy an apprenticeship to a painter. On arrival in London however, Hooke reneged on the apprenticeship, pocketed the money, and somehow managed to be admitted by the master of Westminster School. There Hooke learned Latin, Greek, and mathematics, reading (by his own account) the first six books of Euclid in a week.

With this background Hooke hired out as an assistant to physicians. He was apparently so valued in this capacity that one physician recommended him to Boyle at Oxford. At about this same time however, with the fall of Cromwell and the restoration of the Stuarts, many in the Invisible College circle of students and teachers were dismissed due to their political leanings. A few moved to London, where they reorganized and—after a couple of years—accepted the king's patronage and became the Royal Society. The Royal Society continues to be important in the history of chemistry as a forum for discussion and a vehicle for publication. It is still very important, and membership in this organization, which is by election, is considered England's highest scientific accolade. *Philosophical Transactions*, an instrument of the Royal Society and the world's oldest currently continuing scientific journal, was founded in 1665. Part of the initial success of the Society lay in the fact that it hired Hooke to be the curator of experiments.

In the rarefied intellectual atmosphere of London, Hooke began his independent career of many astoundingly diversified achievements. He enunciated the relationship known as Hooke's law, which states that stretching in an elastic body, such as a spring, is proportional to the force applied. This law was later used to describe the motion of atomic nuclei in molecules. Hooke used a telescope to make several original astronomical observations and a microscope to describe snowflakes, cells (a word he first used), and microscopic fossils. Hooke speculated on using the barometer to predict weather, but he later doubted its efficacy, confounded no doubt by variables that weather forecasters still struggle with today.

Appointed surveyor of London after the Great Fire, Hooke amassed a considerable sum of money, and with this and his new, respected position, he should have been a contented man. But in Hooke's diary we see a lonely person in pain. He never married but had a succession of mistresses, some of whom took monetary advantage of him. His poor health now included a chronic and painful sinus inflammation accompanied by headaches, vomiting, giddiness, sleeplessness, indigestion (possibly worms), and nightmares. These various maladies he occasionally treated with alcoholic overindulgence, and when a rare night passed peacefully, he noted it in his dairy as "Slept well. Deo Gratias."[9]

Hooke was chronically embroiled in various controversies in which he felt that his ideas had been usurped without proper credit. As with most things of this sort, there was probably some basis for his feelings. It has been reported that Newton, due to disagreements with

Hooke, refused to acknowledge Hooke's contributions to the study of light and delayed the publication of *Opticks* for two decades after Hooke's death.[10] When the one true love of his life, Grace, his niece, ward, and later mistress, died, Hooke became even more reclusive and cynical. His illnesses and discomforts must have increased because it was said that in his last 2 years, he neither undressed nor went to bed. But despite whatever flaws of personality, it should be noted that when he died at 68, the entire membership of the Royal Society attended his funeral to mourn his passing.

Hooke contributed substantially to the advancement of chemistry. In addition to his work with Boyle, he pointed to some directions for future chemical research. In his book describing his observations with a microscope, *Micrographia*, Hooke put forth a theory of combustion in which he states that a substance common to both potassium nitrate and air, *nitrous air*, is the agent of combustion. But as with many of his insights, he did not pursue the idea further (a habit that was the source perhaps, of his perception that others usurped his ideas), leaving this to the next person we encounter, John Mayow.

THE QUESTION OF COMBUSTION

We might have expected Boyle's observations on the physical properties of the gas phase to excite a flurry of experimental activity in this direction, but the opposite occurred. Boyle's results seemed to show that all gases behaved alike, so there seemed to be no need for further research. Boyle himself realized there was more to be discovered, as he stated in the late 1670s in *Suspicions about the Hidden Realities of the Air*. But what other investigators found more interesting was Boyle's observation that air had something to do with combustion.

To the delight of children and the despair of insurance agencies, combustion, though commonplace, still remains one of the most spectacular and fascinating chemical reactions. Aristotle believed fire to be one of the four constituents of matter, and he reasoned that during combustion, this element was released. Paracelsus reasoned that all that burned contained sulfur; therefore he considered this substance to be an element and a component of all compound things. The function of air in supporting combustion was seen by the alchemists as vaguely mechanical: a means by which heat and fire were carried off. All these speculations sounded reasonable enough, and they could be envisioned without too much imagination. The only problem was that tin smiths and lead smiths kept pointing out with irritating regularity that the materials they worked with *gained* weight when heated, which should not have been occurred if the element of fire were released.

But this observation did not cause the four-element theory immediate, insurrmountable problems. For one reason the significance of weight gained or lost in chemical processes did not become a central part of chemists' thinking until the late part of the next century, although it was hinted at in the work of Helmont and Angelus Sala. When it was observed that metals heated to a high temperature in air (a process called *calcination*) gained weight, some imagined—as did Cardanus in the mid-1550s—that the element of fire was somehow working against gravity, buoying the metal up while it was still part of it. Others, such as Boyle, attributed the weight gain to the absorption of some part of the heat or light or flame. A few, such as John Mayow and a certain Jean Rey, found a more plausible explanation—and one that won out in the end.

John Mayow

John Mayow was born to an old, established, and respected Cornish family. Though a practicing physician, he dedicated much of his personal time to scientific research, and he was an intimate of the Oxford group and eventually a Fellow of the Royal Society. In the course of his work, Mayow made two important advances in gas-handling techniques: He showed that gas could be transferred from one vessel to another underwater and that volumes of gas could be directly compared if they were at the same pressure. He achieved this equalization of pressure by leveling the water inside and outside a gas-containing vessel with a siphon. Mayow also did some interesting, and (some say) precocious work on the theory of combustion.

The level of understanding about this important reaction is summarized in Boyle's researches on the subject. Boyle knew that air was involved in combustion because if he dropped a combustible material on a red-hot plate under vacuum, nothing happened. If he admitted air, the material burst into flame. Mayow refined this observation by conducting a series of experiments in which he showed that only part of the air was used in combustion and respiration. He inverted a glass vessel over a candle or an animal perched on a pedestal in a tub of water, equalized the water levels inside and outside the glass vessel by means of his siphon, then watched the water level rise as the candle burned or the rodent breathed. Because part of the air was still left when the candle or rodent extinguished, he knew that this second type of air could not support combustion or respiration. He named the first part of the air, the part that supported combustion, *nitro-aerial*. Mayow then made a leap in imagination to conclude that weight gained by metals during calcination was due to absorption of nitroaerial particles.

Mayow came enticingly close to anticipating by a hundred years the oxygen theory of Lavoisier. But before too much is read into his insight, we must note that in a style reminiscent of the alchemists, he overinterpreted his result and began seeing nitro-aerial particles everywhere: Rays of sun were nitro-aerial particles; iron, because it sparked when struck, contained nitro-aerial particles. And Mayow was not the only one to come up with such a notion. Though not so well developed, the idea was earlier expressed by Jean Rey, a French physician, who explained why lead and tin gained weight on heating by saying, "[Weight] comes from the air, which . . . mixes with the calx . . . and becomes attached to its most minute particles . . ."[11] However he did little experimental work. He did point out (with his own leap of intuition and in anticipation of the law of definite proportions) that the increase in weight never exceeds a certain amount: "Nature in her inscrutable wisdom has set limits which she never oversteps."[12] Unfortunately Mayow was confounded by another law of nature: An idea is not recognized until its time has come.

The delay in the development of an accurate theory of combustion however cannot be blamed entirely on timing. Some credit has to go to the rivalry of another theory of combustion that was being developed at the time. It, too, had a very logical derivation and in addition a catchy name—*phlogiston* (flō-jis´-ten). Phlogiston was supposed to be a material that leaves a substance undergoing combustion. Although little more than a rehash of Aristotle's fire element, this time the theory was supported by methodical experimentation. And though this theory was proved incorrect—some say it was a roadblock in the development of chemistry—the debate it inspired was eventually the basis for an new age in chemistry. The story starts with an interesting entrepreneur, promoter, and (some say) opportunist: John Joachim Becher.

John Joachim Becher

Becher, whose work included a book, *Foolish Wisdom and Wise Folly* (in which he described a stone that turned people invisible and a flask that could hold words) and the invention (on commission) of a universal language for which he was never paid, was in most ways a complete throwback to the European middle ages. But he was also in a way the herald of the new age. With the nonexperimental, head-scratching methods of the ancients, he set about trying to determine the basic elements of which all matter is composed, and with a blend of previous ideas rather than anything original, he came up with air, water, and three earths: the fusible, the fatty, and the fluid. He then envisioned the process of combustion as the loss of this fatty, inflammable earth, which later became known as phlogiston. He also envisioned calcination as the same process: "Metals contain an inflammable principle which by the action of fire goes off into the air; a metal calx is left."[13] He knew that the metals gained weight in this process, but with perhaps the first of what would be a long stream of rationalizations in support of the phlogiston theory, he at times explained this as an incorporation of ponderable particles of fire and at other times as the power of levity of the phlogiston: "Something minus another thing that weighs less than nothing, weighs more."[14]

These ideas proposed by Becher were not fully developed by him—or in the 1600s—but more in the 1700s by our next subject, Georg Ernest Stahl. In fact Stahl should be placed chronologically in the next chapter. But standing as he does with one foot in this era and one in the next, Stahl makes an ideal conclusion to this equivocal age.

Figure 6.3. Georg Stahl, popularizer of the phlogiston theory. (Courtesy of the John F. Kennedy Library, California State University, Los Angeles.)

Georg Ernest Stahl

Georg Ernest Stahl, a German physician and teacher of chemistry and medicine, was born in the latter half of the century. Giving full credit to Becher for the idea, Stahl popularized the theory of a fatty earth that leaves a burning material, calling it phlogiston. However leaving Becher behind, Stahl supported the theory on experimental grounds. He reasoned thus: Sulfur burns in a flame and produces sulfuric acid; therefore sulfur consists of sulfuric acid and phlogiston. If phlogiston were restored to sulfuric acid, the original sulfur would be recovered. Because charcoal reduces metal oxides to metals and when burned leaves only a small amount of ash, charcoal must be rich in phlogiston. Therefore charcoal heated with sulfuric acid should restore the phlogiston and yield the original sulfur.

And indeed when Stahl *fixed* sulfuric acid (made it a solid by allowing it to react with potassium hydroxide), then heated it with charcoal, he obtained *liver of sulfur* (a potassium polysulfide). This is the same dark brown amorphous mass obtained when pure sulfur is fused with potassium hydroxide, so he concluded that he had his sulfur back, and the phlogiston theory worked.

Stahl's explanations, based on experiment and couched in a simple, intuitively appealing form, were so seductive that the phlogiston theory remained viable for some 70 years. Many observations were explained by phlogiston: Metals heated in air changed because they lose phlogiston. When the resulting *calxes* (oxides) are heated with charcoal, they change back into the metal because they gain phlogiston from the charcoal. Combustion could not occur in a vacuum because air was needed to absorb phlogiston mechanically. Combustion in a sealed container ceased when the air became saturated with phlogiston.

The primary difficulty however remained: Metal oxides lose weight when reduced to metals, and this was the weakness that led to the eventual downfall of the theory. But the information gained in the effort to prove or disprove the theory was the data base for the next theory—and this one would stand.

So although Europe in the 1600s had both medieval and modern aspects, progress in chemistry was made. To understand this progress, we need compare only the experiment-based phlogiston theory with the conjecture-based theorizing of, say, Paracelsus. This new faith in experimental methods—topped with the successes of the Scientific Revolution—culminated in a new philosophy: Enlightenment. The essence of Enlightenment was a rejection of tradition for the sake of progress and an improvement of the human condition. Politically this meant individual freedom and equality. Scientifically this meant a critical application of scientific methods. Chemically this meant Aristotle had to go. The problem however was finding a replacement.

chapter
SEVEN

ca. 1700: The Search for System and Phlogiston

Chemists of the 1700s were carried along with the spirit of the age: progress, Enlightenment, system. Unfortunately their initial attempt at system—phlogiston—was wrong, though it would make its own, upside-down, contribution: through the battles phlogiston inspired, the right system would eventually be found.

THE WORLD CA. 1700

Though the century ended quite differently, in the beginning of the 1700s, the world found itself an unusual circumstance: a period of relative peace. In the Near East and northern Africa, peace was imposed by the Ottoman Empire. In central Africa harbors of slave trade pockmarked the coast, but the interior remained out of reach of outside adventurers. In the New World inhabitants struggled to recover from

127

the European invasion, and the invaders had not yet turned on each other.

By the 1700s Japan had tossed out the Europeans (with the exception of a few who were willing to defile the cross and promise not to preach), and it enjoyed the great peace of the Tokugawa shogunate. When the British tried to counter an unfavorable trade balance by selling Indian opium to China, Chinese leaders decided the Europeans were not only culturally inferior but also purveyors of drugs, so the Chinese erected barriers, mental and real, and settled behind them in a temporary peace. India of the 1700s found itself caught up in a dark age of religious fatalism, as Europeans had been a few hundred years before, and political shifts meant little to the average Hindu. With this apathy came a kind of peace.

In Europe, the feudal system with its councils of lords and round table discussions, had been efficient enough for wars by knights with spears, but when faced with artilleries and large professional armies, a need for a more responsive system soon became apparent. The solution arrived at temporarily was to concentrate government power in nationalistic, absolute monarchies. Although the primary impetus for absolute monarchies was the ability to wage war more effectively, the actual consequence, at least at the opening of the 1700s, was a power standoff and a period of relative peace.

So despite the chaos of the religious wars of the 1500s and the ambiguities of the 1600s, the 1700s emerged with a clear trend: a rejection of tradition and a new passion for progress. This Enlightenment (as it was called) found a curious partner in the absolute monarchies: The resultant philosophy of government came to be known as Enlightened Despotism. The ideal enlightened despot drained marshes, built roads, codified laws, repressed regionalism, curtailed the church, and administered everything through a professional bureaucracy. The new centralization of power made commerce as well as war more efficient, and for many the standard of living improved. The result was that individual loyalties shifted from the village to the nation, and this new nationalism expressed itself in national rivalries when times were good. When times were bad, it meant a return to war. Because these rivalries and wars of the European nations in large part dictate the course of our chemical history, it behooves us at this point to take a brief look at some of these players as they take their positions on the board.

THE ENLIGHTENED DESPOTS

England

England (or the United Kingdom, after the union of England and Scotland in 1707) had survived a civil war, a Puritan commonwealth, and now had its monarchy restored. But the powers of the restored monarchy were curtailed, and a strong parliament controlled the army. The nation had a Bill of Rights guaranteeing habeas corpus, the right to petition, the right to bear arms, and the right to debate. It had an Act of Toleration granting religious freedom (except to Catholics, Jews, and Unitarians). It remained happily insulated by its still imposing channel of water, and it had little interest in continental political problems.

Spain

At the opening of the 1700s Spain suffered from one of the inherent ills of rule by a monarchical family: Its king was an impotent, imbecilic product of inbreeding in the Hapsburg family. When this king died in 1700, the king of France and the Holy Roman Emperor (each of whom had married one of the sisters of the King of Spain) struggled for the throne. The inevitable war however had a new twist. Because union with Spain would make the winner too strong, none of the noncompeting countries wanted either contestant to win. To prevent this, they invoked a new tactic called the Balance of Power, under whose precepts alliances of convenience were made, broken, and reformed until the war ended with compromise. Spain lost some European possessions, retained its American ones, and gained a French king on the understanding that the same king could never occupy the French and Spanish thrones simultaneously. The Bourbon family reigned with few interruptions until the republican revolution of the 1900s.

Prussia

Frederick William, at the age of 20, succeeded as head of the house of Brandenburg and what was left of Prussia after the devastation of the Thirty Years' War. With a large portion of his population lost by starvation, plague, or murder by marauding armies, he instituted, understandably, a military culture. The first Prussian king to appear always in uniform, he focused on the cultivation of his army. In fact he ruled his country as though it were one large army, disciplining citizens with his walking stick as though they were troops.

During his reign Berlin grew to be a city of 100,000 inhabitants, one out of five of whom were soldiers. By 1740 under his son, Frederick the Great, Prussia had become a great power. This Frederick though was different from his father. He liked to play the flute, correspond with French poets, write prose and verse, and at 18 years of age tried to escape the kingdom and his military father. He was caught, brought back, and forced to watch the execution of a friend who had helped him run away. Though eventually a capable leader, he remained a freethinker like many of his age, rejecting religion and the divine right of monarchs. As he wrote to Voltaire, "My chief occupation is to fight ignorance and prejudices in this country . . . I must enlighten my people, cultivate their manners and morals, and make them as happy as human beings can be. . . ."[1]

Austria

According to Voltaire, by the 1700s the Holy Roman Empire was neither holy, Roman, nor an empire. But the Austrian family of Hapsburg managed to emerge from the ruin to build an empire of their own. By 1740 Austria was a populous empire of great military strength. However when the Hapsburg throne passed to 23-year-old Maria Theresa, the empire and its leader were tested.

Frederick II of Prussia was the first to invade when Maria Theresa was pregnant with her first child. But far from buckling, she used the birth of her son to rally support. Although some territory was lost, the empire survived, and Maria Theresa bore nine more children, one of whom we encounter again as the French queen Marie Antoinette.

Joseph, her son, embraced Enlightenment, but unfortunately a bit too quickly. He abolished serfdom, and he ordered tolerance of all religion to the point of making Jews responsible for military service for the first time in Europe. But the Catholic church opposed him, and some segments of his population rose in revolt. When his brother Leopold succeeded Joseph, he had too much to straighten out in Austria to be able to help his sister in France.

Russia

The leader of Russia at the opening of the 1700s, Peter the Great, saw Enlightenment as westernization, and to this end he built up St. Petersburg and founded the Academy of Sciences. He also made European dress and the use of tobacco compulsory for members of his court. After his death he was succeeded by his wife, Catherine I, because he had put his son to death when the son supposedly voiced plans to revert to the old ways. This first Catherine reigned for only two years, then was followed by a succession of tzars and tzarinas until another Catherine, Catherine the Great, dethroned her half-witted husband and reigned during the latter half of the 1700s. This ruthless woman of German origins and many lovers, took steps to codify Russian laws, restrict torture, and remodel local government, but then responded to a serf rebellion in 1773 with iron-handed repression. Rebel leaders were drawn and quartered, nobles given more power, so that Russian serfdom became virtually the equivalent of slavery. She said to an encyclopedist of the time, on the subject of reforms, "You write only on paper, but I have to write on human skin, which is incomparably more irritable and ticklish."[2]

France

France in 1700 was a wealthy country with three times as many people as England, twice as many as Spain, and a government that was the epitome of absolute monarchy. Louis XIV, the Sun King, inherited his throne at the age of five and reigned until his death at the age of 77. He was able to state, quite correctly, that "the state, that is I." And Louis XIV made war an affair of state: He revamped the military system so that soldiers now fought for him, not regional leaders. He integrated artillery into the army, systematized military ranks, clarified the chain of command, and placed himself at the top.

But the Sun King's successor, Louis XV, inherited the Sun King's two problems: high military costs and an inefficient system of taxation. Neither problem was insurmountable, but Louis XV had no will to tackle them, which points up another weakness of absolute family rule: If a leader chose to be indifferent, there is little that could be done. Little that is short of revolt, which is of course what happened. But before we describe the next reign of violence, let us look at the accomplishments of this period of peace.

THE ENLIGHTENMENT AND SCIENCE

For science the enlightened approach was to reject the mystical in favor of clear, Cartesian logic. Newton and Descartes had made it appear that all systems in the natural world could eventually be described by mathematical formulas, and for such disciplines as physics, the approach brought gratifying results: Newton's gravity and his laws of motion described the behavior of objects on Earth, and Kepler's law described the behavior of the planets. But then physicists were dealing with the macroscopic world—a world they could see or manipulate. Chemists have never had this luxury. The scale of their world is atomic, and the activity of atoms has to be inferred rather than observed. So in the early 1700s chemists did not yet have the tools to reduce their systems to exact mathematical models. But they had the desire.

PHLOGISTON

The 1500s had given a purposeful redirection to chemistry (shifting emphasis from gold to medicine), but the basic alchemical theory remained the same. Materials were made of an admixture of some small number of elements: the air, earth, water, and fire of Aristotle, though perhaps refined—in the case of earths and metals—to include mercury and sulfur (the Arabic notion); mercury, sulfur, and salt (the Paracelsean notion); or the fusible, the fatty, and the fluid (Becher's notion).

ca. 1700: The Search for System and Phlogiston

Chemistry consisted of trying to alter the proportion of these elements in what wasn't wanted to make what was wanted. But by the 1600s the chemists had tried about all the combinations they could, and they found that they still did not have much control over the end product. They were starting to see that the Aristotelean hypothesis would fail. But before it was thrown out completely, it underwent one last resurgence in the theory of phlogiston. This time however hypothesis testing and rejection took about a hundred years instead of 2000.

It is tempting to try to describe phlogiston in the clear, precise terms we have become accustomed to now, but chemists in the early 1700s did not find it necessary to do so. Although it was generally agreed that phlogiston was a material that left burning substances, its characteristics beyond that were difficult to pin down. No one had measured its weight, and some said it was weightless. It was supposed to have some combining power, but no exact proportions were proposed. For some chemists this flexibility in the system amounted to irritating imprecision. But other chemists were quite comfortable because the flexibility allowed them to explain most phenomena in terms of phlogiston. For instance there was the burning question of combustion.

CALCINATION AND COMBUSTION

It is worth a few lines to explain combustion and calcination as they are currently understood and to show how phlogiston was used to explain them.

In ideal combustion, oxygen from the air combines with hydrogen and carbon in the fuel to produce carbon dioxide gas (a carbon–oxygen compound) and water (a hydrogen–oxygen compound). In a real fire however, combustion is not complete, so there are other products formed, such as soot and smoke, which complicated the picture for these early chemists. The phlogiston theory explained combustion by saying that something—namely, phlogiston—left the material that burned. Because smoke and flame do rise from the fire, this is not an illogical conclusion. Chemists had no reason to believe that something (oxygen) was going into the burning material.

In calcination, oxygen from the air combines with a hot metal. Tarnish is caused by oxygen from the air combining with clean metal. So clean copper-bottomed pots eventually tarnish in air, but copper-bottomed pots heated while cooking tarnish more quickly due to calcination. In the 1700s calcination was viewed as a process similar to combustion, which it is. But here chemists had evidence that something was going into the calcined material because the material gained weight. But chemists at the beginning of the 1700s were not yet completely aware of the significance of mass in chemical reactions, and they did not routinely include mass measurements as a part of their investigations. Besides the visual evidence of flames leaving the fire was so compelling that they felt it was easier to adjust the notion of phlogiston than question their own eyes. So when explanations were offered, chemists gave phlogiston either a negative weight or no weight at all. But this was also unsettling because there was nothing else in common experience that behaved this way. Chemists realized more information was needed. As the chemist Carl Wilhelm Scheele said, "I soon realized that it was not possible to form an opinion on the phenomena of fire as long as one did not understand air."[3]

THE PNEUMATICK CHEMISTS

We now know that the air we breathe is a mixture of gases: about three-fourths nitrogen and the rest oxygen with a little bit of argon, carbon dioxide, and water vapor thrown in. In the early 1700s however air was considered one substance (and so it certainly appears to be). Though Helmont's term gas was useful because it emphasized the novelty of this previously undescribed phase of matter, the term was not widely used in the 1700s. Instead gases were called airs and thought of as various states of common air, though later distinguished by adjectives describing their properties. The group of chemists who studied the various airs were later called pneumatic chemists or pneumatick, as it was then written.

Joseph Black

Born in France to a Scottish family, Joseph Black's father was a wine merchant, his mother was the daughter of a wine merchant, and Joseph was the fourth of their 12 children. He was educated by his mother until he was 12, then sent to Belfast and from there to the University of Glasgow. Having to choose a profession (and chemistry not yet being a choice), Black chose medicine, but he enjoyed chemistry most in his course of study and for 3 years assisted William Cullen, the chemistry instructor.

For his MD thesis Black chose to look for a solvent for kidney and gall stones so that they could be removed without surgery—a laudable study in these pre-anesthesia days. At first Black thought to study a solution of quicklime (calcium oxide) because he knew it would dissolve the stones, but unfortunately it also destroyed tissue. Because of this and because it was not at all clear that quicklime from different sources was the same compound (one of Black's professors preferred quicklime made from oyster shells while another preferred the material made from limestone), Black chose to avoid controversy and study magnesia alba (magnesium carbonate) instead. Though Black soon found that magnesia alba did not dissolve kidney stones, he was intrigued by its other properties, such as its effectiveness as a laxative, so he continued his study. From this study came his insights into the nature of air.

First Black found that magnesia alba gave off bubbles when treated with acid, as did other materials then known as mild alkalis. These mild alkalis are in fact carbonate salts, which we now know give off bubbles of carbon dioxide when treated with acid. In continued experiments Black subjected the magnesia alba to strong heating, then treated the residue with acids. He found the residue formed the same salts as the unheated material but gave off no bubbles. On the basis of this, Black decided that the bubble-forming gas was in fact part of the magnesia alba that had been driven off with heating. He repeated the heating, this time planning to trap the gas over water, but he found that the gas was absorbed in the water. Undaunted he set off on a series of methodical experiments that used weight differences to determine the amount of gas in the material.

Black weighed the material before and after heating and assumed the difference was the weight of the gas lost. He weighed portions of magnesia alba and acid, mixed them, and after all the bubbling had ceased, weighed the residue, and assumed the difference was the weight of the gas. Then bringing everything together, he showed that the gas from magnesia alba was the same as the gas given off by mild alkalis. He heated the magnesia alba to drive off the gas, dissolved the residue in acid, added an excess of a mild alkali to replace the gas driven off by heating, and obtained the same weight of magnesia alba—a demonstration of the conservation of mass.

Interestingly in this series of experiments, no weight is attributed to phlogiston, so Black could have explained all of the preceding experiments without resorting to the phlogiston theory. But chameleon that it was, phlogiston was not excluded either. True to the phlogiston theory, Black decided magnesia alba gained phlogiston when heated at the same time it lost the gas that had caused the bubbles.

Black continued his studies but now concentrated on the gas instead of magnesia alba. He found that his gas made limewater cloudy, as did gas given off by burning wood. We now know that this is because they are both carbon dioxide, and carbon dioxide forms a fine suspension of calcium carbonate (or chalk) in solutions of calcium hydroxide (limewater). Black then showed that the gas given off in fermentation also turns limewater milky as does gas that the human body exhales. He concluded that combustion, respiration, and fermentation all produced the same gas and it was the same gas that was trapped—or fixed—in magnesia alba. He called it therefore fixed air.

To demonstrate that fixed air was different from regular air, Black put two jars in an air pump, one containing just water and the other limewater. When a vacuum was created, dissolved air bubbled out of both the jars, but the limewater did not turn milky. By experiments with birds and small animals, he demonstrated that fixed air, unlike common air, did not support life. With these results Black pushed experimental chemistry forward two giant steps: He clearly demonstrated that the identical gaseous compound could be isolated from several different reactions, and he clearly demonstrated the need to account for the mass of gaseous reagents and products in chemical reactions. Black published these momentous results in his MD thesis, then published very little for the rest of his career.

Reasons for his lack of publications were probably both personal and professional. He replaced his teacher, Cullen, when Cullen moved to Edinburgh, and because students paid Black directly by fee, he had to lecture intensively to earn a reasonable income. The university at

Glasgow was also administered by its faculty, so Black had administrative duties, which took up much time. But he did continue to do research, and many of Black's research results were disseminated in his lectures and became widely known even though they were not formally published. In fact his work surfaces in another context soon.

Black was also not a seeker of fame. He was apparently a quiet person of tenuous health, possibly asthmatic. Described as tall, very thin, and very pale, he had large, dark eyes and thin hair powdered and arranged in a long braid. He was also noted for meticulousness with finances, though not for the stinginess that this has sometimes been interpreted to be. Teaching was apparently his forte. His lectures were so popular that students who did not need the course still chose to attend (which as any chemistry teacher will attest is unusual). Black stressed the practical and spent little time on theory unless it were thoroughly upheld by experiment. He rejected the idea of four elements, saying instead that materials should be separated into classes.

Black eventually moved to Edinburgh (replacing his mentor, Cullen, once again) and taught there. He lived out his quiet bachelor's life in the company of his friends—including Adam Smith, the economist, and David Hume, the philosopher—at the Philosophical Society (or the Royal Society of Edinburgh, as it was called after the 1780s) and those informal clubs for which Edinburgh was famous: the Select, the Poker, and the Oyster. Black's contributions to chemistry were substantial and varied, but as with any teacher, his true success is best measured by the success of his students. One of these students was Daniel Rutherford, nephew of Sir Walter Scott, the novelist and poet.

Daniel Rutherford

In his MD thesis Daniel Rutherford discussed Black's fixed air and called it mephitic air—the mephitis was a legendary noxious, pestilential, foul exhalation from the Earth—probably because the gas did not support respiration. Rutherford observed that common air has a "good" part that is able to support respiration; when this is used up, the rest cannot support life. He knew that part of the good air had been converted into fixed air (carbon dioxide) by respiration, so at first Rutherford speculated that the part that would not support combustion was contaminated by fixed air. But after he removed the carbon dioxide by passing the gas through limewater, the remaining gases still did not support combustion or respiration.

The observation that this residue, which we know as nitrogen, was different from common air was made by Priestley, Cavendish, and Scheele (other chemists who are discussed shortly) at about the same time. But Rutherford is generally given credit for the discovery of nitrogen because he was the first to publish it (in his MD thesis). He did not however recognize nitrogen as a distinct chemical species; he believed it to be atmospheric air saturated with phlogiston or phlogisticated air. In combustion it was thought that phlogiston left materials and went into the air, so Rutherford reasoned that if the air over a material were already saturated with phlogiston, combustion could not take place. Phlogiston worked equally well as a theoretical framework for the next worker, Priestley, when he isolated his dephlogisticated air.

Joseph Priestley

An English chemist whose name has become synonymous with political and religious nonconformity, Priestley, the son of a Yorkshire cloth dresser, was raised by his Calvanist maternal grandparents, probably because of his mother's poor health. When his mother died, the boy, only six, was sent to live with an aunt. At his aunt's home he came in contact with the Presbyterian philosophy, and he eventually trained to become a Presbyterian minister.

In his choice of the Presbyterian ministry as a career, we see early evidence of Priestley's capacity for independent action: As a Presbyterian he became a member of a dissenting religion (that is, a religion other than the Church of England), and the ministry involved public speaking, which he took on despite an inherited speech impediment. Beyond this we find that Priestley supported the cause of the colonists over the crown in the U. S. war of independence (a view that was not favored in England) and rejected the Trinity, stating what would come to be known as the Unitarian view that "Jesus was in nature solely and truly a man, however highly exalted by God"[4] (a view that had few sympathizers among Christians at that time).

After completing his studies Priestley went rapidly through two ministerial jobs (his unorthodox views were not very popular with traditional congregations), and while holding the second position, he opened a school. At the time English universities were open only to male students who were congregants of the Church of England. Those who dissented from the Anglican doctrines set up their own schools, called the dissenting academies, and these were often more progressive than the traditional universities. Priestley's school was so successful that it won him a position at a well-known dissenting academy in Warrington. There he was asked to teach languages, history, law, oratory, and anatomy and other science—but not, it may be noted, theology.

Required to teach science, Priestley found he had to learn some. As part of his studies, he wrote a History of Electricity, an effort that was encouraged by Benjamin Franklin, whom he had met in London. With an apparent propensity for meeting famous people, Priestley met and married Mary Wilkinson of a famous family of ironmasters whose work is illustrative of the industrial trend in England. However with his new family responsibilities (and perhaps a bit of his lifelong penchant for rancor), he moved from his teaching position back into the ministry in the town of Leeds.

At Leeds Priestley lived next door to a brewery, which Joseph Black had shown was a good source of his "fixed air," carbon dioxide. Priestley with his ever lively mind began to study this gas. In the course of his studies, he noticed that not only did the gas dissolve in water, but when it did, it produced a pleasant drink. This invention of soda water (the very same soda water that is the basis for today's billion-dollar, international soft-drink industry) won him the prestigious Copley Medal from the Royal Society of London, and his scientific career was off and running. (It is of passing interest to note that Priestley in this same period discovered that rubber from South America, called India rubber, efficiently erased pencil marks—an invention to which these authors owe much.)

Priestley was soon recruited by the Earl of Shelburne to be his secretary, resident intellectual, librarian, and adviser to the household tutor. Priestley did most of his work on gases during the 8 years that he served Shelburne at Shelburne's quiet country house in the southwest of England. Priestley was able to isolate carbon dioxide by bubbling it through a pool of mercury into an inverted glass bulb (the carbon dioxide did not dissolve in mercury as it did in water), and in this manner he was also able to work "nitrous air," which is now known as nitric oxide. Many chemists had noted the red–brown fumes generated when nitric acid attacked metals, but colorless nitric oxide had gone unnoticed.

Priestley found that his nitrous air, on standing over iron filings, forms another gas that can cause a candle to burn more brightly than usual. This gas, nitrous oxide, he called "dephlogisticated nitrous air" because in the phlogiston system, a gas that promoted burning was thought to do so because it was deficient in phlogiston (which gave phlogiston in the burning material somewhere to go). Priestley further reasoned that nitrous air was rich in phlogiston and dephlogisticated nitrous air was formed when iron drew phlogiston from the nitrous air—not unreasonable, given the phlogiston system.

The brown fumes of nitrogen dioxide, Priestley found, were formed when nitric oxide reacts with common air. This brown gas was novel at the time, but it has become all too familiar to everyone in large cities today. High temperatures of combustion in gasoline engines generate nitric oxide from the nitrogen and oxygen in the air. This nitric oxide reacts now as it did then: It combines with oxygen in air to form the familiar brown fumes whose accumulations in atmospheric inversion layers are a large component of smog.

Priestley was a couple of centuries removed from this concern, but he did find a use for the reaction. When nitric oxide reacted with oxygen to form nitrogen *di*oxide (that is, with two oxygen atoms attached to the nitrogen rather than one), it diminished the volume of oxygen in the air. In fact Priestley noted that it diminished the volume of common air by about one-fifth. This gave him a method for measuring what he called the "fitness" or "goodness" of air for respiration (which was in fact its oxygen content). His test was more quantitative than timing how long the air supports respiration in mice (which depended on the size and condition of the mouse) and eventually evolved into an instrument for measuring oxygen content of air called the *eudiometer*.

Priestley went on to isolate and study hydrogen chloride gas (made from sulfuric acid heated with common salt, sodium chloride) and ammonia gas (noting when he sparked a sample that he obtained hydrogen and nitrogen). Then in what must have been a gratifying experiment, he mixed hydrogen chloride and ammonia and saw them come together in a fine white cloud that settled into a soft pile of ammonium chloride crystals.

Then in August 1774 Priestley prepared oxygen by collecting the gas driven off when mercuric oxide is heated. Whether or not he was the first to do so soon became the subject of lively debate, but in August 1744 Priestley certainly believed he was the first. He found that a candle burned brightly in his new gas and that heated charcoal glowed. When quite as a second thought, he put a burning candle into a sample that he had allowed to react with nitric oxide, he saw that it still supported combustion, and he concluded that he had a new gas—one that supported combustion more than ordinary air.

Priestley wrote:

> I cannot, at this distance of time, recollect what it was that I had in view in making this experiment; but I know I had no expectation of the real issue of it. If . . . I had not happened for some other purpose, to have had a lighted candle before me, I should probably never have made the trial; and the whole train of my future experiments relating to this kind of air might have been prevented.[5]

Because his new gas was better than common air for supporting respiration and combustion, he called it "dephlogisticated air." In October Priestley went with Lord Shelburne to Europe to meet and discuss his results with a certain French chemist, Antoine Laurent Lavoisier. (Lavoisier at this time was formulating his own ideas about air and combustion, and his meeting with Priestley had repercussions, as we shall soon see.) By March 1775 convinced of his priority in the discovery of the new gas, Priestley wrote: "Hitherto only two mice and myself have had the privilege of breathing it."[6]

Besides his interest in gas experiments, Priestley, a prodigious worker and writer, produced volumes on theology, history, education, metaphysics, language, aesthetics, and politics. He wrote so voluminously about his gases, that the Royal Society suggested that he use a vehicle for publication other than the Philosophical Transactions.

Between 1774 and 1786 Priestley wrote six volumes of *Experiments and Observations on Different Kinds of Air.*

After 1780 Priestley chose to leave the service of Lord Shelburne (a decision possibly brought on by ill feelings occasioned by Lord Shelburne's second marriage), but the parting must not have been too acrimonious because Priestley retained an annuity from Lord Shelburne for the rest of his life. And the change coincided with other social difficulties. He had never been welcomed by the membership of the Royal Society for political and religious reasons, and it is said that Cavendish, our next subject for discussion, avoided him entirely. So after leaving his patron, Priestley decided to relocate in Birmingham.

In Birmingham he headed the New Meeting House, one of the most liberal congregations in England and became associated with the Lunar Society, an informal collection of provincial intellectuals, scientists, and industrialists who met during the full moon (when it was easier to travel at night). During the years of Priestley's attendance, the Lunar Society included (among others) Erasmus Darwin, Richard Edgeworth, Jonathan Stokes, James Watt, and Josiah Wedgwood. Members supported his research intellectually and financially, and Priestley must have finally felt at home.

Priestley however continued to run counter to the established church and government. On the anniversary of the fall of the Bastille, a "Church and King" riot (not seriously discouraged by the authorities) destroyed the New Meeting House and Priestley's home and laboratory. Priestley retreated to London, but even there he felt increasing threats of political persecution. He finally moved with most of his family to the United States.

He was offered a professorship in chemistry at the University of Pennsylvania, but he opted to settle with other British political refugees in the quiet of rural Pennsylvania. In his new house and laboratory in his new country, he went to work. By passing steam over glowing charcoal, he managed to generate a new gas (now known as carbon monoxide) that burned with a beautiful blue flame.

Priestley died in 1804, and when mourners met at his grave 70 years later to recognize the one-hundredth anniversary of his discovery of oxygen, they decided to form a society—the American Chemical Society—an important force in developments to come.[7] Priestley never abandoned his support of the phlogiston theory, and he continued his political diatribes, too, taking the side of Jefferson against the administration of John Adams. Ironically when Jefferson was elected president and became his personal friend, Priestley was able to die within the current political fold—a rather unfitting end for this cantankerous, antisocial, anti-establishment man.

But then antisocial behavior seems to have been common among pneumatic chemists, as evidenced by our next chemist, Henry Cavendish. Cavendish was the first to study hydrogen, the infamous gas of the Hindenberg dirigible explosion. Cavendish calmly calls it his inflammable air.

Henry Cavendish

Born of aristocratic lineage, Henry Cavendish had no title, but he was independently wealthy all his life. He went to Cambridge but took no degree, a practice not uncommon for a person in his position. After Cambridge he lived with his father in London, where he built a laboratory and workshop, then started on his unique career of unrestricted scientific investigation.

Cavendish seems to be one of the few characters we have encountered (or will encounter) whose scientific pursuits were free of ulterior motivation. A reclusive soul, Cavendish never married, and he avoided frivolous society. His father was a member of the Royal Society, and he helped his son to find a place in London scientific circles. Elected to the Royal Society, Cavendish attended meetings faithfully. He shunned publicity, writing fewer than 20 articles during his 50 years of productivity, but when he did publish, the work was careful and exhaustive. His publications also covered a broad range—electricity, freezing points, the density of the Earth—and of interest to us here, pneumatic chemistry.

Cavendish's first paper in fact was on "factitious" airs, that is, gases fixed in some material but capable of being freed, such as carbon dioxide in carbonates. One of these factitious airs was his inflammable air, hydrogen gas, which he collected from metals treated with acid.

When Cavendish sparked his inflammable air with Priestley's dephlogisticated air (oxygen), he found that water was formed, confirming what had been observed by Priestley, Watt, and others. Cavendish concluded that his inflammable air (hydrogen), soaked up by dephlogisticated air (oxygen), must be pure phlogiston. And hydrogen with its light weight and explosive characteristics certainly had attributes expected of phlogiston.

There was however a hitch: Cavendish found that the water he produced contained a small amount of nitric acid. Cavendish concluded (correctly) that this was caused by Rutherford's phlogisticated air (nitrogen) present as an impurity in his dephlogisticated air. Before he would publish, he carefully confirmed this by continually sparking and removing nitric acid. When no further nitrogen or oxygen could be removed, he found that a small volume of gas still remained. This unreacted volume, about 1 percent of the original volume, went unexplained for over a hundred years until Rayleigh and Ramsay identified it as atmospheric argon.

As combustion has always had its fascination, so have explosions, and this new inflammable gas, hydrogen, was capable of both. At least one daredevil was willing to inhale the gas, then set fire to it as it was exhaled. However when he tried to do the same with a mixture of hydrogen and common air, "the consequence was an explosion so dreadful that I imagined my teeth were all blown out."[8] Dr. Jacques Charles (a name we encounter again) constructed the first hydrogen-filled balloon, and for the entertainment of some 300 thousand spectators, he climbed in and made aerial history.

Cavendish himself was not one to participate in such theatrics and in fact he would not even teach nor hire out his chemical talents. He may best be described as socially clumsy: He dressed 50 years out of style in somewhat soiled, shabby clothes, and he always wore a high-coat collar. He had very rigid, exact habits, he was stingy (with himself), and nonreligious. There are accounts that paint him as a misanthrope, but these accounts may be unfair: He was more blind to the human condition than unsympathetic to it. When his personal librarian fell on hard times after leaving Cavendish's employment, a mutual friend told Cavendish about the librarian's troubles, and Cavendish said he was sorry to hear it. It was only after the friend suggested a small annuity for the fellow that Cavendish finally responded, but then it was with 10,000 pounds (a substantial fortune in those days).

For whatever reasons though, Cavendish was antisocial to the bitter end. On his deathbed he was attended only by his valet, and he even sent him away, asking not to be disturbed until a certain hour. When the valet returned at that hour, he found his employer dead. We begin to suspect that there was something in the air of these pneumatic chemists, because our next subject also seems to have had his own peculiar social habits.

Carl Wilhelm Scheele

Swedish Carl Wilhelm Scheele, the seventh of 11 children, chose pharmacy for a career, but he was more interested in doing research on chemicals than in producing them for sale. After his apprenticeship however, he had to work for 13 years as a production technician before he had an opportunity for independence. This opportunity took the form of a pharmacist's widow looking for someone to manage her husband's business. Scheele applied for the job, but he found the business financially burdened and realized that the widow would soon be forced to sell. Scheele and the widow struck a deal in which Scheele agreed to manage the pharmacy for 1 year and in that time try to position himself financially so that he would be able to negotiate its purchase. The widow held negotiations with another potential buyer unbeknownst to Scheele, and the opportunity was almost lost. Local customers however were so happy with Scheele's work that they demanded that he be allowed to continue, and so he did. The young widow signed the pharmacy over to him and stayed on as his housekeeper.

As a confirmed phlogistonist, Scheele set out to study the effects of phlogiston on gases. Observing that combustion reduced the volume of air and that the unrespirable air remaining was less dense than common air, he reasoned that phlogiston from the combustible material combined with a component of the common air and was dissipated as weightless heat—a reasonable working hypothesis. Scheele called the unknown component of the air *fire-air*, and he was determined to isolate it.

Because nitric acid acts on some metals to produce oxides and the same thing occurs when the metal is heated in air, he reasoned that nitric acid must absorb phlogiston from the metal the way air does when the metal is heated. Therefore if heat is, as he suspected, phlogiston plus fire-air, then contacting heat with nitric acid would draw off the phlogiston and leave the fire-air.

To try this experimentally, he "fixed" the nitric acid (made a solid) by allowing it to react with potassium hydroxide, then heated the resulting potassium nitrate in a retort fitted with a bladder containing a salt that absorbed nitrogen oxides. As he had hoped, the bladder filled with a gas in which a taper burned brightly. Moreover when mixed with depleted air, the air was restored to respirable, common air. Scheele had isolated oxygen and named it fire-air.

But like his taper Scheele burned brightly, but not for very long. He died at age 44, but by that time he had accomplished a great deal. He had also known the warmth of companionship because the widow was at his side. Shortly before his health failed, he willed the pharmacy to her, and 2 days before his death, they sealed the friendship with marriage.

Though not isolated, Scheele was somewhat off the trodden path: Although his laboratory notes show that he probably isolated oxygen earlier than Priestley, when Scheele finally published his results, oxygen was already well-known. But then it is a well-recognized phenomenon that given the same set of data, more than one mind will arrive at the same conclusion. Therefore it is difficult (and really not that enlightening) to try to assign priority to each advance in science. What is interesting is to see how many people came up with the idea and in how many different ways. For instance the work of Mikhail Vasilevich Lomonosov actually anticipated the breakthroughs detailed in the next chapter. This chemist however was Russian, and in those times Russia was also a bit off the beaten path—though not for long. As Lomonosov challenged Russian youth, the time was coming "to show that the Russian land can give birth to its own Platos and quick-witted Newtons."[9]

Mikhail Vasilevich Lomonosov

Gifted with special linguistic abilities, Mikhail Lomonosov learned to read and write at an early age. He went to study in Moscow at the age of 20 against his father's wishes, and on his own slim budget, he managed to travel to Germany and Holland. After that he returned to St. Petersburg to marry and stayed in Russia for the rest of his life.

A man with an unusual mix of abilities, he wrote poetry, a textbook on grammar, and a textbook on rhetoric. But it often takes a creative as well as an analytical mind for insights into science. Lomonosov's insight was that he refused to see phlogiston where there was none. He heated metals in air-tight vessels and found—contrary to Boyle—no increase in the weight of the system. Furthermore Lomonosov did not see the necessity of invoking phlogiston to explain these results. He believed they could be accounted for by assuming that something from the air combined with the metal. Unfortunately Lomonosov did not have the scientific stature, nor did any Russian at the time, for his views to be widely influential. Before such ideas gained acceptance, they would have to be advanced by a nation viewed as more scientifically acceptable: precisely the role written for France in the 1700s.

But wherever they occurred, breakthroughs were inevitable; the theory of phlogiston was clearly an effort to fit a square peg into a round hole. In the 1700s physics and astronomy became so systematic that it seemed only right that chemistry should behave the same. As if by sheer effort of desire, chemists had come up with phlogiston, hoping it would be the unifying system for them. But the theory of phlogiston is more than an historical artifact: It shows that a search for fact can be influenced by a desire to work within an established framework of thought and that it is possible to fall in love with a theory. And the flexibility with which Stahl imbued phlogiston made it possible to use it to describe a number of phenomena: The worst criticism seemed to be that it was unnecessary to invoke it to explain all results. With each success it became more entrenched, and it looked more and more as if it would take a revolution to overthrow it. In the end, it did.

part TWO

chapter
EIGHT

ca. 1700: Révolution!

The remark in the last chapter that France was the scientifically acceptable culture of the 1700s may have caused a few curious second readings because in the last chapter we detailed the work of one Russian, two Scottish, one Swedish, and two English but no French chemists. The reason for this however was not a lack of material but rather such a surplus of material that it warrants a chapter unto itself. France was the center of European Enlightenment, and in France this passion for throwing off tradition took the extreme form: revolution. France in the 1700s saw two revolutions—one political and one chemical.

THE FRENCH REVOLUTION

Politically speaking Enlightenment philosophy valued equality and freedom for individuals. In France there was a keen need for freedom and equality: The country was in financial trouble, and those who were bearing the burden of the debt were powerless to do anything about it. Under the tax system the church and the nobles did not pay taxes, and the *bourgeoisie* (middle class) that did pay had to go through organizations of private tax collectors called tax farms. These tax farms purchased the privilege of collecting taxes by prepaying to the monarchy, then descended on the populace to collect the taxes and whatever profit they could. Toward the close of Louis XV's reign there were more than a million beggars in a French population of around twenty-three million—one out of every 23 people.

Louis XVI, a well-meaning but naive and weak monarch, tried some tax reforms, but he only managed to anger the nobles. Inspired by the Enlightenment and the United States' war of independence, some of these nobles formed a National Assembly to institute government reforms. Despite these efforts toward reform, the majority of the French people remained dissatisfied. In addition to bad harvests, trade slowed when the United States' war of independence ended, and the economy became depressed. Peasants refused to pay manorial dues and taxes. The Bastille was stormed by city dwellers attempting to arm themselves against hungry hordes from the country. In response the National Assembly nationalized church lands, abolished the feudal system, and created a system of centralized representative government. But the government got off to a shaky start.

In the midst of all this political strife, the philosophic debate on phlogiston also heated red hot. Scientists were polarized into phlogiston and antiphlogiston camps, basing their views on nationalistic and personal reasons as much as on experimentation. The debate was finally resolved, but with a resolution radical enough to constitute a revolution of its own.

THE CHEMICAL REVOLUTION

The Chemical Revolution was a revolution against the mysticism of the alchemists, the authority of the Scholastics, and the autocracy of the philosophy of the long-dead Aristotle. The weapons in this war were precise analytical measurement and precision in theoretical thought. The immediate result of the Chemical Revolution was to redefine chemical elements as separate materials with separate identities and properties. The long-term result was to reject chemical theory based on impressionistic speculation and replace it with theory based strictly on verifiable experimental results. Although there were many players in this philosophic drama, there was decidedly one central figure: Antoine Laurent Lavoisier—chemist, tax collector, and believer that the balance sheets in science must add up as well. When Lavoisier did the books, phlogiston did not fit in.

ANTOINE LAURENT LAVOISIER

Lavoisier's family was part of the prosperous French bourgeois, and Lavoisier's youth consequently was comfortable. His father was a lawyer, his mother the well-dowered daughter of a lawyer, and Lavoisier was an adored, indulged, and idolized child. Though he was one of two children, his sister died in her teens, and when his mother died when he was 5 years old, his recently widowed maternal grandmother and a maiden aunt took over his care.

He was given the best possible primary education, and when it came time for him to chose a profession, he started out (naturally) by studying law. But acting on what was probably a suggestion by a family friend, the geologist Guettard, he also attended popular courses in chemistry given in the Jardin du Roi. The main speaker at these lectures was Professor Bourdelain, but apparently more people went to see demonstrations following the lecture than to hear the main lecture itself. Performed by the lively apothecary, Guillaume-François Rouelle, the demonstrations sometimes contradicted the very theory that the main speaker had just proposed.

According to several accounts Rouelle was so unabashedly enthusiastic about his subject that he arrived at the lecture properly attired, but during the course of the demonstration, he stripped off wig and waistcoat and carried on his show in shirtsleeves. Though described by some as untutored and semiliterate, Rouelle's unconventional style achieved its purpose: He captured the interest of his audience, and his audience included Lavoisier.

Though Lavoisier passed the bar examination in the 1760s, his interest in chemistry did not flag. He traveled with the geologist

Guettard on a mineral survey of Alsace and Lorraine, and at the age of 21 he decided to become a member of the Academy of Science. Created in 1666 the French Academy of Science fills the same role as the Royal Society in England: It provides a forum for discussion, a vehicle for publication, and some laboratory facilities for experimentation. Like the Royal Society, membership in the French Academy of Science is by election. While Lavoisier's father and aunt would not have been averse to pulling strings to gain his election, Lavoisier did not need their help. His research into better methods for public street lighting was published by the academy and rewarded by a gold medal from the king. He became an associate member of the academy by the time he was 25 and a regular member when he was 26. Shortly after his election he justified further the academy's decision by performing a fine piece of original research: the water transmutation experiment. With this experiment he demonstrated his intention to take seriously the principle hinted at by Helmont and Black and hedged or ignored by the proponents of phlogiston: the conservation of mass.

Water Transmutation Experiment

It had long been believed and was an article of alchemical dogma that repeated and prolonged heating caused water to transmute into earth. This is because a residue is generally found when water evaporates to dryness (the well-known "water spots" on dishes). In truth water is such an excellent solvent for inorganic salts that it is difficult to obtain absolutely pure water—that is, water that does *not* leave some residue. Because water with dissolved salts may appear completely transparent, there was no reason for alchemists to suspect that deposits found after evaporating water were actually materials dissolved in the water in the first place. But Lavoisier believed that this transmutation should be put to the test. Carefully including the mass measurements that were now beginning to be recognized as important, he weighed some distilled water, sealed it in a glass container, then heated the container for 101 days. When he weighed the sealed system, he found no change in weight, which meant that no material had entered the water from the outside through the glass. Some flecks of material appeared in the water, but when he weighed the water and the dry container separately, he found that the container had lost a weight equal to the weight of material suspended in the water. In other words the material in the water was not formed by transmutation but rather by some sloughing off of bits of material from the walls of the glass container. Thus with one well-planned and well-executed experiment, Lavoisier refuted an established theory of alchemists and demonstrated that valid chemical theory must be experimentally verifiable.

Talented though he was, Lavoisier, like other scientists of the time, had to finance his own experiments. In an effort to achieve financial security, he made one of the best and one of the worst moves of his life: He used his inheritance from his mother to purchase membership in a tax-collecting firm. Why this was one of his worst moves, we shall presently see. It was one of his best moves because he met Jacques Paulze, another tax collector, and his 13-year-old daughter, Marie. Marie Paulze was to became Lavoisier's wife before she turned 14.

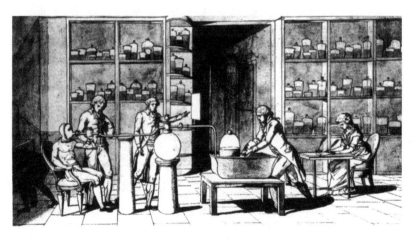

Figure 8.1. Marie and Antoine Lavoisier in the laboratory (drawn by Marie Lavoisier). They are conducting experiments on metabolism. (Courtesy of the John F. Kennedy Library, California State University, Los Angeles.)

Marie Anne Pierrette Paulze Lavoisier

Educated in a convent since the death of her mother when Marie was 3 years old, Marie Anne Pierrette Paulze Lavoisier had talent in both art and languages. She must also have had some innate proclivity for science, because she immediately became involved in Lavoisier's scientific work: translating scientific papers, working in the laboratory, keeping laboratory notebooks, and making illustrations of experimental setups. Lavoisier, now well settled, well financed, and well assisted, started on what was to be his most important work: to use the principle of the conservation of mass to revolutionize chemistry.

When Lavoisier learned of a report by a Paris pharmacist, Pierre Mitouard, that phosphorus seemed to absorb air when it was burned to form acid. Lavoisier repeated the experiment with phosphorus, then with sulfur, and found that the acids produced did indeed weigh more than the starting materials. He wrote down these results in a sealed note and deposited it with the Academy—a practice common for his times and similar to today's practice of notarizing notebooks to ensure credit for originality. (It took several years for articles submitted to the Academy's journal to be published.) Working with colleagues at the Academy, Lavoisier burned a diamond in the focus of a giant burning glass. As a combustion product he found carbon dioxide, the gas described by Black as fixed air. In other experiments he heated metal oxides (calxes) in the presence of charcoal and found fixed air again given off. Ruminating on these results, he wrote in a notebook with characteristic assurance:

> all this work . . . seemed to be destined to bring about a revolution in physics and chemistry. I have felt bound to look upon all that has been done before me as suggestive. I have proposed to repeat it all with new safeguards, in order to link our knowledge of the air that goes into combination or that is liberated from substances, with other acquired knowledge, and to form a theory[1].

Lavoisier soon had both his theory and his revolution, when he identified his air.

Oxygen

In October 1774 Lord Shelburne and his secretary, Joseph Priestley, arrived in Paris. In the course of their visit they dined with Lavoisier and several other French scientists. During this dinner Priestley told the assembled company about the new air he had found that caused a candle to burn brighter than in common air. How much of Lavoisier's subsequent work depended on this revelation by Priestley is debatable, but afterward Priestley was certainly willing to debate it. Lavoisier had already investigated air absorbed by metals on heating and—although he believed it was Black's fixed air, carbon dioxide—with or without Priestley's new information, he probably would have eventually associated it with oxygen. Lavoisier may have already known about Scheele's fire-air, and he probably knew that other investigators, such as Pierre Bayen, had observed a gas being given off when heating mercuric oxide. But by whatever route Lavoisier arrived at his own conclusions, and by March 1775 he was able to announce that the gas liberated when heating mercuric oxide was a gas closely resembling common air.

Priestley, on reading Lavoisier's result, pointed out that the gas had a "goodness"—an ability to support combustion—that was greater than common air, citing his own tests with nitric oxide. Lavoisier tried the test on a sample of what he called "dephlogisticated air of M. Prisley [sic],"[2] and he found what Priestley had said to be true. But on April 25, 1775 he read a paper to the academy on "the nature of the principle which combines with metals during calcination and increases their weight,"[3] he did not mention the contributions of Priestley, Scheele, or Bayen, and he did not invoke phlogiston.

It is clear that Lavoisier believed he deserved some credit for the discovery of oxygen, and while Lavoisier was certainly not the first to isolate this gas physically and explore its properties, he should be credited with being the first to recognize that this air was not just a variety of common air but a pure and independent substance with unique properties. He definitely deserves credit for his strict adherence to the balance sheet of mass, which left no room for phlogiston.

In 1775 Lavoisier was appointed scientific director of the Royal Gunpowder Administration, and he and Marie moved to the Paris Arsenal. Here Lavoisier set up a laboratory and immediately found procedures to improve the quality of the potassium nitrate used for French gunpowder. Here he also carried out experiments, assisted by Marie Lavoisier and others (to be met shortly), on combustion and respiration, and he began to publish papers clarifying his ideas on the role of oxygen. In a paper published in 1777, he concluded that all acids contain a portion of the "purest part of the air."[4] To name this part of the air, he used the term *oxygène* from the Greek, by which he meant to imply *acid former*, but this assumption was wrong. While it is true that oxygen is part of carbonic, sulfuric, nitric, and many organic acids, it is not a component of such acids as hydrochloric acid, and this was pointed out to Lavoisier at the time. Lavoisier countered that this only appeared to be the case, and that the composition of hydrochloric acid had not yet been fully established.

Given our present understanding of chemistry, the idea that the presence of an element in a compound determines the nature of a compound—that is, if it has oxygen, then it must be an acid—is clearly an oversimplification. Sodium chloride, table salt, is present in virtually everyone's diet and generally causes few problems. In fact saltwater, or *normal saline solution*, is given intravenously to prevent dehydration. Sodium however is a metal that reacts violently with water, and chlorine is a highly toxic gas. Clearly sodium chloride does not derive its properties solely from the properties of sodium and chlorine. Lavoisier's conviction that the presence of oxygen gave acids their nature was similar to residual Aristotelian ideas about the qualities associated with elements: Fire is hot and dry, water is cold and moist, and so forth. This thinking had been ingrained as much as the Aristotelian theory of the pervasive nature of the four elements. Whereas Lavoisier had managed to step free of the latter, the former may still have influenced his thinking on acids. However it was this persuasion that led him to the next famous experiment we discuss and to an accusation of plagiarism.

The Composition of Water

Several workers had noticed that water formed when a mixture of hydrogen with oxygen (or common air) was sparked, but they were cautious in their conclusions. Priestley sparked hydrogen with common air and found water, but he did not publish these results. James Watt, of steam engine fame, also tried the experiment, as did Henry Cavendish. Cavendish prepared a large sample of water by the method, but he did not immediately publish his results because he found his water to be slightly acidic. He located his acid source—small amounts of nitrogen in the oxygen that formed nitric oxides, which react with water to form acid—and then published his results, but by then others—among them Lavoisier—were aware of the reaction.

Cavendish's assistant, Charles Blagden (another person to whom Cavendish left a considerable annuity) visited Paris in 1783. There he congenially and benignly conveyed information on the experiments of Cavendish and Watt to Lavoisier. Lavoisier, who had thought that an acid should be formed from the oxygen–hydrogen mixture, verified that water was the only product, then went one step further and decomposed water by running steam through a red-hot gun barrel and collected hydrogen and iron oxide as his products. Lavoisier did not hesitate. He made the pronouncement that water was not an element as previously thought but the combination of oxygen with an inflammable principle, which he named hydrogen, from the Greek for the begetter of water. He claimed priority for this discovery, making only slight reference to the work of others. There was perhaps understandably a furor. Watt felt that Cavendish and Lavoisier had used some of his ideas, but of course all three owed some debt to Priestley. Again it may be asserted that the significance of Lavoisier's work lies not in the timing of his experimental work but in his interpretation of the results. From the same results Cavendish concluded that inflammable air (hydrogen) was really water with added phlogiston. When it combined with dephlogisticated air (oxygen), the result was just water. Lavoisier however saw it as the combination of two elements to form a compound, and he did not mention phlogiston.

Heat and Light

With these results Lavoisier was nearly ready to formulate his promised new theory that he believed was "destined to bring about a revolution in physics and chemistry."[5] But if this theory included an explanation of combustion, he knew he would have to account for the more spectacular products of this particular reaction: heat and light. Lavoisier set out to do just that, working with a soon-to-be-famous younger colleague in the academy, Pierre-Simon, Marquis de Laplace.

Pierre-Simon de Laplace

Besides their interest in science these two men also seemed to have in common a strong belief in their own talents. Laplace considered himself the best mathematician in France, and he did not hesitate to give his opinion on nearly any other subject as well. But in the cases of Laplace and Lavoisier, it is hard to say that their self-assessments—though perhaps irritating—were ill-founded. Laplace, unlike Lavoisier, did not extend his outspokenness to politics, which may have been fortunate for him.

Lavoisier and Laplace developed a calorimeter, a device for measuring heat released during respiration and combustion. The device consisted of a chamber surrounded by an ice-packed jacket. The amount of water collected from melted ice was used to measure the heat evolved in the inner chamber. To improve the accuracy of the measurement, the whole device was insulated with another ice-packed jacket, and experiments were performed only on days when the outside temperature was a few degrees above freezing. With this device Laplace and Lavoisier measured the amount of heat that evolved in the combustion of charcoal to form carbon dioxide. They then measured the amount of heat generated by a guinea pig while collecting the carbon dioxide the animal exhaled. They compared the amount of heat generated by the animal with the amount of heat generated by combustion to produce the same amount of carbon dioxide, and they obtained accurate enough results to conclude that respiration was a form of combustion.

Despite this interesting experimental work, Lavoisier's resulting explanations of the nature of heat and light seem forced. Laplace favored a mechanical explanation of heat as the motion of particles of matter (as it is currently understood), but Lavoisier described heat as a substance. This material he called caloric, the matter of fire, and described it as weightless (or too difficult to weigh), which made it reminiscent of phlogiston. But unlike the phlogistonists he could quantitatively measure his matter of fire. He could not however clearly separate heat from light, and his description of light was not quantitative.

Reflections on Phlogiston

Not feeling that these were fatally weak points in his system, Lavoisier proceeded formally to attack the phlogiston theory. In 1783 he submitted a memoir entitled *Reflections on Phlogiston* to the academy. In *Reflections* he credited Stahl with the discovery that calcination is a form of combustion and with the observation that combustibility could be transmitted from one body to another. But then he proceeded to turn the whole system upside down. Calcination had been viewed as a separation—phlogiston from metal—but Lavoisier asserted it was a combination—oxygen with metal. Reduction had been considered a combination—metal with phlogiston, but Lavoisier made it a separation—oxygen from metal. Oxygen, he said, was not a variety of the element air but a separate substance—an element unto itself. And phlogiston was not just a variety of the element of fire—phlogiston, he said, did not exist.

Lavoisier had by no means arrived at a modern definition of an element. Instead he retained some of the Aristotelian notion that an element must be found everywhere. He wrote

> it is not enough for a substance to be simple, indivisible, or at least undecomposed for us to call it an element. It is also necessary for it to be abundantly distributed in nature and to enter as an essential and constituent principle in the composition of a great number of bodies.[6]

He also retained the idea that the element itself dictated functionality (which is an extension of the Aristotelian notion of fundamental properties of elements). His explanations of heat and light also required something like metaphysical faith for acceptance. His true accomplishments however were that he broke the Aristotelian barrier of four elements, established the conservation of mass as an inviolate law, and confirmed the need for verifiable experimental results as the basis for valid chemical theory. And for one lifetime, perhaps this was enough.

After learning of Lavoisier's exposition, several other chemists, including most notably Claude Louis Berthollet, Louis Bernard Guyton de Morveau, and Antoine François Fourcroy, began to see the utility of his notions. Lavoisier's ideas became known as the theory of the French chemists. Lavoisier, however, said "It is mine."[7]

Claude Louis Berthollet

Though born to French nobility Berthollet did not have the financial advantages of Lavoisier. He chose the profession of physician and studied chemistry in connection with his studies in medicine. Perhaps through family connections he made the acquaintance of the Duke of Orleans, a widower, retired from the king's service, with a passion for the theater, music, and a certain Marquise de Montesson. The duke recommended Berthollet as private physician to Madame de Montesson, and this position provided Berthollet with the support to marry and the leisure time to do chemical experiments in the duke's personal laboratory. Berthollet repeated experiments of Lavoisier, Priestley, and Scheele, and that work won him election to the Academy of Sciences. He then worked with Fourcroy, de Morveau, and the Lavoisiers on the promotion and dissemination of Lavoisier's new system of chemistry.

Bernard Guyton de Morveau

A lawyer from a bourgeois family, Bernard Guyton added de Morveau to his name to designate family property. Following the course of politics during the French revolution, he became Guyton-Morveau, then Guyton, then Guyton-Morveau again. He taught himself chemistry and equipped a laboratory in his house. He chose as a personal project to verify that in fact *every* metal gained weight on heating in air. No one had so carefully and systematically approached the question before, and de Morveau's work had worth. He based his original explanation for the phenomenon on the levitation powers of phlogiston, but Lavoisier convinced him that it was caused by the absorption of part of the air.

De Morveau became a lecturer in chemistry at the Dijon Academy. Like Marie Lavoisier, de Morveau had a talent for languages and translated a number of books and papers, working with the wife of an associate at the Dijon academy, Claudine Picardet (he and Picardet later married after the death of her husband). His understanding of languages also manifested itself in his frustration with the current chemical nomenclature. Based on appearance (oil of vitriol, butter of antimony), or a person (Glauber's salt), or a place (Epsom salt), names of chemicals rarely described their composition or chemical behavior. He proposed some reforms, then worked with Berthollet, Lavoisier, and Fourcroy to develop a systematic method of nomenclature.

Antoine François Fourcroy

The youngest of the group of Lavoisier's close associates was Fourcroy. Fourcroy's family had been nobles, but they now were part of the working class. Fourcroy's father was an apothecary, and his mother died when he was seven. Never described as a particularly brilliant student, Fourcroy was very hard working, and when the secretary of the Royal Society of Medicine met this earnest young man, he managed to have Fourcroy's medical education sponsored by the Society of Medicine. Fourcroy never practiced medicine, but finding he had a talent for chemistry and teaching, he began lecturing in the private laboratory of Bucquet. He eventually became professor of chemistry at the Jardin du Roi, and when Fourcroy married, he was able to use part of his wife's dowry to set up a private laboratory where he gave private instruction in chemistry.

Reportedly an excellent teacher, Fourcroy's approach to research seems a bit mundane. The body of work that this hard-working chemist was able to accomplish is impressive, but there are few remarkable insights. He was important however for his association with, and support of, Lavoisier. This came about through his membership in the Academy of Sciences. Although he first applied to the Academy of Sciences in his early twenties, he did not succeed until he was in his thirties. When he was admitted, he became predictably one of the hardest workers.

Aside from assignments from the Academy of Sciences, Fourcroy pursed his own research interests, which reflected his previous training in medicine. Interested in changes that take place in matter during putrefaction, he took advantage of the opportunity to examine

1000 corpses exhumed from the Cemetery of the Innocents. He carefully described the effect of heat, air, water, alcohol, acids, alkalis, and other solvents on decayed material. Fourcroy also investigated the effect of injecting putrefied blood into animals (there was a theory at the time that scurvy was caused by blood putrefying in the veins), but his subjects did not survive to develop symptoms. Fourcroy and an associate, Vauquelin (whose story more rightly belongs in the next chapter, and that is where we have placed it) also examined chemically the urine of such diverse species as birds, lions, tigers, and snakes.

Fourcroy was not particularly political (although it was an age that forced politics on everyone), but he favored reforms proposed by the revolution. After the revolution had begun, he taught revolutionary courses (short intensive courses on the manufacture of niter, gunpowder, and cannon, etc.) and helped write instructions intended to guide the general public in the production of such practical items as soap. In another support of the revolutionary effort, he developed a process to separate copper from the copper–tin alloy of church bells.

In addition to his voluminous work, Fourcroy was also a voluminous writer. Though described as verbose his writings were accurate and understandable. Several women attended his lectures (including Marie Roland, destined to meet the guillotine), and Fourcroy wrote some works directed specifically to this audience. He also wrote a textbook, *Elements of Natural History and Chemistry*, the second edition of which presented both the phlogiston and oxygen theories. When Fourcroy presented the second edition to the Academy of Sciences, Lavoisier was on the committee appointed to examine it. Perhaps because Lavoisier was aware of how important the full endorsement of such an influential person in French chemical education would be, the publication of the book was held up until Fourcroy became convinced of the validity of Lavoisier's theory. It happened at about the same time as the conversion of Berthollet.

Nomenclature

These three chemists—de Morveau, Berthollet, and Fourcroy—then joined the Lavoisiers in establishing Lavoisier's theory. One major step in this direction was the proposal and publication of a new system of nomenclature that agreed with the theory. The new system (which is still in use) named compounds according to elements they contained and their functionality. For instance a compound of a metal with a nonmetal had the suffix *ide* (for example tin oxide); acids were named for their nonoxygen component (for example sulfuric acid). Salts of acids had different suffixes to distinguish them from acids (for example a salt of sulfuric acid was a sulfate). Acids with the same nonoxygen component but different amounts of oxygen were named differently (for example acid formed from sulfur with less oxygen was called sulfurous acid, and the salt of this acid was a sulfite). Significantly the nomenclature text included tables of such elements as oxygen, nitrogen, hydrogen, carbon, sulfur, and phosphorus (Lavoisier also listed light and caloric, which he postulated to be the element of heat), a list of organic radicals, and a list of compounds of alkaline earths and alkali metals.

In other efforts to disseminate their views Lavoisier, Berthollet, Fourcroy, de Morveau, and others established a new journal, *Annales de Chimie,* and Lavoisier published his important work, *Traité élémentaire de chimie.* In it he listed 33 elements and proposed the term "gas" for materials that he and others had been calling "airs." He included his theory of heat and light as elements—a decided weakness in the system—but his explanation of calcination, combustion, and reduction in terms of oxygen rather than phlogiston had such a ring of truth that the system had the strength to stand.

Not all chemists were immediately convinced. Scheele died believing in the truth of phlogiston, and Cavendish believed that Lavoisier's explanations worked about as well as phlogiston and turned to studies in physics. The irascible Priestley wrote two of his last publications in defense of phlogiston, but the newer generation of chemists were generally convinced. Elizabeth Fulhame, a chemist in the newly formed United States, published an *Essay on Combustion* in 1794 that accepted Lavoisier's nomenclature. The Lavoisiers, with a rather dramatic touch, staged a celebration in which Marie, dressed as a priestess, burned on an altar the writings of Stahl and Becher. But then many things were being burned in those days. The *Traité élémentaire de chimie* was published in 1789, the first year of the French Revolution.

The momentum of the revolution gathered rapidly. Marie Antoinette appealed to the Austrian monarchy for intervention. The fear of foreign invasion enabled more zealous revolutionaries to gain control of the Legislative Assembly. Recruits from Marseilles brought to Paris the marching song known as the Marseillaise and working-class mobs seized and imprisoned the king and the royal family. Refractory priests and counterrevolutionaries were dragged out of prison, tried, and executed. Instead of enacting the constitution and disbanding as originally planned, the National Assembly extended its life from year to year, assigning its executive powers to a group known as the Committee of Public Safety. The role of the committee was to control anarchy and counterrevolution so that the nation's resources could be concentrated on the international war. The committee proceeded to carry out its work so enthusiastically that this period became known as the Reign of Terror.

The Academy of Sciences in this phase of the revolution was seen more and more as the bastion of the elite and privileged. Moreover Lavoisier, as a member of one of the tax-collecting firms, was a special target for such militants as Maximilien Robespierre and Jean Marat, a disgruntled would-be member of the Academy of Sciences whose publication on the element of fire had earned disparaging comments from Lavoisier. Although Lavoisier was a member of the important committee on weights and measures formed for the establishment of the metric system (a stated revolutionary goal) and had greatly improved the quality of the gunpowder used by revolutionary armies, he was arrested and imprisoned. Pluvenet, an obscure druggist who sold chemicals to Lavoisier, got the prosecutor to agree to move Lavoisier to a better prison if Marie Lavoisier would make the request in person. She went to the appointment, but apparently every bit as arrogant as her husband, she did not plead but used the opportunity to heap abuse on the people who had arrested her husband. Lavoisier was not transferred, and on the morning of May 8, 1794, he was tried. Later that day he followed his father-in-law to the guillotine.

There has been continued debate in history as to why Lavoisier's associates, Fourcroy and de Morveau, respected leaders in the revolutionary movement, did not rise to Lavoisier's defense. Apparently there is some evidence that a day or two before the execution, Fourcroy did go to a meeting of the Committee of Public Safety and spoke in Lavoisier's defense. Lazare Carnot (whose son is prominent in this history) confirmed the account that Fourcroy's plea was met with stony silence by Robespierre. Fourcroy must have reasoned that his hands were tied—in the figurative sense—and if he pursued his protest, the sense could become literal.[8]

Of the group around him though, Lavoisier suffered the worst. Fourcroy, always philosophically in support of the revolution, after Marat's assassination, became an active supporter of the revolution. He survived the Terror and even participated in the regime of Napoleon. He had long been in love with a married cousin, and when her husband died, he divorced his wife and married the cousin. From 1800 he resided in the Museum of Natural History and with his associate, Vauquelin, spent productive years plodding along with his accumulation of data.

Fourcroy was ultimately disappointed in his ambitions, not achieving the academic rank to which he aspired (a decision based on politics rather than qualifications), and his health failed. He died at the age of 54, probably from a stroke. De Morveau remained an active revolutionary and voted for the execution of Louis XVI. In fact he served as president of the first Committee of Public Safety, but he was too moderate by some standards and was replaced. Away fighting the Austrians at the time of Lavoisier's trial (de Morveau helped to organize the first military air force in the form of crew-carrying reconnaissance balloons), he may not have known of Lavoisier's arrest until too late, or had he known he may not have been able to return in time to protest. After the Terror his scientific career, like Fourcroy's, continued to prosper. De Morveau also worked well with Napoleon and under Napoleon was administrator of the mints, retiring finally only after Waterloo. Even when the Bourbon monarchy was restored and many of those responsible for the execution of Louis XVI were exiled, de Morveau was allowed to remain in France where he died 6 months later.

Perhaps the last and the most loyal of the entourage surrounding Lavoisier, Marie Lavoisier was to survive her husband by 42 years, living to be nearly 80 years old. Although she was also arrested during the Terror, she was held only briefly, and most of the family's confiscated possessions were returned to her after her husband's execution. Shortly after Lavoisier's death, Marie edited and published a collection of his scientific work. In this she was initially assisted by Armand Séguin, who had been Lavoisier's assistant near the time of his demise. But the collaboration between Séguin and Marie Lavoisier did not continue long. Séguin hesitated to commit his name to a document denouncing Lavoisier's still politically active executioners, and he believed he should be given credit for work Marie Lavoisier did not believe he deserved. Though it cannot be doubted that he put much of his person into his researches (part of Lavoisier's experiments on respiration included enclosing Séguin in a varnished silk bag with a brass face mask so that he breathed oxygen from a reservoir and the exhaled products could be collected), there is some question as to how much of the interpretation of results was his.

Marie Lavoisier remarried some 10 years after her husband's death, and her new husband, Count Rumford, figures prominently in chemical history in the 1800s. However the marriage was not a happy one. One story has her ordering her husband out of their house, and another has her pouring boiling water over his favorite flowers. Marie insisted on keeping the Lavoisier name, calling herself Countess Lavoisier of Rumford, and reportedly she became increasingly arrogant with time. The marriage lasted only 4 years.

But then all things change with time, including revolutions. Robespierre met his fate on the guillotine, and the young army officer, Napoleon, decided that the fight for freedom could be furthered by conscription, national war, and a permanent dictatorship with himself at the head. But though perverted in the end, the revolution in France shook the foundations of national monarchies and left a Europe never to be the same, just as Lavoisier's revolution broke the stranglehold of Aristotle on European thought, so that chemistry would never be the same—Lavoisier managed this feat because he was exceptionally brilliant but also because he was a product of his times. In another less revolutionary age, he might not have felt free to question traditional assumptions as he did.

Lavoisier's success may have had another more subtle source. Lavoisier was not a humble man: He was at times annoyingly arrogant, but then he was born to the privileged class, wealthy nearly to the point of independence, and married to a beautiful, devoted wife, so there was little in his life to convince him he should be otherwise. This may have been a source of his strength; maybe Lavoisier succeeded because he was just arrogant enough to put his ideas up against those of Aristotle.

But the ideas of Aristotle had served their purpose as a starting place for chemical theory, and as an early chemical theory with an empirical foundation, phlogiston served chemistry well, too. It provided a theoretical framework for interpreting experimental results, and it seemed to have some predictive powers (witness Scheele's prediction of fire-air, which he subsequently isolated). In fact the problem was that it worked *too* well; it clouded interpretations and crowded out alternative explanations. Lavoisier was not the first to challenge phlogiston, nor the first to propose conservation of mass, nor the first to discover oxygen, but he did have the insight to realize that oxygen could be used to defeat phlogiston finally and firmly. Once this crack in the Aristotelian edifice had been found, it quickly crumbled, and real theoretical progress began. The next generation of chemists started discovering new elements at an amazing rate, and with hardly a backward glance at the first famous four: earth, water, air, and fire.

chapter
NINE

ca. 1800–1848: Après Le Déluge

Louis XV, whose reign preceded the French Revolution, is said to have commented, "Après moi le déluge": After me, the deluge. In the history of chemistry we might say, "After the deluge, the deluge": After the chemical revolution of Lavoisier and the political revolution of France, the number of chemical discoveries—new elements and new laws to describe their interactions—increased dramatically. The French Revolution was a setback for the French Academy of Sciences, but it did little to slow the growth of scientific societies in general. In the 1800s they sprang up in England, France, Prussia, Sweden, Italy, Russia, Spain, Mexico, and the newly united United States. The French Academy itself was reorganized as part of the Institute of France, and it soon regained its former strength. The French Revolution also cut short Lavoisier's career, but it did little to slow down French science. In fact the revolution promoted chemistry in an unexpected way by furthering the career of a new patron of the sciences: a short, crass-mannered, ill-tempered army general who cheated at cards, Napoleon Bonaparte.

NAPOLEON

At the beginning of the 1800s there was a general desire in France for peace and calm. As the seedling French government, now in the hands of a directory (governance by committee) struggled to root itself, it relied heavily on Napoleon for defense. Bonaparte was so skillful in this defense that his armies became self-sustaining, and he functioned more or less independently. Ordered by the directory to invade England, he decided instead to cut off England's Mediterranean trade first by invading Egypt. Fortunately for our story he also decided to take a troop of scholars with him. These scholars discovered the Rosetta Stone for Egyptology, and for chemistry the first hints at what would later be known as the law of mass action.

BERTHOLLET, BONAPARTE, AND THE BEGINNINGS OF THE LAW OF MASS ACTION

One of the scholars who accompanied Napoleon to Egypt was Lavoisier's contemporary and collaborator, Berthollet. In Egypt Berthollet had the opportunity to study inland salt lakes that had at their edge a crust of sodium carbonate, known in ancient times as natron and used (among other things) for embalming. The sodium carbonate formed slowly as sodium chloride in the salt lakes reacted with the limestone (calcium carbonate) in their shores and bottoms: calcium carbonate and sodium chloride coming together to form sodium carbonate and calcium chloride. In his laboratory the only reaction Berthollet had seen occur spontaneously was the opposite reaction: calcium chloride solutions reacting with sodium carbonate solutions to give calcium carbonate, an insoluble precipitate. Berthollet reasoned correctly that the presence of excess sodium chloride in the waters of Lake Natron was responsible for the opposite reaction.

Consider the generic chemical reagents: A, B, C, and D. Chemical reactions are not so simple as A and B coming together to form C and D: A and B come together to form C and D until the reactants and products reach *equilibrium*—the point at which there is no further change in the amount of material. At equilibrium there is some C and D, but there is still some A and B left over, too. This is because chemical reactions are *reversible*: If A and B can come together to form C and D, then C and D can come together to form A and B. What Berthollet saw in his beaker was A and B coming together to form C and D. On the shores of Lake Natron there was already so much C and D (relative to A and B) that A and B were formed. The reaction proceeded in the opposite direction.

Berthollet stayed in Egypt for 2 years writing on the phenomenon he had observed and that 50 years later became known as the law of mass action. The French government though was in trouble again, and Berthollet's patron found it necessary to leave Egypt. Though successful in land warfare against the Egyptians and Ottoman Turks, Napoleon had not defeated the British at sea, and he had to slip through a British naval blockade to return. When he got to Paris he found a Directory suffering from corruption and profiteering, and famine and inflation in the Parisian streets. One of the directors decided the solution lay in overthrowing the directory and that Napoleon, a popular military hero, could help him. The coup was successful. A new government was installed with executive power ostensibly shared by a three-body consulate. Bonaparte became a consul, then the most powerful consul, then consul for life, and finally in 1804 Napoleon I, emperor of France, king of Italy, and protector of the Confederation of the Rhine.

Napoleon may be considered one of the enlightened despots. He drained marshes, enlarged harbors, built bridges, developed the Napoleonic legal code, made the metric system mandatory (using any other system became a penal offense). He also at last revised the tax system, making it uniform and efficient. He continued education reforms begun by the National Assembly early in the Revolution, establishing local high schools, military and technical schools, and a national university. He supported the École Polytechnique, which became a model for training in science and engineering, and its staff and students soon included France's best.

On Berthollet's return to France Napoleon made him a member of his newly established senate. The purpose of the senate was to safeguard the constitution, but in a dictatorship such duties are light, so Berthollet used the salary from the position to buy a country house just outside Paris at Arcueil and made it a center for work and discussion for the new chemists of France.

Berthollet performed an extensive experimental program to test his new hypothesis that the relative mass of reactants and products determined the direction of the reaction by allowing mixtures in various proportions to react under various conditions and analyzing the products. He found ample evidence for the reversibility of reactions, but unfortunately he also came to another conclusion that was false. Solutions could be made with varying concentrations, so he believed that products could have varying composition. For instance sodium chloride could be 1 part sodium, 1 part chlorine; 2 parts sodium, 3 parts chlorine; or any fraction in between. In the absence of an atomic theory, this was not unreasonable. Berthollet based his conclusion in part on crude analyses of impure products, which *did* give him varying results, and on analyses of salts that crystallize with different amounts of water. Other analysts of the day also published widely discrepant analyses of supposedly similar compounds, so the conclusion that materials could combine in almost any proportion was not unreasonable.

When this aspect of his theory was challenged, he answered the challenge with true Bertholletian diplomacy and reserve. But then this was the same person who had managed to get along with Lavoisier, Napoleon, and the Duke of Orléans. It is understandable that Berthollet saw that chemical reactions could shift under the influence of external forces—his own career required considerable bending with the wind. This compatibility was not manipulative but stemmed from a sincere love and appreciation of humanity, as can be seen from the following account of Berthollet's bleaching process.

In Berthollet's day the love of color and personal adornment was as strong as ever, and the demand for dyed textiles came from those who could afford it. Before dyeing textiles however they had to be bleached, which required spreading the cloth on the ground. This was not only labor intensive, but it kept fertile fields from being tilled. Berthollet developed a chemical bleaching process that used chlorine absorbed in sodium hydroxide solution—essentially the bleach sold in supermarkets today—and published the technique immediately without patenting it, so that it could be used without delay. He also found that potassium chlorate produced a more explosive powder than potassium nitrate (gunpowder), but when a public demonstration resulted in deaths, Berthollet pursued the project no further. But then fire power was not an area in which Napoleon needed help. Lavoisier had made French gunpowder the best in Europe, and Napoleon had an excellent army. Napoleon defeated the Austrians, then divorced his wife (Joséphine) and married Marie Louise, daughter of the emperor of Austria. He forced Spain to cede Louisiana to France, defeated the Prussians and took Berlin, and organized a Confederation of the Rhine, finally bringing the Holy Roman Empire to an end. England however remained out of his reach—and continental Europe did not take well to the Napoleonic yoke. Nationalism ran high. Within each country there were those who supported the goals of the French Revolution, but even they did not appreciate being subjects of an imperial France. Prussia and Austria repeatedly rebelled, and Britain was always ready to aid Napoleon's enemies. When Spain rebelled, Napoleon felt compelled to crush the rebellion ruthlessly, antagonizing one chemist who should have been solidly in his ranks. But this chemist was not one for falling in line. He was the same man who defied the prominent Berthollet, Joseph Louis Proust.

LAW OF DEFINITE PROPORTIONS

Joseph Louis Proust

Born in France, Joseph Louis Proust was the second son of an apothecary. (We have seen—and will continue to see—that apothecaries produce fine chemists.) Proust was initially apprenticed to his father, but when he was 20, despite his father's objection, he moved to Paris. He continued his education in chemistry and physics and participated in some of the first piloted balloon experiments. He eventually found a permanent teaching position in Spain, though he was reportedly an indifferent teacher. He married a Spanish woman of French descent, received an appointment to head a well-equipped chemical laboratory in Madrid, and settled.

After many experiments in which he perfected his analytical techniques, Proust became convinced that each chemical compound has a fixed and invariable composition by weight; that is, each compound has a *formula*. Whereas Berthollet had seen a blurred average (similar to the infamous statistic that the average family has 2.6 children), Proust found that his chemical children could not be divided into tenths. The observation, eventually known as the law of definite proportions, is best stated in Proust's own words:

> A compound is a substance to which Nature assigns fixed ratios . . . Nature never creates other than balance in hand . . .
> Between pole and pole . . . No differences have yet been observed between the oxides of iron from the South and those from the North. The cinnabar of Japan is constituted according to the same ratio as that of Almaden. Silver is not differently oxidized . . . in the . . . [chloride] of Peru than in that of Siberia. . . .[1]

Although Berthollet did not withdraw his theory, the controversy was carried on with courtesy, and in the end the predominance of the experimental evidence was on Proust's side, and an important physical law was established. (As an instance of historical irony however, we note that some of the compounds analyzed by Proust in support of the law of definite proportions—some metal oxides and sulfides—*do* have compositions that vary over a narrow range. In modern materials science and inorganic chemistry, these are called *berthollides*. Later in this history we see an interesting example.) But in the 1800s Berthollet's idea about varying composition was discredited, and in the process his ideas about the salt lakes of Egypt were discredited, too, which is why the law of mass action would take 50 years to surface again.

Proust's fate was no more kind. His work was suspended when French troops marched into Madrid and swept away Proust's laboratory in their wave of destruction. By this time the European diet had become dependent on refined sugar introduced from the New World, and Napoleon's supplies were cut off by Britain's naval blockade. Napoleon tried to use Proust's skills to come up with another source, but Proust, though reduced to poverty, refused Napoleon's offer to become supervisor of the manufacture of grape sugar, a substance Proust had discovered in grape juice. Napoleon was able to enlist other French chemists to refine the production of sugar from beets, but sugar alone could not save him. Napoleon's military genius met its match when he decided to invade Russia. Tsar Alexander lured the French army to the outskirts of Moscow, burned all supplies and shelter, then let the Russian winter do its work. Napoleon's prestige plummeted, the Bourbon dynasty was restored, but the influence of the revolution remained. The monarchy recognized the new legal codes and made no attempt to restore the system of feudalism and privilege. Internationally, too, the revolution left a legacy of bitter national rivalry felt especially between the British and French who had tangled at every turn. John Dalton, who capped Proust's law of definite proportions with a systematic atomic theory, was an insightful and able chemical problem solver, but nationalism, on occasion, clouded his solutions, too.

ATOMIC THEORY AND THE LAW OF MULTIPLE PROPORTIONS

John Dalton

The second son of a modest Quaker weaver in England's Lake District, John Dalton's contribution to chemistry was to reintroduce a systematic atomic theory based on the elements of Lavoisier. We say *reintroduce* because the concept of atoms was certainly nothing new: Democritus postulated atoms in pre-Aristotelian Greek philosophy, and atoms were proposed by Descartes and Hooke. In 1738 Daniel Bernoulli correctly derived Boyle's law by assuming gases consisted of collections of particles that continuously collided with the container walls. But Dalton did not propose atoms as an abstraction or mathematical device; Dalton's atoms were physical. They had characteristic masses (atomic weights) and combinations of these atoms in fixed ratios made up the range of chemical compounds.

A forerunner of Dalton, Jeremias Benjamin Richter (a porcelain chemist who died of tuberculosis at the age of 45) had proposed that chemical processes are based on mathematical laws, and he had coined

Figure 9.1. John Dalton, Quaker schoolteacher and 1800s champion of the atomic theory. (Courtesy Chemical Heritage Foundation.)

the word *stoichiometry*, to describe mass ratios of chemical elements in reactions. Dalton built on this insight by experimentally comparing mass ratios of elements in compounds and importantly between compounds.

To understand how these measurements led Dalton to his atomic theory, imagine a box of Velcro covered atoms: green atoms covered with "hook" Velcro and red atoms covered with "eye" Velcro. Only green atoms and red atoms can combine, but sometimes one red can combine with two green atoms, or two red atoms with one green, and so forth. If a sample of two-reds-with-one-green is removed and analyzed, the ratio of the red mass to the green mass will be two to one. This information alone doesn't prove the material is made from individual atoms. But if a sample of two-greens-with-one-red is taken out and analyzed, and the mass of red in this sample is compared to the mass of red in the last sample, then the fact that this ratio is also a whole number ratio (1 to 2) *does* say that the red must come in discrete units. This is the comparison Dalton made with chemical compounds.

Carbon and oxygen can form carbon monoxide (one carbon atom with one oxygen atom) or carbon dioxide (one carbon atom with two oxygen atoms). When Dalton compared the masses of carbon and oxygen between the monoxide and the dioxide, he found simple whole number ratios. He compared nitrogen monoxide and nitrogen dioxide and found the same. This first direct evidence for atoms became known as the law of multiple proportions. He observed "the elements of oxygen may combine with a certain portion of nitrous gas or with twice that portion, but with no intermediate quantity."[2]

Dalton also derived a scale of atomic weights. He arbitrarily chose the weight of hydrogen to be 1, and based on a belief that like atoms repel each other, he postulated that the most stable compound of two elements must contain only one atom of each. Because water was the only compound of hydrogen and oxygen known in Dalton's time, Dalton believed its formula had to be HO (today we know it is H_2O). Lavoisier found the mass ratio of oxygen to hydrogen in water to be 17 to 3 (today we know it to be 16 to 2), so Dalton assigned to oxygen the atomic weight of 17 divided by 3, or 5.5 (today we know oxygen has a mass of 16 compared to a hydrogen weight of 1).

Despite its foundation in dubious hypothesis and its erroneous initial results, Dalton's theory was just the breakthrough that was needed. For the first time it allowed chemists to interpret mass relationships rationally. In 1808 he published his findings in his book *New System of Chemical Philosophy*, at the age of 42.

Though Dalton's atoms helped explain many observations, such as the tendency of materials to combine in simple whole number ratios, the atomic theory brought to the fore more questions, such as the nature of the force that held atoms together: the hazy concept of chemical affinity. This question was addressed by Swedish chemist, Jäns Jakob Berzelius, using a new spice in the chemist's rack: electricity.

ELECTRICITY

Electricity itself, like atomic theory, was nothing new. The Greeks knew how to generate static electricity by rubbing amber with wool (the word electricity is derived from *elektron,* the Greek word for amber), and Otto von Guericke of air-pump fame made a machine for generating a high-potential electric charge in the 1500s. The Leiden jar for storing static charge was invented in 1745 by Pieter van Musschenbroek of Leiden, who stumbled across the method while trying to preserve electrical charge in an empty glass bottle. Unknowingly he built up considerable static charge on the surface of the bottle, which he discovered when he touched the bottle,"the arm and the body was affected in a terrible manner which I cannot express; in a word, I thought it was all up with me."[3] In the 1750s Benjamin Franklin carried out his famous kite experiment in which he collected a charge from a thunder cloud in a Leiden jar. He was fortunate in surviving this experiment; others who attempted to duplicate it did not. Franklin performed many revealing experiments with the Leiden jar, but these experiments were limited because the Leiden jar provided only one jolt of electricity at a time. The study of phenomena caused by a continuous charge was sparked by an invention of Alessandro Giuseppe Antonio Anastasio Volta.

Alessandro Volta

Volta's father was a Jesuit for 11 years but left the priesthood and married when it became clear that the family line would otherwise die out. The marriage produced three nuns and three males who went into the church. Volta however was withdrawn from the local Jesuit college when a philosophy professor tried to recruit him into the Jesuits with gifts of chocolates, bonbons, and secret communications. Volta probably would have not succeeded as a Jesuit in any case, because he was a man who "understood a lot about the electricity of women,"[4] and for many years he enjoyed the companionship of the singer Marianna Paris. He was drawn to natural philosophy, and he became interested in the Leiden jar, lightning, and static electricity.

Luigi Galvani

About this time Luigi Galvani, an anatomist, inserted a copper hook into a frog's leg and hung it on an iron fence to dry. The frog's leg, though completely disembodied, twitched. Galvani published his observation and named the new phenomenon "animal electricity," believing it to be present only in animal tissue. When Volta read of the observation of Galvani, he began his own experimentation and found that he could induce the same response with dissimilar metals. After thorough testing on several insects and animals—"it is very amusing to make a [headless] grasshopper sing"[5]—he concluded that the source of the response was not the animal but an electrical charge somehow created by joining different metals. The frog, he concluded, just served to complete the circuit.

Volta's researches were then slowed by two events: his marriage at the age of 49, which brought him three sons in as many years, and the French invasion of Italy. When he returned in earnest to his researches, he decided to mimic the animal conductor with paper soaked in brine. It worked, and he found that he could produce electric current on demand. He connected a series of cups with metal connectors and found that he could amplify the charge. In 1800 he demonstrated his invention for Napoleon, and Napoleon awarded him with a gold metal and authorized an annual prize for experiments on electricity.

Volta also submitted his results to the Royal Society in a letter to the president, Sir John Banks, dated March 20, 1800. Banks showed the letter to Sir Anthony Carlisle, a socially prominent physician, and within days Carlisle had made a "pile consisting of 17 half crowns [of silver], with a like number of pieces of zinc, and of pasteboard, soaked in salt water..."[6] He found that this pile likewise generated electricity. He attached it to an electroscope: two sheets of gold leaf that separate on being charged. This new device for measuring electric charge was more accurate (and less discouraging) than estimating the charge by feeling the shock. He began experiments with a friend, William Nicholson.

Nicholson was an entrepreneur who had shipped out with the East India Company; worked as commercial agent for Josiah Wedgwood of china tea-service fame; and served as master of a school for mathematics, a patent agent, and water engineer. He patented several of his own inventions, collaborated on at least one novel, had time to get married and have at least one son, wrote a chemical dictionary, and most important for our story, published a monthly science journal. In this journal Nicholson was able to report that he and

"Mr. Carlisle observed a disengagement of gas" where "a drop of water upon the upper plate" completed a circuit.[7] This was the first observed *electrolysis*, the decomposition of materials by electricity. The journal immediately became the forum for new investigations of electricity. Ultimately the journal was a commercial failure, Nicholson spent time in debtor's prison, and died in poverty after a lingering illness. But he set the stage for a spectacle of discovery, and onto this stage stepped Berzelius in the summer of 1800.

THE DUALISTIC THEORY OF CHEMICAL AFFINITY

Jöns Jakob Berzelius

Berzelius' father died when he was 4 years old, and his mother remarried a man who had five children of his own. Accounts vary as to the difficulty of this situation, but when his mother died when he was nine, he and his sister were sent to the home of a maternal uncle. At 15 Berzelius went to tutor on a nearby farm and became interested in natural science and medicine. At 19 he earned a three-year scholarship and was able to concentrate his efforts on study. Two years later it was 1800, the year Volta introduced his electricity-generating pile.

For his doctoral dissertation Berzelius built a voltaic pile and studied the effects of galvanic current on patients. He found no effects (and gained no new patients), but this started a chain of thought that culminated 11 years later in a dualistic theory of chemical affinity. Berzelius followed up the experiments of Nicholson and Carlisle to find that not only did electricity split water, but it also split salts. Simultaneously with Davy, who we encounter shortly, he used electrolysis to isolate such alkaline earth metals as calcium and barium. He then proposed a dualistic theory of chemical affinity based on electrical attraction:

> in every chemical combination there is a neutralization of opposite electricities, and this neutralization produces fire in the same way as it is produced in the discharges of the electric bottle, the electric pile, and lightning....[8]

This was intuitively pleasing because it held a germ of truth (the full picture would slowly evolve from this point). His ideas were well received, and Berzelius became a respected and successful chemist.

Berzelius also thought in terms of atoms, and he began the practice of writing chemical formulas using the first letter of the element name to stand for the atom of the element (adding a second letter when necessary to distinguish between two elements beginning with the same first letter). He also used superscripts to indicate their relative number in the compound. This system proved so serviceable that it is still used (except that our familiar H_2O would have been written by Berzelius as H^2O). Actually Berzelius would not have written the formula for water with either subscript or superscipt 2 because he initially accepted the Dalton's premise that a molecule of water was formed from one hydrogen and one oxygen atom (a *molecule*, formed from atoms, is the smallest particle of a compound substance that retains all the properties of the substance). From this assumption—and discoveries of several other chemists—he built a table of atomic weights.

Berzelius, his fame and fortune secured, was finally able to consider his personal life. He married at the age of 56, the 24-year-old daughter of an old friend. He wrote to an associate (who will soon be of our acquaintance, too), "Yes, my dear Wöhler, I have now been a benedict for six weeks. I have learned to know a side of life of which I formerly had a false conception or none at all."[9] When Berzelius entered his bride's home just before the ceremony, his future father-in-law handed him a letter from the king of Sweden with directions that it be read before the assembled guests. The letter announced Berzelius's new title of Baron.

The marriage was by all reports very happy, but even so a certain sadness hung over Berzelius's final years. He had worked exceptionally hard to make up for his financially poor start, and at one point he had been injured in a chemical explosion and recovered only after spending several months in darkness. He became subject to headaches and depression as he grew older. He became rigid in his thinking and refused to accept newer developments in chemistry. In his final years he was respected, but his opinions were not given much weight.

But this in no way negates a career that was otherwise long and productive. In his prime Berzelius was regarded by the chemical community as its leader. For 20 years he published his annual review of all significant developments in chemistry, and his pointed critique went a long way toward enhancing the perception of chemistry as a rigorous science. Among his other achievements was the recognition that the elements chlorine, bromine, and iodine all belonged to the same chemical family, which he called the *halogens*. He introduced the use of the terms *isomerism* to indicate substances with the same chemical composition but different physical properties (the concept of complex molecules was just beginning to emerge), *catalysis* to describe the ability of some materials to speed up reactions without being consumed, and *allotropy* to describe the fact that certain elements exist in different solid forms with different properties (such as diamond and graphite: both solid carbon, but with widely divergent properties and prices).

In the process of determining his accurate atomic weights, Berzelius designed many items of chemical apparatus that are now commonly used, including glass funnels, beakers, washing bottles, filter paper, and rubber tubes (sewn leather tubing had formerly been used). He used an alcohol burner known as the Berzelius lamp, which was replaced in laboratories by Bunsen burners only with the coming of gas lighting and piped gas supplies. He also used this new laboratory equipment to isolate and discover several new elements, which we will now discover, too.

In the early 1800s, as though now freed to look elsewhere for explanations, post-Lavoisier chemists began to see an underlying system in chemical reactions. During the early 1800s a series of laws were discovered that established the atomic theory, provided methods for determining atomic weights, and proposed theories to explain chemical affinity (though debate over atoms and affinities would continue into the next century). Although Berzelius's dualistic theory of affinity was too simplistic, the basic tenet that chemical affinity is based on electrical attraction was, after a hundred years or so, shown to be true. But before this point could be reached, there was more ground work to be done—literally. Now that Lavoisier had shown that there could be many elements, chemists set out to find them, and they did—mostly by digging in the dirt.

chapter
TEN

ca. 1800–1848:
The Professional Chemist

Between 1790 and 1848 it was open season on elements, and chemists added some 29 to Lavoisier's list—an average of about one every 2 years. These new elements had interesting new properties, and many had an immediate useful purpose. With the increased success of chemistry came the increased success of the chemist. No longer alchemists hiding in cellars—and more than artisans perpetuating a craft—chemists now were members of Societies, founding industries, and getting rich. Chemists gained respectability, and more people wanted to study this respectable skill. When industrialists saw the profits from improved processes, they wanted to hire chemists, too. There was now interest, income, and impetus to educate and support a new social class: the professional chemist. But professional chemists have always been more than practitioners of a trade; they are part artist and philosopher, too. This may stem from the fact that these first professional chemists were the product of two influences: the Industrial Revolution and Romanticism.

THE INDUSTRIAL REVOLUTION AND ROMANTICISM

Beginning in England in the mid-1700s technical advances and overseas trade expansion created a demand for products that could no longer be met by piecemeal output from homes. To make production more efficient, workers were brought to centrally located factories. The resulting collection of social and technical changes—known as the Industrial Revolution—benefited production greatly, though it is not clear that it did the same for society. When Dickens wrote about the Industrial Revolution, he wrote about Scrooge, Oliver, hard times, and elusive expectations. The Romantic movement was in some ways a reaction to the Industrial Revolution and the Enlightenment with its steel-edged rational approach. The Romantic thinker was able to accept some truths not necessarily proved by rational or scientific examination. The Romantic movement had as a basic doctrine harmony with nature and the belief that simplicity was the true character of the natural world. The professional chemist was therefore driven to technical advances by the Industrial Revolution but guided in this quest by Romanticism. The combination turned out to be very productive.

Berzelius, caught up in the Romantic spirit, went on long hikes to collect mineral and plant samples. Another such nature enthusiast, Friedrich Strohmeyer, made it a habit to go to apothecary shops as well as on hikes looking for new samples. In one of these shops he picked up a sample of contaminated zinc oxide, isolated the contaminant, and found a new element, cadmium. Abbé René Just Haüy liked to collect gemstones, and he gave one to Nicholas Louis Vauquelin, Fourcroy's assistant. Vauquelin analyzed the gemstone beryl and found the new element beryllium. To read of their successes, it would seem as though chemists of the day had only to shake a rock to have a new element fall out, but each discovery was actually the product of careful, painstaking work. The tools and techniques of the analysts and assayists—those who determined the identity of salts, amount of metal in ores, or the purity of materials—were employed and improved.

Those who excelled in these techniques soon found a ready market for their skills and from them evolved a type of specialist: the analytical chemist. One important representative of this class is Martin Henrich Klaproth.

Martin Henrich Klaproth

Martin Henrich Klaproth was born in the same year as Lavoisier, and he was in large part responsible for introducing Lavoisier's new chemistry into the German states (no minor task because Stahl, Lavoisier's antithesis, was German). Klaproth started out as a pharmacist's apprentice and eventually became the manager of an apothecary shop owned by Valentin Rose, who was to found a long line of chemists and apothecaries. The elder Rose died a short time after engaging Klaproth, who was then left with the task of educating his son, Valentin the younger. Klaproth discharged his duty faithfully. The son became a pharmacist and chemist, as did his sons in their turn.

Klaproth also collected and analyzed minerals, and like other chemists, there were instances when his analytical results did not add up to one hundred percent. But contrary to the habits of some other chemists, he did not assume that these differences were due to errors nor did he doctor his data to match expectation. He carefully repeated analyses, and when he confirmed his results, he started looking for the missing pieces. Sometimes the missing pieces turned out to be new elements. Klaproth was the first to isolate tellurium, naming it after the Latin for Earth. Klaproth also analyzed black magnetic sand and found titanium, a new metal named after the Titans, the first sons of the Earth.

Klaproth married well, which allowed him to set up his own laboratory. He purchased a mineral collection that included a specimen of pitchblende, a well-known ore. He attacked it with acid and heat and finally isolated large, clear yellow, four-sided crystals of a new salt. There are conflicting reports as to whether or not he isolated the new metallic element present in the salt, but he probably did not. He did however realize that the crystals were not compounds of any known element, so he made the following announcement to the Royal Prussian Academy of Science in Berlin:

chapter TEN

> The number of known metals had been increased by one—from 17 to 18. . . . A few years ago we thrilled to hear of the discovery of the final planet by Sir William Herschel. He called the new member of our solar system Uranus. I propose to borrow from the honor of that great discovery and call this new element Uranium.[1]

Though once believed rare, uranium is more common in the Earth's crust than mercury, antimony, silver, or cadmium, and its salts were used as coloring agents as early as 80 CE. The silvery white metal was later isolated when the French chemist Eugène-Melchior Peligot reduced the chloride salt with potassium.

Klaproth was well respected as a chemist, and he became the first professor of chemistry at the newly organized University of Berlin, though he was 67 at the time. He died, still teaching and still collecting rocks, at the age of 74.

There are other wonderful stories to accompany the discovery of other elements in this era, including the discovery of niobium in a specimen from the collection of John Winthrop the Younger, first governor of Massachusetts Bay Colony, alchemist, physician, and rock collector; the discovery of selenium by Berzelius while analyzing the residue from the floor of his sulfuric acid factory; the discovery of vanadium by Andrés Manuel del Río, professor at the School of Mines in Mexico City; and the discovery of palladium announced anonymously in a handbill by William Hyde Wollaston.[2] With the number of discoveries of new elements in the early 1800s, it would seem as though each announcement would be met with great jubilation. But one of the precepts of Romanticism was that nature is simple, implying there should be *fewer* elements, not more. Therefore each announced new element was met with some suspicion. In fact our first professional chemist, who by modern reckoning has as many as eight new elements credited to his name, actually resisted the idea that they were really elements, feeling it contrary to the unity of nature. He distrusted abstract and mechanical theories that attempted to rationalize nature (including Dalton's atomic theory). But then he, too, was the product of conflicting influences. His birthplace was in the romantic, natural setting of the wind-swept moorlands of Cornwall, with ancient fishing villages, Celtic monoliths, and the sheer cliffs of Land's End—but a romantic setting into which the steam engine had just been introduced by the Cornish tin miners. Throughout his life he would always be torn between romance and progress and his two great loves: chemistry and fly fishing.

Humphry Davy

Davy's ancestors were at times laborers, at times landed gentry who lived in Cornwall before records were kept on tombstones. Raised by a rich uncle, Davy's father was to be an heir, but the will was unsigned, so he was left with only a small farm, and he had to supplement his income with wood carving. After risking money in farming ventures and tin, he died in debt when Davy was still in his teens.

The United States' war of independence was being fought when Davy was born and the French Revolution spanned his early youth. His formal education ended when he was 15, and a year later Lavoisier ascended the steps of the guillotine. Davy's mother left the farm to set up a millinery business with a young female refugee from France, and Davy was apprenticed to John Borlase, surgeon and apothecary. Here he read romantic literature and wrote his own romantic poems. Once while delivering medicine to a patient in the country, he was so carried away by poetic inspiration he flung open his arms and let the medicine bottle fly into a hayfield beside the path. It was not found until the next day. It is easy to see why grinding powders and filling prescriptions would seem mundane to such a spirit, and although Davy liked the patients—and was liked in return—he disliked surgery and felt he was destined for better things. Apparently the Fates agreed, because they so arranged it.

As part of his mother's livelihood, she took in boarders, and one of these happened to be Gregory Watt, the son of James Watt, who we meet again in connection with his improvements to the steam engine. The son had been sent to Cornwall because his sister had died of tuberculosis, and his parents felt that he, too, was threatened. Watt had been educated at Glasgow in geology and chemistry, and he shared his knowledge with Davy. Through Watt, Davy met Davies Gilbert (née Giddy; on his marriage he took the name and arms of his wife's family) and through Gilbert, Dr. Thomas Beddoes.

Dr. Beddoes had lost his lectureship at Oxford because of his writings on the French Revolution, and he was now attempting to set up a Pneumatic Medical Institution in which he would investigate the use of factitious airs (synthetic gases, such as carbon dioxide) in treating disease. There were several precedents for this, including an investigation by English physicians into the use of carbon dioxide administered rectally to treat a condition described only as "putrid fever." Beddoes decided that gaseous emissions of cow barns were useful in treating consumption, and he prescribed residence in the hayloft to his patients. It is interesting that the title of his publication describing this treatment changed from *Speedy and Certain Cure for Pulmonary Consumption* to *Speedy Relief and Probable Cure* to *Probable Relief and Possible Cure* between the first and final drafts.

Davy's neighbors thought Beddoes was an agitator and a quack, but Davy got on well with the Beddoes family:

> Dr. Beddoes . . . one of the most original men I ever saw—uncommonly short and fat, with little elegance of manners, and nothing characteristic externally of genius or science; extremely silent, and, in a few words, a very bad companion. . . . Mrs. Beddoes is the reverse . . . extremely cheerful gay, and witty . . . We are already great friends. She had taken me to see all the fine scenery about Clifton; for the doctor, from his occupations and his bulk, is unable to walk much.[3]

Beddoes hired Davy to work at his institution, experimenting with various factitious gases. Davy's first experimental efforts were of a speculative nature, and, because he was essentially self-taught in the experimental arts, he lacked a systematic and critical approach. However he wrote about them in a series of essays, which were published by Beddoes in a book entitled *Contributions to Physical and Medical Knowledge, Principally from the West of England*. In some of the essays Davy attempted to refute the notion of caloric to show that heat was not a material but a mode of motion. This is the currently accepted view, but Davy's methods for demonstrating it were highly questionable. Interestingly he accepted the idea that light was a substance and proposed that electric fluid was condensed light. He also speculated that just as terrestrial plants renew the oxygen in the air, marine plants do so in the sea.

Partly because of Davy's youthful self-confidence and partly because of his association with Beddoes, the scientific establishment took it all as a grand joke. Davy angrily wrote in a notebook:

> These critics perhaps do not understand that these experiments were made at a time when I had studied Chemistry but four months when I have never seen a single experiment executed.... They do not perhaps consider that my apparatus could not be made more perfect and that infinite labour was required in performing every experiment . . . their inaccuracy have been determined by . . . eminent Chemists [of which] there are few enough God knows in England. . . .[4]

In the United States Joseph Priestley wrote in an appendix to his work *The Doctrine of Phlogiston Established* that "Mr. H. Davy's Essays . . . impressed me with a high opinion of his philosophical acumen,"[5] but Davy never quite recovered from the criticism of his homeland colleagues.

Fortunately for science though his ego was somewhat assuaged when he defied the notion that nitrous oxide was the very principle of contagion and inhaled it. He found it produced a significant physiological activity. He reported

> A thrilling, extending from the chest to the extremities, was almost immediately produced. I felt a sense of tangible extension highly pleasurable in every limb; my visual impressions were dazzling, and apparently magnified.... By degrees, as the pleasurable sensation increased, I lost all connection with external things . . . I existed in a world of newly connected and newly modified ideas. I theorized; I imagined that I made discoveries. I was awakened from this semi-delirious trance by Dr. Kinglake, who took the bag from my mouth, indignation and pride were the first feelings produced by the sight of the persons about me.[6]

Although Davy proposed the use of nitrous oxide, now sometimes known as laughing gas, in minor surgical operations, it achieved its first success as a recreational drug. Davy himself employed it in this capacity, though after being introduced to recreational drinking by

Gregory Watt, he also found it a good medicine for a hangover. The poet Coleridge, an acquaintance through Mrs. Beddoes and eventually a great admirer of Davy, inhaled the gas as did Dr. Peter Roget of thesaurus fame, and Thomas Wedgwood, son of Josiah, of china tea-service fame.

Buoyed by his success, Davy tried other gases. He knew nitric oxide formed nitric acid on contact with moist air, so he expelled as much air as he could from his lungs before inhaling it, but even so burned his mouth, larynx, and probably his lungs. He inhaled water gas, a combination of carbon monoxide and hydrogen, and nearly ended his experimental career and life.

He published *Researches, Chemical and Philosophical; Chiefly concerning Nitrous Oxide, or Dephlogisticated Nitrous Air, and Its Respiration*, a 580-page work, that was well received. It established Davy's reputation as a chemist and earlier criticisms were forgotten by everyone but Davy.

Davy was partly right in his sense of injury. He was working without the benefit of a formal chemical education, and in this, he was not unique in England. France was still the leader in science, having been stirred to action by the French Revolution. The German states were close behind, gaining energy in 1871 when the loose collection of German states was finally made into a nation. But in England there was no systematic effort to teach science in universities until University College, London, was founded in 1826. The Dissenting Academies did teach science, but they were not an option for the majority of students. A ray of light was shed on this dark situation when Benjamin Thompson, Count of Rumford, the second husband of Marie Lavoisier, decided that England would be wise to emulate France and established the Royal Institution.

THE ROYAL INSTITUTION

The French institution that Thompson sought to emulate was the Conservatory of Arts and Trades, founded to teach industrial skills to the general populace. He formed the Society for Encouraging Industry and Promoting the Welfare of the Poor and then submitted to this society a proposal to establish a

Public Institution for diffusing the knowledge, and facilitating the general introduction, of useful mechanical inventions and improvements, and for teaching by courses of philosophical lectures and experiments the application of science to the common purposes of life.[7]

After a general fund-raising, the Royal Institution of Great Britain was set up in London. Thompson invited Davy to become director of the laboratory at the newly founded Royal Institution and assistant professor of chemistry. Davy was at the time in his early twenties.

Thompson's dream of providing practical education for the masses was confounded by the fact that such efforts require funding, and donations were not forthcoming. Davy sensed this and with economic as well as scientific acumen, designed entertaining lectures to attract wealthy patrons. This in essence changed the mission of the institution and Thompson objected, but other more practical directors prevailed. Thompson responded by leaving England and spending the rest of his life in France.

It was hard however to argue with Davy's success. When Dalton was slated to deliver a series of lectures at the Royal Institution, Davy coached him on delivery and told him to concentrate on his first lecture because that was what people would remember. Dalton reported that the audience of some 200 people seemed pleased with the result. He described Davy as a "very agreeable and intelligent young man . . . the principal failing in his character as a philosopher is that he does not smoke."[8]

Nitrous oxide was naturally among the subjects Davy chose for his lectures, and a popular cartoon of the day (see Fig. 10.1) may have depicted the administration of a dose of it to a volunteer from the audience. But he also chose to lecture on another hot topic of research: electricity.

Davy was in his early twenties and working on the composition of laughing gas when Volta invented his pile. Davy was immediately attracted to this area of research and improved Volta's design by putting zinc and silver plates directly into acid and connecting them with wires. He used this more efficient battery to show that passing current through a salt or acid broke it down, moving part of the material toward the negative pole and part toward the positive pole. The substance of this work was delivered in a lecture to the Royal Society in 1806. Even though France and England were at war at the time, Napoleon awarded Davy the 3000-franc prize promised for "the best experiment . . . on the galvanic fluid."[9]

Davy went on to use his battery to attack caustic potash and soda, two materials that had defied analysis. Edmund Davy, his cousin and assistant at the time, said:

> When he saw the minute globules of potassium burst through the crust of potash, and take fire . . . he could not contain his joy—he actually danced about the room in ecstatic delight; some little time was required for him to compose himself to continue the experiment.[10]

Figure 10.1. A James Gillray caricature of the Royal Institution. The lecturer is Thomas Garrett, the assistant with the bellows is supposed to be Davy, and Rumford is standing off to the right, smiling on. (Courtesy of the John F. Kennedy Library, California State University, Los Angeles).

In his laboratory notebook Davy wrote in large letters "Capital experiment . . ."[11]

Compounds of potassium had been known since ancient times; potassium is actually the seventh most abundant element in the Earth's crust. But this was the first time the metal had been seen and understandably so. Potassium metal reacts with moisture to form hydrogen gas, which catches fire from the heat of the reaction. Davy stored his reactive potassium under naphtha (today used as lighter fluid) and proceeded 2 days later to isolate sodium from caustic soda. But as often happens, Davy's solution of one puzzle provided another. *Bases* (materials referred to as caustic or alkali) react with acids to form salts and water, which makes them, so to speak, the opposites of acids. When Davy electrolyzed his bases, he found that oxygen was one of the products. Lavoisier had said that oxygen was the acidifying principle, but Davy now showed that it could equally well be considered the principle of bases. Davy dedicated the next 4 years to demonstrating that elements do not behave as "principles" in the manner Lavoisier had said.

In his early thirties at the pinnacle of success, after having resisted the advances of women in his audiences for years, Davy abruptly fell in love. The object of his affection was a rich Scottish widow, Mrs. Apreece: a small, dark patron of the women's discussion groups then popular and called (derisively by a male-dominated intellectual mainstream) blue-stocking clubs.

Although most of the members tended to be intellectual dilettantes, women of serious intellectual ability, such as Mary Sommerville, the Scottish writer on mathematics and physical science, also associated with the clubs. Mrs. Apreece herself was more interested in the social than the intellectual value of the clubs, but Davy saw in Mrs. Apreece what he chose to see. He associated her Scottish background with his own rural origins, and he imagined that her interest in science meant that she would be supportive of his efforts. He lent her his favorite copy of *The Compleat Angler* and asked her to marry him. On April 8, 1812 he was made *Sir* Humphry Davy, and on April 11 he married Mrs. Apreece.

By all accounts the marriage was unhappy. Lady Davy did however accompany her husband to France to receive his prize from Napoleon, which did take some spirit because as mentioned, the countries were at war. Of interest to this history however was not the companionship of Lady Davy, but the companionship of Humphry Davy's "valet," Michael Faraday. Faraday was destined to make fundamental discoveries about the interaction of electricity with matter—foundations of the field of *electrochemistry*.

Michael Faraday

Michael Faraday was born in the hilly country of Surrey a little over a decade after the birth of his mentor, Davy. Faraday's father was a blacksmith in chronic ill health, but the family had the community support of their small dissenting Christian sect, the Sandemanians. His father eventually gave up his smithy and moved with his wife and four children to London to seek work. The father died when Faraday was 18, and his mother took in lodgers. His older brother became a blacksmith like his father, and Faraday became apprenticed to a bookbinder.

It is always difficult to say when any other than extreme circumstances change a person's life, but Michael Faraday's life certainly would have been different had he not become a bookbinder's apprentice. The job gave him an opportunity that other circumstances had not, because Michael Faraday not only bound books, he read them.

He read *The Improvement of the Mind*, which suggested keeping a notebook of ideas and observations. He began one. He read an article on electricity in *Encyclopedia Britannica* and confirmed what he could using a small electrostatic generator. He read Jane Marcet's *Conversations on Chemistry*, intended "more particularly [for] the female sex"[12] and decided to become a chemist.

Writing a scientific work for women was not unprecedented. Fourcroy had written a chemistry text for women and Euler's *Letters to a German Princess* on physics was very popular. Marcet's book was unusual in that it was written by a woman for women:

> In venturing to offer to the public, and more particularly to the female sex, an Introduction to Chemistry, the author, herself a woman, conceives that some explanation may be required; and she feels it the more necessary to apologize for the present undertaking, as her knowledge of the subject is but recent, and as she can have no real claims to the title of chemist.[13]

Marcet had become knowledgeable in the subject by the same method she used in teaching: conversation—conversations with her husband, a London physician and chemist, and friends, Berzelius, Davy, and other notables of the scientific community. In Marcet's book conversations were among a fictitious Mrs. B and her pupils Emily and Caroline. This style was evidently effective because the book went through 16 editions and an estimated 160,000 copies were sold in the United States before 1853.[14] Its influence on Michael Faraday can be read in his own words: "Mrs. Marcet's *Conversations on Chemistry* . . . gave me my foundation in that science."[15]

Faraday decided that science was the career for him, so he wrote to Joseph Banks, president of the Royal Society, asking him for a job and letting him know he would take anything. When he received no reply, he went to the headquarters of the society and requested a reply—repeatedly. Finally he was told that Banks had said that "the letter required no answer,"[16] but fate and Faraday defied Joseph Banks. A customer of the bookbinder's gave Faraday tickets to one of Davy's lectures, and he went and took notes. He then bound the notes with accompanying illustrations into a 386-page manuscript. He sent the manuscript, misspellings and all, to Davy with a request for a job. Davy did not have immediate work for him, but Faraday did not have long to wait. Davy, ever interested in chemical affinities, had heard that another chemist had made a compound of nitrogen and chlorine that exploded with the heat of the hand. Although Davy was aware that the other chemist had been injured in the synthesis, he decided to investigate the compound, too. His sample also exploded, wounding his eye. Still desiring to continue his work, he remembered Faraday. Faraday jumped at the chance to work for Davy part-time, taking dictation, even though Davy advised him not to give up his bookbinding job. When Davy fired a laboratory assistant for sloppiness, lax work, and brawling, he thought of Faraday again.

Davy and Faraday were physically mismatched. Davy was described as slight, round-shouldered, and a dandy, whereas Faraday described himself as having a brown beard, a large mouth, and a great nose. Portraits of Davy always show a perfectly poised, precisely groomed man, whereas photographs of Faraday show an unmanaged shock of hair. But these two men worked together well.

When Davy prepared to go to France to receive a prize for his work on electricity, he asked Faraday to accompany him as assistant and secretary. When Davy's valet got cold feet and refused to go, Davy asked Faraday if he would mind serving in that capacity. Though the traditional duties of a valet are to clean and care for a man's clothing and grooming, which at the time probably meant holding the towel at bath time and rinsing out socks, Faraday, if he did mind, kept it to himself

When they returned to London however, Faraday was promoted from valet. He received a room of his own at the Royal Institution and a salary of 30 shillings a week. Within 6 months he began giving lectures at the City Philosophical Society, and within a year he published his first paper, "The Analysis of Caustic Lime of Tuscany."[17] As he acknowledged, "Sir Humphry Davy gave me the analysis to make as a first attempt in chemistry, at a time when my fear was greater than my confidence, and both far greater than my knowledge."[18]

Faraday's experience grew. He studied chlorine and its reactions and isolated the first two compounds of carbon and chlorine. While serving as an expert witness in a court case, he investigated the ignition point of heated oil vapor, which led to the discovery of benzene. When he was 30 years old he married Sarah Barnard, the sister of one of his friends at the City Philosophical Society; when he was about 40 he began investigating connections between electricity and chemistry—the work that would be his greatest claim to chemical fame.

ELECTROCHEMISTRY

Faraday found two fundamental laws governing the action of electric current on solutions. The first was that the amount of chemical decomposition in a solution is proportional to the amount of electricity passing through it (in fact the unit used to measure the amount of electricity transferred in electrolysis is called a *faraday* in his honor). The second was that the mass of a material deposited at an electrode by a given amount of current is proportional to the atomic mass of the material divided by its charge (or *equivalent weight*). In essence these two laws imply that electricity is a chemical reagent and takes part in chemical reactions in definite proportions. Much was built on this foundation.

Faraday also concluded that *electrolysis* (the decomposition of material by electricity) occurred when electricity flowed through solutions, and he chose to call the electrically charged fragments *ions*, from the Greek word for wanderer. He theorized that these *ions* wandered between the poles of the battery (which he called *electrodes*), carrying the electrical current. Faraday then wished to find a prefix to combine with the words ion and electrode to denote positive or negative, and here he consulted with the Cambridge scholar William Whewell.[19] Faraday proposed possible prefixes such as *dexio* (right) and *scaio* (left), but Whewell advised against these terms, suggesting instead *anode, cathode, anion,* and *cation:* "If, however, you still adhere to dexio and scaio, I am puzzled to combine these with ion without so much coalition of vowels as will startle your readers . . ."[20]

What startled the readers however was not Faraday's terminology but another discovery that rocked the theoretical world: Hans Christian Oersted showed an interaction between electricity and magnetism when a wire carrying an electric current caused a nearby magnetic needle to move. His first experiment was made during a lecture, and he reported that the experiment made little impression on the audience. But news of the experiment made a strong impression on Faraday, and he began experiments on the effect. Unfortunately in the process he incurred the displeasure and censure of Davy—the person he respected most.

William Hyde Wollaston (of palladium fame) was probably the first to decide that the magnetic field induced by a current was not directed just toward or away from the wire but encircled the wire. He wagered that a wire carrying a current could be made to revolve around another current-carrying wire, and the wager was open to all comers. In April 1821 Wollaston and Davy tried the experiment, but it failed. Faraday heard them discussing the results, but he did not see the apparatus they used. During the summer those who could afford to generally left London, and so it was with Wollaston and Davy. Faraday remained to work on an historical sketch he had been asked to write on electromagnetism. While writing it he confirmed for himself the results of published experiments and tried some on his own, including a demonstration of the rotation of a current-carrying wire around a magnet and of a magnet round a current-carrying wire. When he published his results, he did not mention Wollaston. In October Davy and Wollaston returned.

Friends of Wollaston protested to Davy about Faraday's supposed usurpation of Wollaston's research. In subsequent publications Faraday took pains to point out Wollaston's anticipation of the phenomena, and this seemed to smooth all the ruffled feathers but one: A few years later when Faraday's friends posted a petition for Faraday's election to the Royal Society, Davy ordered Faraday to take it down. Faraday replied that he had not put it up and that his friends would not withdraw it. Davy then said that he would take it down himself, to which Faraday replied that he would no doubt do what he thought was for the good of the Society. (This, we must note, was the only instance we found of Faraday making a curt comment to Davy.)

This was an unfortunate set of circumstances all around. Faraday was so moral a person that it is impossible to believe that he purposefully did anything unethical, and the intimation that he did so must have been devastating. With the benefit of hindsight we can accept that several people presented with the same evidence may very well come to the same conclusion, and priority in execution does not necessarily mean exclusivity in inspiration. In fact across the Atlantic in what was viewed by most Europeans as the scientific backwater of the United States, Joseph Henry also demonstrated the interaction of electricity and magnetism, and in the absence of input from Wollaston, Davy, or Faraday. But Faraday suffered from the incident, unfortunately he was also being attacked on other fronts. Berzelius modeled atoms as dipoles, that is, having both positive and negative charges present but separated. This did not agree with Faraday's view of charge being carried by distinct positive and negative ions, so Berzelius declared Faraday's results incorrect. When Faraday proposed that the electric force was a field (that is, a force felt through space, not needing physical contact), no one took him seriously. In 1839 possibly also suffering from mercury poisoning, he had a nervous breakdown. For the next 5 years he spent more time working on administrative tasks and less on experimental work.

In 1845 William Thomson, the future Lord Kelvin, developed a mathematical treatment of Faraday's intuitive, nonmaterial lines of force. This regenerated Faraday's vigor and led him to investigate the effect of an electromagnetic field on light, the study of paramagnetic and diamagnetic substances, and other physical effects. James Clerk Maxwell eventually built a rigorous field theory founded on the work of Faraday, but Faraday could not follow Maxwell's mathematical descriptions because he was by this time in his creative decline. He concentrated his efforts on teaching and Christmas lectures for children. He died at the age of 71 in a house that had been provided for him by a grateful British state.

During the period of Faraday's greatest achievements, Davy was less active, seemingly content to play the role of the grand patron of chemistry. He served with Wollaston as co-secretary of the Royal Society and after the death of Joseph Banks, its president (though Banks had always thought Davy "rather too lively" for the presidency.[21]) Though he occasionally carried out some research, inventing for instance a miner's lamp that would not cause the gas explosions blamed for frequent and tragic mining disasters of the time, his scientific career was on the wane. Davy's marriage remained disappointing, too, though he and Lady Davy found they could get along passably well if they minimized their time together. Davy fell into the habit of going on extended hunting and fishing trips by himself, and Lady Davy nursed herself with "a morning draft of a small quantity of old rum in new milk."[22]

In his last days Davy wrote two meditative, nonchemical works entitled *Salmonia, or Days of Fly Fishing* and *Consolations in Travel*. And although Davy's life has been characterized as "tragic," a better description might be "slightly sad." After all in his life he had accomplished much, and when he died it was at night in bed, with his brother beside him. He had lived some 50 years.

Of the many achievements usually accredited to Davy, there are two in particular—the characterization of chlorine and iodine as elements—that we have not included because credit for these achievements is controversial. Accounts of these events often vary, depending on which side of the Channel the commentator stood. The other contender for the honor of these discoveries was France's premier chemist of the time, Joseph Louis Gay-Lussac.

Joseph Louis Gay-Lussac

Born in the same year as Davy, Gay-Lussac was a product of the Revolutionary era. His father, a lawyer and public prosecutor, had hoped that reform would be possible under the monarchy. Though originally considered a fairly liberal stance, this hope came to be viewed as reactionary as the Revolution progressed. The father was eventually arrested and not released until the fall of Robespierre. The family fell on hard times but managed to educate Joseph Louis, the eldest of their five children. Just before he turned sixteen, about a year after they released his father from jail, he received a government grant to attend the École Polytechnique.

Like Davy, Gay-Lussac was a country boy. Constantly in need of money, Gay-Lussac barely managed to hang on to what little he had. At the École he attended lectures in chemistry and physics as well as mining, public works, mechanics, mathematics, and drawing. He probably heard lectures by Fourcroy, Vauquelin, Guyton de Morveau, and perhaps Berthollet, though Berthollet was on the Egyptian expedition for at least part of this time. On Berthollet's return he asked a friend to recommend a student from the École Polytechnique to work as an assistant in his laboratory at the École, and the friend recommended Gay-Lussac. So for his last year at the École, Gay-Lussac worked for Berthollet and used Berthollet's prestige to soften the normal attendance requirements.

Impressed with the young man's chemical talents, Berthollet asked Gay-Lussac to work in his country house at Arcueil. This must have been an interesting and inspiring experience for the young Gay-Lussac. Laplace had property next door, and many chemists came by to work and hold discussions. It was also during his tenure at Arcueil that Gay-Lussac had his first hot-air ballooning adventures, an experience destined to be significant for his later work.

Gay-Lussac worked at Arcueil for Berthollet during the Napoleonic Empire, but when Napoleon suffered a final defeat, Berthollet retired.

Although Gay-Lussac had benefited indirectly from Napoleon's patronage, he had never been directly associated with the imperial court, and therefore he fit easily into the restored Bourbon monarchy. He assumed Berthollet's position at the École Polytechnique. He resisted his father's plans for an arranged marriage and instead married a shop assistant he found reading a chemistry book she kept hidden under the counter. She was well-educated, able to read English and German, and the marriage was happy and produced five children. From time to time though the relationship suffered from the monetary ills characteristic of the families of many early chemists: Some money always seemed to be spent on equipment and experiments when it was needed elsewhere.

Gay-Lussac had a fine reputation as a teacher at the École Polytechnique, giving clear presentations and demonstrations. A syllabus of one of his lecture series preserved by a visitor from the United States reads almost like a modern first-year general chemistry course syllabus, listing such topics as the law of definite proportions, acids and bases, principal metals, salts, and organic chemistry.

Like Davy, Gay-Lussac was also one of the new breed of professional chemists, but unlike Davy, Gay-Lussac crossed gracefully into the new era of collaborative work. Chemistry was getting to be so broad a body of knowledge that one person could hardly expect to be experienced in it all. Although chemists had worked in teams before, Gay-Lussac made the practice routine. When the news was received of Davy's isolation of potassium and sodium with a voltaic pile, Napoleon ordered a bigger voltaic pile to be built at the École Polytechnique. Gay-Lussac and his coworker, Louis Jacques Thenard, were put in charge.

Louis Jacques Thenard

Louis Jacques Thenard, the son of peasant farmers, received his first education from a local priest. It was realized that he had talent, so efforts were made to secure further education for him, and he was eventually sent to Paris to take advantage of educational opportunities there. He attended public lectures by Vauquelin and Fourcroy, and he managed to be hired by the Vauquelin household as a bottle washer and scullery boy. He must have performed well in this capacity, because Vauquelin soon had Thenard assisting at his lectures. Thenard was eventually appointed demonstrator at the École Polytechnique, and he was on his way. When Vauquelin retired from the College de France, Thenard was nominated to succeed him. He married and continued to rise in the academic–scientific community. (After his death the village he was born in, La Louptière, was renamed La Louptière-Thenard.) He was an able experimenter, preparing hydrogen peroxide for the first time, and he wrote a textbook that went through six editions and was translated into German, Italian, Spanish; sections on chemical analysis were also translated into English.

So Gay-Lussac and Thenard, like Davy, were knowledgeable and accomplished. And they, like him, believed they were the first discoverers of chlorine and iodine.

Chlorine, a greenish yellow, poisonous gas, was first prepared by the Swedish chemist Scheele. It was called oxymuriatic acid because it can be prepared from a reaction between manganese dioxide and muriatic (hydrochloric) acid. Following Lavoisier's dictum, hydrochloric acid should also have contained oxygen, because it was an acid. Therefore it was believed that oxymuriatic acid must contain a lot of oxygen. But oxygen in almost any compound combines with red-hot charcoal to produce carbon dioxide, and when Gay-Lussac and Thenard passed oxymuriatic acid gas over red-hot charcoal, they collected no carbon dioxide. This led them to suspect that oxymuriatic acid did not contain oxygen after all and that it might be an element. Berthollet at this time however was still their patron, and he advised against any radical conclusions. As a result they made the following report:

> Oxygenated muriatic acid is not decomposed by charcoal, and it might be supposed from this fact that . . . this gas is a simple body. The phenomena which it presents can be explained well enough on this hypothesis; [but] we shall not seek to defend it . . .[23]

Davy had independently done many experiments attempting to extract oxygen from oxymuriatic acid, and in the report of Gay-Lussac and Thenard, he found reinforcement for his tentative conclusions. He reported in no uncertain terms the elemental nature of chlorine. Davy was generally credited with the discovery, which may explain why feelings were a bit touchy when the English citizen Davy landed on French shores to collect a reward from a French emperor and to usurp the research of French chemists on another substance newly discovered in France, iodine. Ironically for all the fuss that followed, the true discoverer of iodine was neither Davy nor Gay-Lussac but the son of a saltpeter manufacturer, Bernard Courtois.

Bernard Courtois

Working in his father's business, which produced niter (saltpeter, potassium nitrate) from seaweed by treating it with strong acids, Courtois noticed that when he added excess sulfuric acid, clouds of violet vapor rose over the solution, then condensed into dark, shiny crystals. Courtois investigated the chemical properties of the new material during the next few months and prepared some of its compounds. However Courtois was busy with the war effort (niter is a component of gunpowder), and he may have been feeling the financial strain of his research, so he told two other chemists about his crystals and asked them to continue the work.

They did, and one used it as the basis for an application to the Institute of France. Gay-Lussac, on the application-review committee, was given a sample, and a sample was also given to Davy. Although Davy was aware of the prior work in France, where the substance was known as *iode* because of the violet color of its vapor, he announced that the new substance and its compounds would provide enough work to occupy several chemists and at any rate French chemists could benefit from another point of view. Working in his hotel with a portable laboratory that he had brought with him, Davy satisfied himself that the new material was a element with chemical properties similar to those of chlorine, and he announced the element iodine.

Davy's actions might have been ethically questionable in his day because at that time (and for many years after) it was considered bad form to intrude on another chemist's area of research. But such territorial boundaries were only policed by reciprocal respect, and once breached, these were impossible to restore. Davy's actions would not warrant a second glance in today's competitive research climate. It can be argued that today's attitude is healthier for science because competition fosters hard work. But we also see instances later on when unrestrained competition forces premature publication and the release of incomplete, unverified data.

Gay-Lussac by this time, perhaps feeling the hound at his heels, had also decided iodine was an element, and he announced his opinion in a newspaper article on Sunday, December 12, 1814. Davy's results were announced in a publicly read letter on Monday, December 13. This would seem to give Gay-Lussac one day's priority, except that Davy's letter was dated December 11. Gay-Lussac however said that he overheard the recipient apologizing to Davy for having read the letter on Monday but that he had received it only late that day.

Courtois in the end was given a prize of 6000 francs by the Institute of France. He gave up the saltpeter business and tried to make a living selling iodine and its compounds. He failed and died in poverty at the age of 61. The candidate for the vacancy in the Institute of France who gave the samples of iodine to Davy and Gay-Lussac came in third; Davy was the committee's first choice.

And so in the early 1800s the success of the atomic theory and the cloudburst of new elements made it seem as though scientific enlightenment now rained down as manna from above. But with the rains came dark clouds—clouds that hinted that nature was perhaps not so simple after all.

PROBLEMS WITH THE ATOM

Not content to experiment with just voltaic piles and iodine, Gay-Lussac studied gases, too, and he took his interest to great heights. Gay-Lussac once set an altitude record for piloted balloons at over 4 miles above the Earth. In another high-altitude attempt, Gay-Lussac reportedly jettisoned some equipment, including a chair, to lighten the balloon.

Locals in the French village where the chair landed were not aware of the latest fad and decided that it was a sign from heaven but puzzled over the meaning.[24] One of the earliest collaborations between Gay-Lussac and Thenard took place when Gay-Lussac brought back a flask of air collected at high altitude, and Thenard analyzed it for comparison with ground-level Paris air.

Another French balloon aficionado was Jacques Alexandre Cesar Charles, a self-taught, free-lance teacher of physics. Charles had the idea of increasing the lifting power of the balloon by filling it with hydrogen instead of hot air and coating his balloon with rubber to contain the hydrogen. He used hydrogen generated by acid reacting with iron, and it reportedly required a quarter of a ton of acid and half a ton of iron to generate enough gas. The balloon stayed aloft for almost 45 minutes and traveled some distance from Paris. However at this distance from Paris, people did not know what a balloon was, so when it landed, they attacked and destroyed it.

Gay-Lussac however also did some terrestrial gas experiments. In one of these he found that the volume of a gas expands linearly with increasing temperature. Dalton had also observed this, and the English sometimes refer to the observation as Dalton's law or Dalton and Gay-Lussac's law. Charles had observed gases expanding, too, but he had not taken care to dry his gases, so he had found inconsistent results. He did not publish, but Gay-Lussac, on publishing his own results, dutifully noted the work of his ballooning friend. An English author, Tait, came across this reference and decided that it gave priority to Charles, so he christened the observation Charles' law (even though Charles thought the relationship false). Today in the United States the law is fairly uniformly known as Charles' law, though it also referred to as the Charles–Gay-Lussac law.

During a session at Arcueil, Gay-Lussac met a visiting German scientist, Alexander von Humboldt, who had studied volumes of gases that combined in reactions. They collaborated on an investigation of the hydrogen–oxygen reaction and found that exactly 2 volumes of hydrogen combined with 1 volume of oxygen to contract into 1 volume of water, measured as a gas. Gay-Lussac went on to do further experiments on combining gas volumes and to recalculate other already published results. In every case he observed simple integral combining ratios. Then in one troublesome experiment, Gay-Lussac found that 1 volume of nitrogen and 1 volume of oxygen combined to form 2 volumes of nitric oxide—the total volume stayed the same.

To understand why this might be a problem, consider the following. Dalton's views about the mutual repulsions of atoms of the same kind led him to believe that atoms either combined with other kinds of atoms or remained alone; that is, atoms of like elements did not pair up. If this were pictured as a party, at Dalton's party, girls come alone and disperse themselves around the room and boys come alone and do likewise. At the sound of a chime partners are found, and the occupied volume of the room contracts. But at Gay-Lussac's party the volume did not contract.

There were a couple of ways of explaining the discrepancy between Dalton's theoretical prediction and Gay-Lussac's experimental result: Gay-Lussac was wrong and measurements were faulty—the explanation some English chemists chose—or Dalton was wrong and had a faulty theory—the explanation some French chemists chose. But the answer as it turned out was none of the above. It was however right under their nose, but not an English nose nor a French nose. It was an Italian nose, the nose of Lorenzo Romano Amedeo Carlo Avogadro di Quaregua e di Cerreto.

Amedeo Avogadro

Born to a line of ecclesiastical lawyers (the name Avogadro probably came from the Italian for advocate), Avogadro first took a degree in law but then became caught up in the excitement over Volta's pile and began to do research with his brother. This was in the early 1800s, and lots of people were working on Volta's pile and publishing in better known journals, so Avogadro's work did not receive much notice. When he proposed what is now known as Avogadro's law—that equal volumes of *all* gases contained an equal number of molecules (under like conditions of pressure and temperature)—this did not receive much notice either.

The implication of Avogadro's law is best understood in terms of *moles*, though the word mole was not used until the early 1900s. A

ca. 1800–1848: The Professional Chemist

mole (from molecule) is the unit of measurement for atomic-sized particles. Just as a dozen is 12, and a gross is 144, a mole is about six hundred billion trillion. It is a countable number of things, but because molecules are so small, a mole has to be a very large number to represent a reasonably sized sample. A dozen copper atoms are too small to see, but a mole of copper fits in a cube a bit less than an inch on each side. A mole of gas at ordinary room temperature and ordinary atmospheric pressure is about 22 liters or about 5 gallons. According to Avogadro's law a mole of nitrogen molecules (formed from two atoms of nitrogen) combines with a mole of oxygen molecules (formed from two atoms of oxygen) to give a mole of nitric oxide molecules (formed from one nitrogen atom and one oxygen atom).

The concept was the missing piece to the puzzle. At Avogadro's party, girls came in pairs and boys came in pairs. When the chime sounded they all exchanged partners for one of the opposite sex, but because the *number of pairs* did not change, the occupied volume stayed the same. But Avogadro's hypothesis contradicted the electrical theory of affinity proposed by the well-respected Berzelius. In Berzelius' theory only opposites should attract—not atoms of the same element. It also contradicted Dalton's ideal of simplest formulas. In short it was not well received. So there was Dalton with his atoms, Gay-Lussac with his combining volumes, and Avogadro with his molecules and they could not get together. But then it was not a synergistic age.

It was the age of nationalism and individual rights, and those forces were found to be as powerful as any religion had ever been. Aggressions and atrocities committed in the name of religion were now committed for flags and freedom as well. People demanded to determine their own destinies and to be citizens in their own nations. In 1848 there were revolutions in France, Vienna, Venice, Berlin, Milan, Parma, the Czech state, and Rome. The Balkan nations pushed back the Ottoman Empire, unleashing hideous, wholesale persecution of Christian Armenians. Marx and Engels issued the *Communist Manifesto*. In the United States Elizabeth Cady Stanton demanded woman's right to vote and Mexico gave Los Angeles to the United States.

But after the political reshuffling had quieted, there was a period of stability, prosperity, and economic growth. There were railroads, steamships, and telegraphs, and in 1860 an international conference was organized in an attempt to unify and organize the discrepant chemical concepts of atom, molecule, alkalinity, and affinity. The conference was mostly a failure, with little being resolved, but there was one notable exception: the contribution of a rebel, Stanislao Cannizzarro.

Stanislao Cannizzaro

The youngest of 10 children, Cannizzaro grew up in a family of Sicilian nobles who supported the Bourbon monarchy in Naples. When there was rebellion against the monarchy, however, Cannizzaro joined the rebelling side. When the revolution turned against the rebels, he fled to France. There he continued his work in chemistry, teaching in Alexandria, Genoa, Palermo, and finally in Rome. Cannizzaro wanted to present to his students a clear and logical development of the subject from first principles. He found that by accepting Avogadro's law, all the ideas about atoms, molecules, and combining volumes fell into place. In 1860 he attended the international Karlsruhe Conference in Germany, where he argued for the concepts of his compatriot, Avogadro. His arguments fell mostly on deaf ears, but on the last day, one of the Italian participants distributed copies of a pamphlet Cannizzaro had written outlining his course and its basis in Avogadro's law. The clear, stepwise, logical presentation was able to make headway at least in the minds of some who were ready for a change. Lothar Meyer, one of our future heroes, read the pamphlet on his way home and expressed the experience as scales falling from his eyes. Another convinced attendee at the conference was Mendeleev, who derived a periodic table for the elements. The world of chemistry was in for a change.

In the early 1800s chemists built on foundations laid down by Lavoisier. They found new theories, new laws, and new elements, right and left. When they were done, they had the tools—elements, atoms, molecules—and they took off in every direction. From this point the field of chemistry grew so quickly and so widely that chemists had to narrow their focus to limited, specialized areas of study to keep their work productive. These areas became known as subdisciplines of organic chemistry, inorganic chemistry, physical chemistry, biochemistry, analytical chemistry, and we will follow each divergent stream. We begin however, after all the wonderful additions to Lavoisier's list, with one much needed subtraction: caloric.

chapter
ELEVEN

ca. 1848–1914: Thermodynamics— The Heat of the Matter

It has always been evident that heat is involved in chemical reactions as a product (as in combustion) or as an ingredient (as in cooking). But what is the nature of heat? How can it be measured? Interesting questions, but prior to 1800s chemists had plenty of other fish to fry, and they did not expend much effort in finding the answers. With the Industrial Revolution however, it was found that heat from combustion could produce work—a lot of work—and the question of heat moved to the front burner.

HEAT

Lavoisier accepted the idea that heat was a (possibly unweighable) material substance (not unlike phlogiston) that flowed from hot materials to cold, and he called this substance caloric. With his associate the young Laplace, he carried out measurements of heat flow in chemical reactions and respiration. Dalton found that the material explanation worked well enough for him, too, but there was another school of thought. Some chemists considered heat a form of motion. The motion theory has deep roots, having been discussed by Boyle and Newton and described rather poetically by Francis Bacon, who wrote, "heat . . . is motion and nothing else . . . perpetually quivering, striving, and struggling . . . whence springs the fury of fire . . ."[1]

Those who supported the motion theory sought to show systematically that mechanical motion could be converted into heat. One of the first to tackle the problem was the youthful Humphry Davy. In the early 1800s he described an experiment in which water was formed when a clockwork device rubbed two ice cubes together in an insulated box. The water, he said, was evidence that motion was converted into heat, which melted some of the ice. The validity of this experiment however has been questioned. The heat to melt the ice could as easily have come from the surroundings. And if Davy had been clever enough to keep the whole system at the freezing point of water, the water formed from rubbing would have refrozen. Another more convincing demonstration came from Davy's eventual employer: the founder of the Royal Institution, Benjamin Thompson, Count Rumford.

Benjamin Thompson, Count Rumford

Rumford was born in the colony of Massachusetts, and he began his career as a storekeeper's apprentice. But he was well-read, attended Harvard College, and married a wealthy widow 14 years his senior. The couple had one child, but they separated a year before the signing of the United State's Declaration of Independence. The quarrel between husband and wife may have been political because Rumford took the loyalist side in the U.S. war of independence (some say he spied for the British), and after the Revolution he emigrated to Europe. There he enjoyed a successful military career and went on to run a munitions works in Bavaria. The Bavarian government showed their appreciation by making him a count, and for reasons unknown—but perhaps nostalgic—he took the name Count Rumford, Rumford being the former name of Concord, New Hampshire.

In the process of impressing the Bavarians with his skill at munitions, Rumford himself became impressed with their cannons. Specifically he found that boring a cannon produced enormous, seemingly endless, amounts of heat. He reasoned that if the heat were material—caloric—flowing from the boring drill into cannon metal, then it ought to stop when the drill ran out of caloric. Heat, he concluded, was not a material substance, but motion transmitted to the cannon by the motion of the bore.

The calorists of course opposed this, and Rumford provided no careful experimental verification. But argumentative and tenacious by nature, he stoutly defended his beliefs and said that he would "live a sufficiently long time to have the satisfaction of seeing caloric interred with phlogiston in the same tomb."[2] But if caloric were interred, it had a restless ghost. It was summoned again to explain that most important agent of the Industrial Revolution, the steam engine.

The simplest steam engine is a cylinder with a movable piston. Water in the piston is heated by an external flame until it turns to steam, which expands and moves the piston out. When the heat source is removed, steam condenses, and the piston moves back into the vacuum. This simple design was improved by Thomas Newcomen, who added a jet of cold water to cool the steam. The Newcomen engine became the standard model until it was improved in the 1700s by a Scottish instrument maker, James Watt.

James Watt

Friends helped Watt obtain his first job as scientific instrument maker at the University of Glasgow when guild restrictions prevented him from getting a job elsewhere. As the story goes, Watt had to repair a model of a Newcomen engine needed for a lecture demonstration. He realized that some of the heat used to boil the water in the Newcomen engine was wasted in heating the cylinder. He added a separate condenser so that the steam could be cooled without having to cool the whole cylinder, which considerably improved the efficiency of the engine.

At the time it was thought (without any real basis in experiment) that 1 ounce of steam mixed with 1 ounce of cold water gave 2 ounces of water with a temperature about half-way between that of the water and the steam. Watt however knew from experience that condensing steam required more cold water than this simple theory predicted. Joseph Black was also at the University of Glasgow, and he may have been the lecturer for whom Watt repaired the engine. Black also noticed that not all of the heat absorbed going from one phase to another (solid to liquid, water to steam) went into raising the temperature. For example heat applied to an ice cube does not raise its temperature until enough heat has been added to melt the ice; then the temperature of the water goes up. This quantity of heat that goes into only changing the ice to water (not raising the temperature) Black called the *latent heat*.

With our current understanding of molecules, latent heat can be readily explained. Molecules are weakly attracted to one another . . . like frogs covered with sticky tape. When cold, these frogs huddle together and stick to each other through the sticky tape. But as the frogs warm up they begin to move around and break free of their sticky tape connections. In this analogy the "frogs breaking free" represents a phase change: going from a solid lump of frogs to a fluid moving mass. So it is with solids and liquids: The first heat absorbed by the liquid or solid goes into speeding up vibrations and rotations of the molecules that make up the solid or liquid, but they remain stuck together. Only when enough energy has been added to break them free of their sticky connections does a solid turn into a liquid or a liquid turn into a gas. Temperature does not increase during a phase change because temperature is a measure of *translational* energy—the energy with which moving particles strike a thermometer. During a phase change energy added to the system goes into breaking the connections between particles, not increasing the energy with which molecules hit the thermometer; therefore this *latent* heat does not affect the thermometer. But all this is hindsight. At the time Black measured the effect, he had no complete explanation of it.

Watt and Black worked in the same institution and talked to each other, but there is no reason to believe, as has been suggested, that Watt got all his ideas from Black any more than Black got all his ideas from Watt. What probably happened was that Black taught Watt a bit of chemistry and physics, and Watt taught Black a bit about instrument making. It should be noted that Johan Carl Wilcke also noticed and reported similar phenomena at about the same time. Wilcke saw that the final temperature of a mixture of hot water and snow was less that the mean between the temperature of the snow and the temperature of the water. Black and Wilcke however did not profit financially from their observations; Watt did.

In addition to the separate condenser Watt made several other improvements to the steam engine, including reciprocating pistons, and in general he brought it from the status of a clumsy, though interesting device, to a reasonably efficient working machine. He joined forces with a Midlands entrepreneur, Matthew Boulton, and the Boulton–Watt engine became the prime mover of the Industrial Revolution. With the growing use of steam engines, theorists began in earnest to look for theories of heat. One of these was an engineering analyst and Napoleonic soldier, Marie François Sadi Carnot.

Sadi Carnot

Carnot's father was the engineer Lazare Nicolas Carnot, who had been a war minister under Napoleon and was known as the Organizer of Victory. Sadi Carnot was educated by his father until he entered the Polytechnique. During the Napoleonic wars he volunteered to fight, though he was exempt as a student. When the Restoration exiled his father, Sadi found his military career hampered by politics. He went on half-pay and resumed his engineering studies.

Carnot's father had written a book in which he analyzed the workings of waterwheels, finding that the quantity of work produced depended on the quantity of water and the height that the water fell. In 1824 Sadi also wrote a book in which he adopted the accepted view that heat was a fluid (caloric) analogous to water. The work derived from a heat engine, he said, depended on the quantity of caloric and the temperature difference between the heat source and condenser—analogous to the height that water fell when driving a waterwheel. The theory provided a good starting point for understanding the relationship between heat and work because it recognized that work is done when steam expands against a piston, and it is also done when the steam cools and the piston moves back into the vacuum created. In addition Carnot pointed out that the efficiency of all perfect heat engines had to be equal, for if it were not, then the more efficient engine could drive the less efficient machine backward, store up heat for use, and create a perpetual motion machine. Carnot rejected the possibility of a perpetual motion machine because he had never seen one, and this also proved an important step forward in the understanding of heat, as we will presently see.

Carnot's 118-page *Reflections on the Motive Power of Fire* received favorable reviews, then fell into obscurity. This was partly because the world was not ready to understand it, and partly because the theory's champion, Carnot, died in a cholera epidemic when he was 36 (and his personal effects, including many of his papers, were burned). And it was partly because the water analogy had limitations.

In a waterfall all the water falls in one direction, and all the water does work. When energy is taken up by a gas however, molecules go flying off in all directions. The group of molecules that go flying in one particular direction, such as against a piston, do work. But energy taken up by molecules going in other directions shows up as heat, not work. Therefore energy added to a gas can produce some work, but it always produces some nonworking heat, too.

In a few loose sheets that escaped the fire, it can be seen that Carnot was beginning to recognize the imperfections in his water analogy and moving toward the idea of heat as motion. He probably would have changed his approach had he lived. As it was, it remained for another adventuresome French civil engineer and graduate of the Polytechnique, Emile Clapeyron, to see the value of Carnot's work and raise it from obscurity.

Emile Clapeyron

In 1820 Clapeyron went to Russia to do construction work, but as a French liberal in repressive tzarist Russia, he found it uneasy going, and he had to return. While he was there however, he saw Watt's steam engines in operation, and he may have "borrowed" a closely guarded industrial secret to employ in the study of heat. Watt was above all a practical person, and one of the practical problems he faced was matching the power output of a particular steam engine to the power requirements of the machine it was to drive. When an engineer in his employ, John Southern, found a method for measuring this power output, Watt realized the advantage it gave him, and he kept the method a secret. A business rival wrote: "I am like a man parch'd with thirst in the Expectation of relief, or a woman dying to hear (or tell) a Secret—to know Southern's mode of determining Power . . ."[3]

The device was actually simple enough. It consisted of a sheet of paper attached to the piston of the engine and a pencil attached to a pressure gauge on the cylinder. As the volume in the cylinder increased, the piston and paper moved forward and pressure decreased, dragging the pencil down. The area under the resulting curve was the pressure–volume product, or power. In the 1830s Clapeyron used the pressure–volume curve and the symbolism of calculus to put Carnot's ideas into a format that could be used as a basis for a mathematically rigorous theory of heat. It is to his credit that he did not appropriate Southern's invention to go into steam engine manufacturing.

Thus the relationship between heat and motion became increasingly evident. But it remained to measure exactly how much mechanical motion produced how much heat. This was subsequently attempted by a German ship's doctor and an English brewer's son.

Julius Robert von Mayer

The son of a German apothecary, Mayer studied medicine, and chemistry in the process, but he was a mediocre student. Arrested once and expelled for joining a clandestine student society, he was allowed to return the next year to meet his graduation requirements. Having done this with a dissertation on worms in children, he signed on as doctor on a ship bound for the East Indies. On arrival he performed a routine bleeding, but he was alarmed by the bright red color of the blood, thinking he had opened an artery by mistake. Local physicians told him that this was normal for people living in the tropics. At the time (mid–1800s) it was understood that blood obtains its red color from oxygen and that the body burns oxygen for heat (Lavoisier, we recall, did work along these lines). But a person living in the tropics requires less heat, and therefore less oxygen, so blood on its way back to the lungs, richer in oxygen, looks like blood fresh from the lungs. Mayer, considering this, began to wonder if oxygen in the body might also produce *work*. On his return to Germany he wrote up his thoughts in a paper and submitted it to a German scientific journal.

The paper contained no experimental results, and Mayer's discussion of energy, work, and force was confused. The journal's editor rejected the paper, and Mayer's letters of inquiry went unanswered. On a second rendering he clarified his ideas and wrote a new paper that was accepted for publication. In this new paper Mayer suggested (based on experimental results of others) that "the fall of a weight from a height of about 365 metres corresponds to the warming of an equal weight of water from 0 to 1°C"[4]—a quantitative equivalence between mechanical motion and heat. The accepted value today is a fall from a 418.4-meter height.

Mayer did not have the acceptance of the scientific community. In the nationalistic 1800s—with its reverence for cults, cliques, and societies—it was difficult for individuals to be accepted by the scientific establishment if they had not been trained into it. This rejection was a source of suffering for Mayer, and unfortunately his personal life was painful, too. Although he built up a prosperous medical practice on returning to Germany from the East Indies, five of his seven children died in infancy, and as a conservative, he was arrested during the revolutions of 1848 and became permanently at odds with his brother. Mayer became understandably bitter, and when Liebig (the first to publish Mayer's work and whose work we discuss in another context) wrote that he believed muscle force came from chemical force, Mayer thought the theory sounded suspiciously like his own and cried foul. Liebig however was a chemist with a substantial reputation, and Mayer was not, so not much attention was paid to his complaint. When James Joule in England presented his own computation of the mechanical equivalence of heat, Mayer in Germany found himself in a priority dispute involving geographic considerations as much as philosophic. Peter Tait, a Scottish scientist and avid golfer (perhaps used to creative score keeping)—the same person who gave Charles credit for elucidating

Gay-Lussac's law—sided with Joule and wrote a text to that effect. Mayer, despondent, attempted suicide at the age of 36 and suffered mental breakdowns severe enough to require hospitalization. Although Mayer finally won some recognition before he died of tuberculosis at 64, the English won the war of words: The accepted unit of energy (mechanical motion or heat) is now called the joule.

James Prescott Joule

The brewery founded in Manchester, England, by Joule's grandfather became so successful that Joule could pursue his scientific interests freely and at least initially without concern for finances. He was educated at home and by tutors, one of whom was John Dalton. He began research at the age of 19.

Joule studied electricity with an eye to impressing the scientific community. He carefully measured the temperature increase in water caused by current passing through an immersed wire, but his report was rejected by the professionals. He then measured heat produced in water agitated by a paddle wheel and found a mechanical equivalent of heat that was only off by a few tenths of a percent from the value accepted today. The Royal Society rejected this report, too. Joule became so obsessed by this measurement that he took his wife to a waterfall on their honeymoon in Switzerland so that he could measure the temperature difference between the water at the top and at the base of the fall, caused by the water falling from the height of the falls. He could not obtain a good measurement however because of spray and because he was trying to measure less than a degree difference. In the mid-1800s the value of his work was finally recognized and brought to general attention by William Thomson, a 22-year-old, 7 years Joule's junior but an accepted member of the scientific community.

William Thomson, Lord Kelvin

Although Kelvin started life as William Thomson, he is more commonly known as Kelvin. In the mid-1850, however, still in his twenties, he was William Thomson, born and bred in the scientific community. His father had been a professor of engineering and the author of calculus textbooks, and Kelvin had been educated by him at home. Kelvin had learned in Paris of Carnot's analysis of heat engines, so when he heard Joule's ideas he was prepared to appreciate their value. He was also interested in the measurement of heat, but his focus was on instrumentation. He realized that the liquid-in-glass thermometers of the day, which employed mercury, alcohol, or kerosene as thermometric fluids, had varying responses and nonlinear scales. They were not adequate instruments for the emerging science of heat. There was also a need to devise a temperature scale based on a natural law rather than the expansion of one fluid or another, and Kelvin found one.

Kelvin observed a striking regularity in the behavior of fixed amounts of different gases. When the pressure was plotted versus the temperature for various gases, the lines, extrapolated back to zero pressure, all converged at the same temperature, $-273.15°C$. Kelvin suggested taking this temperature as an absolute zero for temperature and making all other temperatures positive with respect to this zero. This was an important refinement because in calculating gas properties, it is sometimes necessary to divide by temperature and the result is meaningless if one uses zero or negative degrees. Kelvin then hit on the idea of making the temperature increments correspond to a fixed amount of work during the expansion of what he termed an *ideal gas*: an imaginary model gas in which there are no attractions nor repulsions between gas particles and in which gas particles themselves occupy no volume—a behavior approached by real gases at low pressures. Based on the behavior of this ideal gas, the size of the degree no longer varied with temperature. He found that a real gas thermometer could be made to approximate his ideal scale fairly well, and the thermodynamic temperature scale, with degrees now called *kelvins* in his honor, came into use.

Kelvin also made another important contribution to the field: the word *thermodynamics*. He had used the term *thermo-dynamic* to describe heat engines, and the word eventually evolved to describe all studies of the transformations of energy, including heat and work.

FIRST, SECOND, AND THIRD LAWS OF THERMODYNAMICS

The German physician and de facto chemist, Hermann von Helmholtz, summarized the findings of Joule, Mayer, Kelvin, and Claperyon into the first law of thermodynamics: In a closed system (in which there is no change in material content), any change in energy shows up as either heat or work, that is, energy is conserved. Saying that there is a first law implies that there is a second, and this second law was arrived at by Kelvin in Glasgow, contemplating the work of Carnot and Joule—and also by Rudolf Clausius in Berlin, contemplating the work of Mayer. Both these theorists believed that in a perfect heat engine, the heat given off to the cold sink at the end of the cycle should equal heat taken from the hot sink minus the amount used to perform work. But they both knew that in real machines, this was not necessarily so. In real engines heat given off at the end of the cycle is lower than predicted—as though there were some wasted heat that was not going into work—and so there was.

Consider a pile of feathers. If a pole is poked through a pile of feathers, feathers go flying. But it can also be imagined that if a pole is poked through a pile of feathers infinitely slowly and carefully, each feather displaced by the pole falls back exactly to the same spot when the pole is pulled out, and the feathers remain in exactly the same position at the end of the process as they were at the start. This infinitely slow and careful process is called a *reversible process*, because at any point, the process can be reversed without causing a change. In a reversible pole-poking, all the energy goes into moving the pole and none is wasted rearranging feathers. So it is with steam engines: If they could be run reversibly—infinitely slowly and carefully—then none of the energy would be wasted. But if real steam engines worked infinitely slowly, the Industrial Revolution would have ground to a halt before it got started. In real, *spontaneous* pole–feather situations, feathers fly irreversibly, and in real machines, some of the energy always goes into just knocking gas particles around without doing any work. Kelvin called this feature of real engines the universal tendency to the "dissipation of energy."[5] Clausius called it entropy.

ENTROPY

Rudolf Clausius

Clausius summarized the second law of thermodynamics by saying, "The entropy of the world strives toward a maximum."[6] There is a universal tendency for feathers to fly.

His justification was similar to Carnot's. Clausius had never seen entropy decrease (scattered feathers do not spontaneously reassemble into a neat stack), so it must not happen. He offered no explanation on a molecular level, and it was rather curious that he did not. A creative, insightful worker, Clausius modeled gases as collections of particles in motion and showed that an analysis of the energy of these particles and their impact on their container predicted Boyle's pressure–volume law and Gay-Lussac's temperature–volume law. Now known as the kinetic theory of gases, this analysis was one of the first clear-cut pieces of evidence for the existence of atoms. (It should be noted that Clausius was not the first with this idea, but an earlier proponent, John James Waterston, had been too far outside the established scientific community to have his ideas accepted. When Waterston submitted his analysis to the Royal Society, a reviewer pronounced that "The paper is nothing but nonsense, unfit even for reading before the Society."[7] His nephew recalled that after that Waterston "rather avoided the society of scientific men. . . . We could never understand the way in which he talked of the learned societies, but any mention of them generally brought out considerable abuse . . ."[8])

In the matter of entropy the challenge to Clausius's priority was more immediate, but then it was a general time of challenge in the European world. In the 1870s France and Germany were at war, and England, though neutral, had close ties to France. Kelvin was a loyal subject of Great Britain and Clausius, born in Prussia, had been wounded while leading a student ambulance corps in the Franco-Prussian War. When the battle moved from the fields to the corridors of learned societies, Clausius and Mayer found themselves pitted against Kelvin and Joule. Clausius, after participating in the battle briefly, seemed to lose interest—perhaps not in science but in the scientific process. He still felt pains from his old wounds incurred as a wartime ambulance driver, and having raised his family alone after his wife died, he chose at the end of his career to remarry and retire—and not take up the battle of a microscopic explanation for entropy. The battle though was taken up by a citizen of Austria, a soon-to-be ally of Germany, when the quarrels got further out of hand.

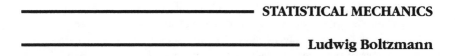

STATISTICAL MECHANICS

Ludwig Boltzmann

Another reason for Clausius's withdrawal from the foray may have been that many people still openly questioned the real existence of atoms, and in a molecular rationalization of entropy, atoms could not be ignored. Boltzmann however was ready to take the heat. He wrote, "I am conscious of being only an individual struggling weakly against the stream of time."⁹ Considering a gas as a collection of particles, Boltzmann used statistical analysis to show that entropy is a manifestation of the natural tendency of a system to seek the state of maximum disorder.

Disorder does seem to be a common phenomenon; we all know what happens when a stack of papers is dropped, and we can imagine what happens when a pile of feathers is poked, but it is not obvious why disorder should be a "natural tendency." Boltzmann showed that the disordered state was the natural state because it is the state with *the highest probability*.

Consider a deck of cards. A new deck of cards, neatly arranged by suit and number, becomes disordered almost as soon as its wrapper is removed. Once shuffled, disorder increases with each shuffle. This is not because it is *impossible* for the deck of cards to end up neatly ordered by suit and color after repeated shuffling, it is just that the probability of this occurring is about one in ten million trillion—not very likely. In contrast the odds that the deck will be in *some* random order (including neatly arranged by number and suit) are one in one (every time).

Boltzmann's insight was that he saw this worked for atoms as well. In a cooled system there is some order. In the limit of absolute zero, in a perfect crystal, everything is frozen in place, and the entropy is zero. This is what Walter Nernst recognized, giving chemists a zero from which to calculate thermodynamic properties and a third law of thermodynamics. But if any energy is added, some entropy is introduced. Specifically Boltzmann showed that if W were the number of ways a collection could be distributed, then the entropy S would be proportional to the logarithm of this number of ways: $S = k \log W$. And while it is one premise of this book that the essence of chemical history can be appreciated without mathematical treatments, we insert this one bit of mathematics here for a good historical reason: $S = k \log W$ is carved on Boltzmann's grave.

Many of the old guard thought Boltzmann's analysis was an amusing mathematical manipulation but not much more. Some historians have ascribed Boltzmann's suicide at the age of 62 to disappointment over this rejection of his work. But Boltzmann had long been subject to moods that swung from elation to depression, which he attributed to being born during the dying hours of a Mardi Gras ball. Had he lived a few years longer, he would have seen his ideas accepted, and as it was, he received support and acceptance from younger scientists of his day. Nowadays the statistical treatment of thermodynamics—known as statistical mechanics—is studied by every chemistry student.

Had he lived Boltzman would also have had the pleasure of seeing someone take a track parallel to his, though this time it would be someone on a frontier other than the frontier of science. Europeans still regarded the United States as inhabited mostly by bears and bear skinners (Boltzmann himself spoke disparagingly of the United States as El Dorado), but while Europeans were puzzling over what all this business of steam engines and gases had to do with chemistry (as is perhaps the reader), Josiah Willard Gibbs in the Unites States showed them.

Josiah Willard Gibbs

In the formulation of statistical mechanics by Boltzmann, he made the assumption that molecules are distributed in all available energy states (such as high energy, low energy, average energy), then calculated properties of gases (such as pressure and entropy) based on the average distribution over time. This can be compared to placing three marbles in an egg carton, shaking it up, noting the distribution of marbles in cups, then shaking it again, and after a trillion or so trials finding the average configuration. Gibbs however said that this should be the same as the instantaneous average over many systems, which can be compared to taking a trillion egg cartons with marbles, shaking them all at one time, and then finding the average distribution. This turned out to be a very useful mathematical device and aided tremendously in the computations of statistical mechanics. (Although computations remain challenging—hence the reference to S&M by less reverent students.) Gibbs called his egg cartons *ensembles*—canonical ensembles, petite canonical ensembles, and grand canonical ensembles—terms that sound more like angelic choirs than atoms or molecules, and his austere language and mathematical format befuddled the scientific community for a long time. In large part Gibbs's work went unappreciated and unnoticed. Gibbs was also a naturally quiet person, so he did not spend a lot of time trumpeting his ideas. Given his family background, it can be imagined why Gibbs opted for a more quiet life.

Gibbs's forefather, Simon Willard, migrated from England to New England in the mid-1600s to be a fur trader and fighter. His son, the Reverend Samuel Willard, spoke out against the Salem Witchcraft trials and ended up being accused as a witch himself. In Gibbs's family tree are signers of the Declaration of Independence, physicians, professors, lawyers, and chemists. His father was a Yale professor and cosigner of a protest sent to President Buchanan over the proposition to use troops to help proslavery Kansas settlers. His mother was also quietly rebellious, making a systematic study of ornithology, not limiting herself to bird watching, as was the usual extent of the pursuit for a woman in her day.

The United States Civil War began while Gibbs was in graduate school, but his number did not come up for the draft, and he did not volunteer. Although his mother died while he was in college and his father while he was a graduate student, his unmarried sister continued to live with him and keep his house.

Gibbs's doctorate was the second in science to be given in the United States and the first in engineering. (Because his career at Yale spanned the next 40-some years, he probably crossed paths with another of Yale's firsts: in 1876 the first American of African heritage to receive a Ph.D. in physics, Edward Bouchet, did so at Yale, where he was elected Phi Beta Kappa.[10]) Wishing to stay at Yale, Gibbs accepted a 3-year tutorship. At the end of this time he and two sisters rented out the family home to finance a trip to Europe, the primary purpose of which was study for Gibbs. On his return Gibbs was able to secure a position at Yale as professor of mathematical physics but only because he was willing to do so without pay. But in the 1870s this was not unusual. He did receive fees from students, but he had few of them at first. His experimental work on soap bubbles, surface tension, and capillary action was done at home in his kitchen sink. He chose to lecture from the works of Clausius, and he seemed to have an intuitive grasp of the concepts of thermodynamics, including entropy.

Previously it had been believed that only reactions that produced heat were spontaneous. But Gibbs understood that the change that produced the maximum entropy in the system and surroundings was the change that occurred spontaneously (heat flows spontaneously from hot to cold, not visa versa, because this is the change that produces the most entropy).

In other words Gibbs saw that the calculated change in entropy for the system and surroundings predicted the direction of spontaneous change in any chemical reaction. Thus with pencil and paper—and not one drop of solution or sweat—we can calculate whether a laboratory or industrial reaction should occur.

For the important conditions of constant pressure and constant temperature (such as reactions carried out in an open beaker on the laboratory bench), a sum known as the Gibbs free energy G totals these entropy changes. And as every good chemistry student knows, reactions proceed spontaneously only when G for the initial state is greater than G for the final, and at equilibrium the change in G is zero.

As though this were not good enough, Gibbs also extended the dimensionality of Watt's pressure–volume diagram to a three-dimensional surface so that the phase of a system at a given temperature and pressure could be determined at a glance. He derived a *phase rule* for determining whether or not a chemical mixture were in equilibrium and what components or conditions could be varied without pushing it into another phase. In many practical situations, such as in the metallurgy of alloys, it is very important to control the composition of the final material. Information on the composition of a mixture could now be summarized on *phase diagrams*, and these diagrams could be used along with Gibbs's phase rule to predict how the mixture would behave under various conditions.

The result of all this work was the production of a new repertoire of tools for chemists, engineers, and theorists alike. The only problem was that it did not immediately reach the people who could make use of it. The work was presented in such a cryptic mathematical format that people did not understand what they were reading, and it was also published in an obscure journal, *Transactions of the Connecticut Academy*, which reached Europe by slow boat, if at all. However it did reach an important few, one of whom was the productive and respected James Clerk Maxwell, who we meet now and will meet again.

James Clerk Maxwell

The Clerk family was of landed Scottish gentry, and when in the 1700s two of the family married Maxwell ladies, illegitimate offspring of the eighth Lord Maxwell, they took the name Maxwell to gain a legal title to some estates. James Clerk Maxwell, born in the 1830s, lived a comfortable life supported by income from these estates, so he was able to indulge his scientific curiosity. Trained as a physicist, his interests were diverse. While speculating on Saturn's rings, he started thinking about the physics of collisions between a large number of bodies. When he read Clausius's paper that connected pressure with the collisions of gas molecules on the sides of their container, he played with the theory and found that it predicted a remarkably nonintuitive effect: Gas viscosity should be independent of its density. Viscosity is a measure of how much a gas slows down when flowing through a tube. Particles along the wall experience drag from collisions with the wall and slow down interior particles as they collide with them. At low densities there are fewer gas molecules in a given volume, so there are fewer drag collisions; but there are also fewer collisions between wall particles and interior particles, so the net drag is the same as at higher densities. When he and his wife, Katherine Mary Dewar, confirmed this experimentally, this also confirmed for them the kinetic theory of gases, and Maxwell revisited Clausius's work.

He noted that Clausius made the simplifying assumption that all gas particles were traveling at the same velocity. Maxwell refined the theory by removing this simplification and using the statistical distribution of velocities that a real gas would have. Maxwell also postulated that Clausius's indication of the direction of spontaneity, entropy, might be defeated in some systems that are completely isolated from the outside world. Specifically he said that heat might be induced to flow from cold gases to hot ones if a demon sat at a gate over the conduit between the gases and allowed through only hot (fast-moving) particles to the hot side, and cold particles to the cold side. (This was answered by showing that "Maxwell's demon" could not stay isolated forever—even demons have to eat—so such a system would require the input of energy.)

Maxwell also grasped the usefulness of Gibbs's analyses and relationships. He even made a model of the three-dimensional thermodynamic surface, which he called a Willard Gibbs surface, as a present for Gibbs, but 2 weeks after finishing it, Maxwell died of abdominal cancer at the age of 48. But by this time Gibbs, through Maxwell, had gained some recognition in Europe.

In his own country however Gibbs remained obscure. When a new university was founded in the United States, its president was sent to Europe to find an expert in molecular physics for the faculty. When he was told at Cambridge in England that the best person for the job was right there in his own United States, he said, "I'd like you to give me another name. Willard Gibbs cannot be a man of much personal magnetism or I should have heard of him."[11]

Personal magnetism was definitely something Gibbs did not have. He had a naturally high-pitched voice—an English chemist, watching Gibbs lecture, described two students seated, scribbling in notebooks, and Gibbs standing at the blackboard, covering it with tiny characters and whining over them. The times he steeled himself to ask questions in seminar were rare enough to be noteworthy. In one faculty debate over the relative merits of teaching language versus mathematics, Gibbs rose to say, "Mathematics is a language," then sat down.[12] He ended up spending 10 years at Yale without salary, receiving only fees from his students. In 1880 when Gibbs briefly considered going to Johns Hopkins, university authorities finally gave him a $2000 yearly salary.

It is doubtful that Gibbs would have gone to Johns Hopkins. The house that he lived in since he was 10 years old was half a block from the elementary school, one block from Yale, two blocks from his office, and two blocks from the cemetery where he is buried. But through Maxwell, Gibbs's work did gain recognition among a few highly influential chemists. One of these was a French chemist who translated Gibbs's work into French and did much to bring the work to the attention of French scientists. His name was Le Châtelier, a name that became synonymous with chemical equilibrium.

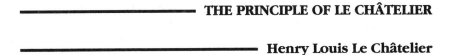

THE PRINCIPLE OF LE CHÂTELIER

Henry Louis Le Châtelier

In the course of investigating some high-temperatures problems, such as mining disasters and cement preparation, Le Châtelier found that equilibria could be shifted by adding or subtracting heat. When Le Châtelier was in his early thirties in the early 1880s, he summarized these observations in what is now known as the principle of Le Châtelier: When a strain is placed on a system in equilibrium, there is readjustment in the direction that most effectively relieves the strain. This strain could be in the form of material, as originally noticed by Berthollet on the shores of Egypt's salt lakes, or in the form of heat, as now shown by Le Châtelier.

The utility of this principle in chemistry cannot be overestimated. The chemist could now show via the Gibbs function whether or not a reaction would occur spontaneously and if it did not, how it could be coaxed in the desired direction using Le Châtelier's principle.

This principle can be illustrated by two jugglers balanced on a the ends of a seesaw. If one juggler drops some balls, then the other must throw some over to keep the seesaw even. A chemical reaction at equilibrium is also like a seesaw: Removing a product causes the reaction to shift and produce more product. If the reaction is highly *exothermic* (a word invented in this era to describe reactions that produce heat), cooling it removes this heat and causes the reaction to produce more product and heat.

Le Châtelier found Gibbs's phase rule especially useful in his work on metallurgy (Le Châtelier had himself worked out parts of the phase rule using his own principle of equilibrium), and much of this work appeared in the metallurgy journal that he and his daughters published. Le Châtelier also used these principles in his work as a consultant on shell casings—a position he held during the spontaneous eruption of entropy and wasted energy known as World War I.

In the midst of this war and the growing hostilities between French and German nationals, Gibbs's work also had to be translated into German for its general acceptance. Luckily Gibbs had his German champion, too: Friedrich Wilhelm Ostwald.

Friedrich Wilhelm Ostwald

Friedrich Wilhelm Ostwald, eventually one of the most influential scientists of his day, was an indifferent student at first, preferring to spend his time on bug collecting, photography, woodworking, and making fireworks. In college his interests included music, painting, socializing with other students, and a general celebration of Romanticism. After a warning by his father, his schoolwork improved, and in the early 1880s he became a chemistry professor.

Ostwald saw the usefulness of thermodynamics and the phase rule in chemistry and taught them to his students. In fact he became so convinced of the productivity in chemical theory promised by the marriage of physics with chemistry that he effectively founded the new discipline of physical chemistry, starting the first journal in the field, the German *Journal of Physical Chemistry* in the late 1880s. For a long time he rejected the atomic theory—referring to it as "the atomic hypothesis"¹³—and the debate between Ostwald and Boltzmann followers raged. But there were also many chemists at the time who saw the debate as pointless. For them modeling their material as atoms and their heat as motion produced the desired product, and this was the bottom line. Not waiting for physical chemists to make up their minds, the organic chemists carried on.

chapter
TWELVE

ca. 1830–1914: Organic Chemistry—Up from the Ooze

Chemists have always been fascinated with materials from living systems—the stuff of life. It may be recalled that Fourcroy in the 1700s counted it a great opportunity to be able to examine decaying material from exhumations—a task not to be undertaken by those given to olfactory delicacy. In fact the field of organic chemistry started as the study of the soup of loosely connected materials extracted from animals and plants, but by the 1800s organic chemists offered a set of theories to explain and systematize their science—and a valuable system it was. Their insights not only advanced the understanding of organic chemistry, but these served as a basis for a new understanding of chemistry as a whole.

Virtually all materials extracted from plants and animals fall under the heading of *organic* compounds—that is, compounds that include some combination of carbon and hydrogen. In the chemist's quest for materials, plants and animals have always been exploited as much as the stones we have focused on so far. Extracts of herbs have been used for medicines and ethanol for mood control. Soaps and many dyes have their origins in plant and animal products, as do many acids, including acetic acid from sour wine, lactic acid from milk, citric acid from lemons, and formic acid from the dry distillation of ants (alchemists tended to throw anything in the pot that was not faster than they were).

Some ninety-eight percent of the dry weight of living tissue is made up of carbon and hydrogen, with some oxygen and nitrogen thrown in for variety. And variety is what is achieved. It has been estimated that a single bacterium contains some 5000 different organic compounds, and the human body as many as five million. The base for all these compounds is carbon, and the basis for organic chemistry is carbon chemistry.

The question is, Why carbon? It occupies a single space on the Periodic Table like all the rest of the elements, and its mineral form, carbonate—as in calcium carbonate (chalk or lime)—does not display any obvious distinction. However in compounds formed from hydrogen and carbon, *hydrocarbons*, carbon has a unique property: Carbon atoms form chains—long chains, short chains, branching chains, circling chains, looping chains, and complex three-dimensional structures. (The only other element that comes close to carbon's chain-forming ability is silicon, but silicon–silicon bonds are less stable than carbon–carbon bonds, and these oxidize in air to the silicon–oxygen–silicon bonds that make up such materials as sand.)[1] The ability to form chains allows carbon to make up the complex materials of life.

It is now known that within all the complexity of organic chemistry there are patterns. For instance many organic acids display similar chemical behavior because they contain the same group—the COOH or carboxylic acid group—and they can be classed together as carboxylic acids. Alcohols are another class, characterized by the OH group. There are classes known as alkanes (for example, methane), alkenes (for example, ethylene), and alkynes (for example, acetylene).

But at the time of Berzelius the term "organic" covered a poorly defined hodgepodge of materials, with the different classes yet to be sorted out. Lavoisier developed a technique for analyzing organic compounds in which they were burned in oxygen: Carbon was collected as carbon dioxide and the hydrogen as water. But just as more precise analyses looked as though they were going to clear the mire of organic chemistry, these same analyses clouded the issue. In the 1820s the German chemist Friedrich Wöhler analyzed silver cyanate and found it to be 77.23% silver oxide and 22.77% cyanic acid. The German chemist Justus Liebig analyzed silver fulminate and found *it* to be 77.53% silver oxide and 22.47% cyanic acid. From analyses, the compounds seemed to be the same, but they had markedly different properties: Silver fulminate is an explosive; silver cyanate does not explode. At the time it was firmly believed that different properties of different compounds resulted from differences in the ratios of constituent elements. Wöhler was curious and baffled, but for Liebig the explanation was simple: Wöhler was wrong. As it turned out, both were right.

ISOMERISM
Friedrich Wöhler and Justus Liebig

Even though chemists as a group are not noted for their great beauty, Friedrich Wöhler was an outstandingly homely man. However this did not hamper him from learning chemistry from his father's books and building a laboratory in his father's home. When he went to Heidelberg to study, he already knew so much chemistry that he did not have to attend lectures, and he probably never did. When he began teaching in Berlin, his laboratory was the former rooms of Count Ruggiero, who had been hanged as an alchemical swindler. It is not clear which if any of these circumstances were responsible for Wöhler's pleasant, relaxed approach to life, but he was able to go benignly through some stormy times, well buffered by his humor. It is probably for the good of chemistry that Wöhler was able to maintain the attitude he did, otherwise his eventual friendship and collaboration with Liebig would not have been possible. Liebig's attitude toward life was considerably more combative than Wöhler's.

Figure 12.1. Friedrich Wöhler, the chemist who showed that an organic compound could be synthesized from inorganic starting materials. (Courtesy E. F. Smith Collection, Special Collections, University of Pennsylvania.)

Justus Liebig was the son of a dealer in drugs, dyes, oils, and chemicals in western Germany. He did poorly in school (when he told a teacher he hoped to be a chemist, the class and the teacher laughed). But he had become interested in chemistry while assisting his father, and Liebig was fascinated after he learned how to make the explosive silver fulminate from an entertainer in a traveling show. He served an apprenticeship with a pharmacist, which terminated when his experiments with silver fulminate caused structural damage to the pharmacy and his father, looking leeward, sent him to a university to study chemistry. During his time at the university, his mentor moved to Bavaria, and Liebig followed. There he became involved in a student political organization, was arrested, and had to return home. Through the intercession of some friends, he eventually earned his doctorate, but frustrated by the lack of opportunity to learn modern chemistry in Germany, he went to Paris where he worked with Gay-Lussac. In the 1820s he accepted a position at a small German university at Giessen. Taking over an unused barracks for a chemical laboratory, he worked relentlessly over the next 25 years to make this small university a center for chemistry education that would attract students from all over Europe, the United States, and Mexico. Liebig in the process became a respected authority, though this reputation may have been based as much on his success as an administrator as on his ability as a chemist. Perhaps Liebig's greatest discovery was the efficacy of the research group as a way of advancing science. Research groups headed by a senior scientist as mentor and junior scientists as lieutenants and students are a Liebig creation.

The role of authority suited Liebig well. He was arrogant and pugnacious, and as Wöhler, his soon-to-be life-long friend, pointed out

> You merely consume yourself, get angry, and ruin your liver and your nerves . . . Imagine yourself in the year 1900, when we are both dissolved into carbonic acid, water and ammonia, and our ashes . . . are part . . . of some dog that has despoiled our graves.[2]

Wöhler was able to say this after they had become friends. Their first encounter was not so amiable: Liebig simply refused to believe Wöhler's results. A careful recheck however revealed that the two chemists were indeed working with two different chemical substances that had the same elements present in the same ratios, and a collaboration resulted. They ended up not only showing that cyanic acid and fulminic acid have the same elemental composition, but that there was a third compound (isocyanic acid) of identical composition.

It is now accepted that the difference in chemical behavior of these compounds arises from the different arrangement of elements in the compounds: Cyanic acid and isocyanic acid both have carbon, hydrogen, nitrogen, and oxygen in a one-to-one ratio, but cyanic acid has hydrogen–oxygen–carbon–nitrogen bonds (HOCN), whereas isocyanic acid has hydrogen–carbon–nitrogen–oxygen bonds (HCNO). Although Gay-Lussac made this suggestion and Berzelius proposed the name *isomerism* to describe materials with the same chemical composition but different properties prior to the Wöhler–Liebig results, most chemists had remained unconvinced. With the Wöhler–Liebig results however there appeared to be clear evidence that arrangement had something to do with chemical reactivity. Arrangement of what? Atoms. An intuitive picture of compounds formed from specific arrangements of atoms was beginning to emerge.

Wöhler, not resting on his laurels, continued to experiment with cyanate. In an attempted synthesis of ammonium cyanate, he noticed something peculiar about his product. It had the same chemical composition as a natural product—urea from urine—and it had the same reactivity, too. They were, he found, identical in every way. Wöhler had for the first time made a natural product from inorganic chemicals, outside a living body. He wrote to Berzelius:[3]

> I can no longer, as it were, hold back my chemical urine; and I have to let out that I can make urea without needing a kidney, whether of man or dog; the ammonium salt of cyanic acid is urea.[3]

Up until this point organic chemicals had only been isolated from plant or animal material. In fact some chemists theorized that the synthesis of organic chemicals *required* some sort of vital force found only in living organisms. This theory, called *vitalism*, had few serious adherents in Wöhler's time, but it had not been disproved either. Wöhler's discovery was a nail in the coffin of vitalism. The discovery gave organic chemists new confidence, and they approached their work with vigor. Wöhler and Liebig found themselves rather taken with the topic, as did the French chemist, Jean Baptiste André Dumas.

Jean Baptiste André Dumas

Of virtually the same age, background, and interests, Dumas, Liebig, and Wöhler should have been close colleagues. But the one small difference between them became a big one. Dumas was born in France and Liebig and Wöhler were born in German states. Still groping toward national status, the Germans found themselves the nouveau nationalists in a neighborhood of grand, old nations. They tended to take a defensive stance in all areas, science not excepted. German scientists resented the dominance of French scientists, and it seemed that every French discovery was met with challenge, and challenge was Liebig's talent as much as science. But even he had to admit Dumas's skill: "it always annoys me that this fellow, in spite of his unclean, impossible and bad way of working . . . with the devil's help fetches masterpieces out of his sleeve."[4] Dumas was more magnanimous, but then, as a French chemist in a world dominated by French chemists, he could afford to be.

Born in 1800, Dumas at first planned a naval career, but given the political atmosphere of France in the early 1800s, he decided he would be better off in science or—considering the fate of Lavoisier—at least no worse off. Dumas served the requisite apprenticeship to an apothecary, then set off on foot to study at the academic centers of Geneva. He eventually served as lecture assistant to Louis Thenard, then succeeded Thenard at the Polytechnique and Gay-Lussac at the Sorbonne. With this pedigree and his considerable talent, he quickly became the most noted French chemist of his day.

He was familiar with all branches of chemistry, and as more organic compounds were isolated and analyzed, he began to see patterns. In inorganic chemistry (the field that deals with compounds not classified as organic), the dualistic theory of Berzelius (which stated that the attraction between positive and negative electrical charges held compounds together) had been productive, and Lavoisier in his systemization of acids had used the idea of radicals: groups of elements within a compound that functioned together as a unit. Dumas put these two ideas together to come up with the concept of an organic radical: a group of elements within an organic compound that functioned as a unit.

RADICAL THEORY

Dumas, working with the French apothecary Pierre François Guillaume Boullay, first proposed that the ethylene group (a two-carbon chain) could serve as the radical base for alcohol and related compounds. The concept of radicals received further support when Liebig and Wöhler published their work on oil of bitter almonds. They found a family of compounds—benzoic acid, benzoin, benzaldehyde, benzoyl chloride, benzoyl bromide, benzoyl iodide, benzoyl cyanide, benzoyl amide, and the benzoyl ethyl ester—all had the benzoyl radical in common. The discovery was met by enthusiasm by Berzelius, though he had been lukewarm to the concept of radicals when first presented by Dumas, because now there was a radical containing oxygen, an element of negative electrical character that could combine with a radical with a positive electrical character and preserve his theory of dualism.

The search was now on to identify different radicals. Dumas found the methyl radical. Robert Wilhelm Bunsen investigated compounds of the cacodyl radical (a smelly, toxic, sometimes explosive set of compounds containing arsenic) and isolated what appeared to be a free radical. It was actually a compound made of two radicals joined together, but it supported the idea that radicals were stable, isolatable entities, which could be treated like organic "elements." Bunsen did not carry this work further because an explosion of cacodyl cyanide cost him an eye and several weeks of illness (and thereafter he steered clear of organic chemistry in general). He did go on to have a successful career in other chemical endeavors, inventing for instance a gas burner called the Bunsen burner, which is still standard laboratory equipment.

With the discovery of Bunsen's free radical, the radical theory seemed set and discussion merely ranged over what different groups were really radicals. But the peace would not last.

SUBSTITUTION THEORY

One evening at a royal ball at the Tuileries palace in Paris, the aristocratic guests were thrown into decidedly common coughing fits by acrid fumes given off by candles. The ball's organizers, no doubt a bit piqued, asked the supplier of candles, who happened to be Dumas's father-in-law, to find the cause. He gave the problem to Dumas, and Dumas found that the irritant was hydrogen chloride, an acidic gas. The vendor of the candle wax had apparently hit on a novel way of whitening a particularly yellow batch: heating it with chlorine gas. When the candles bleached with chlorine were burned, hydrogen chloride was produced. In the course of his investigation, Dumas found that when many organic compounds are treated with chlorine, chlorine replaces some of the hydrogen.

Berzelius, an aging chemist by now, did not like the idea. It did not fit with his dualistic theory that a negative element, such as chlorine, could substitute for a positive one, such as hydrogen. But Faraday had carried out such reactions, as had Liebig and Wöhler when they prepared benzoyl chloride from benzaldehyde. It was even suggested that carbon might be replaced by chlorine without greatly changing the properties of the compound. Believing this was going too far, Wöhler, the perennial jokester, wrote to Berzelius and Liebig about a bogus discovery in which *all* the elements of manganous acetate had been systematically replaced by chlorine:

> I have found that it was formed from 24 atoms of chlorine and 1 atom of water. . . . Although I know that in the bleaching action of chlorine there is replacement of hydrogen by chlorine, and that the fabrics, which are bleached in England . . . conserve their type, I believe nevertheless that the replacement of carbon by chlorine, atom for atom, is a discovery which belongs to me. Please make note in your journal . . .[5]

Liebig went one better. He published the letter in the journal he edited, signing the name S. C. H. Windler (Schwindler is German for swindler) and added an unsigned footnote:

> I have just learned that in the shops of London there are already fabrics of spun chlorine, very much in demand in the hospitals and preferred over all others for night caps, [under] drawers, etc.[6]

Berzelius continued the attack on the theory of substitution. He proffered the copula theory, which proposed that chlorine caused a drastic rearrangement of the radical when it joined (copulated) with the hydrocarbon. This notion had no intuitive appeal because compounds did not show drastic differences in chemical properties, so this theory was generally rejected. Berzelius protested loudly, but few listened. Younger chemists, such as Auguste Laurent, were listening to Dumas.

NUCLEUS THEORY AND THEORY OF TYPES

Auguste Laurent

Born in a small village in France, the son of a wine merchant, Laurent studied under Dumas, then pieced together a patchwork chemistry career from porcelain manufacturing, perfume manufacturing, industrial ventures, professorships, private instruction in chemistry, and independent research. Frequently clashing with the chemical establishment, and never accepted by it, Laurent did not obtain the professional posts his talents warranted. Stricken early by tuberculosis, he died young and probably disappointed. His research in chlorine substitution however led him to the nucleus theory, a modification of the radical theory. According to the nucleus theory, compounds were made of nuclei (radicals) in which substitution could take place. The original nuclei was changed into derived nuclei that retained many of the properties of the original. With this seemingly minor modification of the radical theory, Laurent stirred up quite a storm.

Liebig said it was unscientific. Berzelius, mistaking Dumas for the originator of the theory, said the whole idea did not even deserve comment. Criticized by the lawgiver of chemistry, Dumas quickly disclaimed responsibility, saying that he had only observed that compounds that take up chlorine seem to lose an equal amount of hydrogen, and "I am not responsible for the gross exaggeration with which Laurent has invested my theory; his analyses moreover do not merit any confidence."[7] Later when Dumas prepared acetic acid with three chlorine atoms substituting for hydrogen (trichloroacetic acid), and noticed its similarity in properties and reactivity to acetic acid, he came up with his own type theory wherein *types* were now the building blocks of organic compounds, and in these types, chlorine could substitute for hydrogen.

The similarity to the ideas of Laurent is as obvious now as it was then to Laurent. He said

> I have not been able to dismiss an emotion of indignation in seeing certain chemists first call my theory absurd, then much later when they have seen that the facts are in agreement with my the-

ory . . . pretend that I have taken some ideas from M. Dumas. If it fails, I shall be the author, if it succeeds another will have proposed it. . . .[8]

Dumas however was decidedly the senior chemist, so though disgruntled, Laurent had no choice but to carry on his experiments using his own meager funds, his ideas being largely unrecognized. He did however have one champion: Charles Frédéric Gerhardt.

THEORY OF RESIDUES
Charles Frédéric Gerhardt

Gerhardt's father was a manufacturer of white lead in northern France, and though Gerhardt tried working with his father, their frequent disagreements would not allow it. Gerhardt did not get along well with very many people—the one exception perhaps was Laurent. Gerhardt's clashes with Dumas prevented him from securing work in Paris, and he relied on tutoring, writing organic chemistry texts, and donations from his wife's relatives while he was there. For his textbooks he developed a theory of *residues*. According to this theory, when certain organic compounds react, other very stable inorganic compounds, such as water and carbon dioxide, are formed as products. Leftover organic fragments minus the water and carbon dioxide—residues—combine to form new organic compounds. The theory bore a similarity to the theory of types, but with the significant differences that residues were not considered to be discreet, indivisible units in compounds and they had neither a positive nor a negative electrical character.

For his organic texts Gerhardt devised a classification scheme that grouped compounds with similar reactivity, such as alcohols, but differed by the number of carbon atoms that formed them: *homologous series*, a term and concept still used today. There arose another tempest because Dumas had done the same for fatty acids and therefore claimed priority, even though other chemists, including Laurent, had used the same idea and were ignored by both Gerhardt and Dumas. The classification scheme, whatever its origins, strengthened the theory of types, though the concept of what constituted a type was now more flexible. New types were identified that could serve as the bases of homologous series. For instance ammonia—formed from a central nitrogen atom surrounded by three hydrogen atoms—was identified as a type. When a hydrocarbon radical is substituted for one or more of the original hydrogen atoms, the compound formed is an *amine* (a member of the class of compounds responsible for the fishy smell of decomposing seafood), and amines—with a characteristic NR_3 group (R is any hydrocarbon radical)—were considered to be of the ammonia type. Water is formed from a central oxygen atom with two hydrogen atoms, and in the water type one of the hydrogen atoms is replaced by a hydrocarbon group. The resulting compound is an *alcohol*, with a characteristic OH group. These characteristic groups are now called *functional groups*, and this same sort of scheme in which compounds are characterized by functional groups is used in organic chemistry today.

Recognition of these various types was an important advance in understanding the reactivity of organic compounds, but the next important advance came from a chemist trying to discredit the first type theory of Dumas, Edward Frankland.

VALENCY

Edward Frankand

The English chemist Edward Frankland started life in an unconventional way, without a legitimate father, but he started his chemistry career in a conventional way, apprenticed to a pharmacist. He might have remained a pharmacist had two doctors not helped him secure a position in a laboratory at the Museum of Economic Geology. There he met Adolf Wilhelm Hermann Kolbe, an outstanding German chemist of the period. Frankland learned enough from Kolbe to study under Bunsen, earn his doctorate, and study with Liebig. In the course of experiments performed right around 1850, meant to show that Berzelius's copula theory was valid, Frankland examined reactions between metallic zinc and ethyl iodide in sealed tubes exposed to sunlight or immersed in an oil bath and heated. He found that some of the organic material had combined with the zinc. Intrigued Frankland extended his studies to compounds of mercury, antimony, and other metallic elements. Studying this new family of compounds—organic materials combined with metals, or *organometallics*—he noticed that there seemed to be "a fixity in the maximum combining value . . . in the metallic elements which had not before been suspected."[9] This fixed maximum combining value, or *valence*, was to become a vital factor in understanding carbon chemistry and indeed all chemistry.

To understand valence, or combining power, consider the saying "I have only two hands." This could be restated as humans have a valence of 2. Monkeys on the other hand can grasp four things, so monkeys have a valence of 4. Sulfur, it was found, had a tendency to combine with two other atoms to form compounds, so it was assigned a valence of 2. Chlorine was found to combine with only one other atom, so it had a valence of 1. Carbon, it was found, had a valence of 4. This last observation caused organic chemists to take note.

TETRAVALENT CARBON

Friedrich August Kekulé and Archibald Scott Couper

Friedrich Kekulé was a student of architecture who was so impressed by Liebig's lectures that he decided to study chemistry. Couper was a Glasgow philosophy student in his late twenties when he became interested in chemistry and went to Paris to study. He worked in the laboratory of Charles Adolphe Wurtz, an eminent chemist and early investigator of the ammonia type. While at Wurtz's laboratory early in 1858 Couper wrote a paper entitled "On a New Chemical Theory" that probably contained the first statement about the tetravalence of carbon and its chain-forming ability. Couper asked Wurtz to help him have his paper read at the French Academy of Sciences, but Wurtz was not a member of the academy (all papers had to be sponsored by a member), so there was a delay while he secured the cooperation of a member. A few months later a paper by Kekulé appeared in print stating the same ideas. Couper was so upset that he raged at Wurtz and ended up being dismissed from the laboratory. Dumas was persuaded to sponsor Couper's paper, and it was finally read on June 14, 1858.

Kekulé's paper was printed on May 19, 1858. The following year it became evident that Couper suffered from a mental illness (which may have been evidencing itself while he was still in Wurtz's laboratory), and he was hospitalized. He was released, hospitalized again, and eventually allowed to remain in his mother's care. He lived another 30-some years, but was never again intellectually able. Described as "a wreck"[10] by an acquaintance, he lived out his life tending flowers.

Whatever the priority, Kekulé's story clearly wins the award for romance. Reportedly the idea of the chain formation by carbon came to him in a dream:

> During [a] stay in London, I lived for a considerable time in Clapham Road near the Common. But I often spent my evenings with my friend . . . at the other end of the great town. We spoke of many things, but chiefly of our beloved chemistry. One fine summer evening, I was returning by the last omnibus, "outside" as usual, through the deserted streets, at other times so full of life. I fell into a reverie. There before my eyes gamboled the atoms. I had often seen them moving before, each tiny being, but I had never succeeded in discerning the nature of their motion. This time I saw how frequently two smaller atoms united to form a pair; how a larger one embraced two smaller ones; how still larger ones kept hold of three or four of the smaller, whilst the whole kept whirling in a giddy dance. I saw how the larger ones formed a chain dragging the smaller ones after them. . . . The cry of the conductor "Clapham Road" awoke me from my dream, but I spent a part of the night putting down on paper the sketches at least of these dream-forms. Thus began the structure theory.[11]

After this breakthrough, progress in the structural theory of chemical compounds was rapid. Kekulé with his great imagination clearly visualized the chemical bond, but he represented it pictorially as a lumpy cloud joining atoms. Though actually close to our present concept, these bonds were hard to draw and hard to set in type. They were referred to as "Kekulé's sausages" by other chemists who preferred to represent bonds by a straight line.

Tetravalent means that carbon has four things attached to it. The simplest hydrocarbon chains are formed from carbon atoms linked together with a single bond and hydrogen atoms making up the rest of the required attachments. Therefore *ethane*, the simplest two-carbon chain, is formed from two carbon atoms linked together and three hydrogen atoms at either end: H_3C—CH_3. However a double bond can also form between two carbon atoms, which means fewer hydrogen atoms are needed to fill up the valence of 4. *Ethylene* is formed from two carbon atoms linked by a double bond and two hydrogen atoms at either end: H_2C=CH_2. *Acetylene* is formed from two carbon atoms linked with a triple bond and one hydrogen atom at either end: HC≡CH. Hydrocarbons that need fewer hydrogen atoms because of multiple bonds are referred to as *unsaturated*—a *saturated* hydrocarbon is made from the maximum number of hydrogen atoms—and unsaturated hydrocarbons tend to be more reactive than their saturated equivalents. The notions of double bonds, triple bonds, and saturation quickly gained acceptance after Kekulé introduced tetravalent carbon, but benzene remained a mystery for a while.

Formed from six carbon atoms and six hydrogen atoms, chemists puzzled over how to arrange a chain of six carbon atoms with six hydrogen atoms and still preserve the tetravalence of carbon. Kekulé came up with the name *aromatic* for the generally pleasant smelling class of benzene-based compounds, and then he came up with the structure—again reportedly in a dream:[12]

> Again the atoms were gamboling before my eyes. This time the smaller groups kept modestly in the background. My mental eye, rendered more acute by repeated visions of the kind, could now distinguish larger structures, of manifold conformation: long rows, sometimes more closely fitted together; all twining and twisting in snake-like motion. But look! What was that? One of the snakes had seized hold of its own tail, and the form whirled mockingly before my eyes. As if by a flash of lightning I awoke; and this time also I spent the rest of the night in working out the consequences of the hypothesis.[12]

And so with the ring structure of benzene (see Fig. 12.2), Kekulé scored another triumph, though the next important discovery eluded him. Interestingly enough for all his imagination and architectural train-

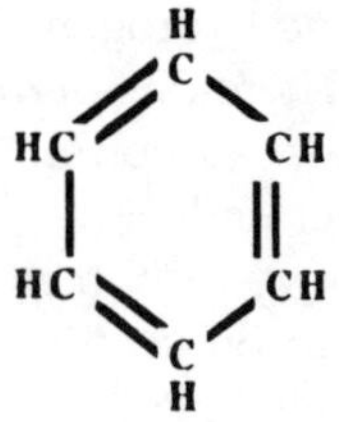

```
        H
        C
    HC /  \ CH
     |      ||
    HC \  / CH
        C
        H
```

Figure 12.2. Kekulé's ring structure for benzene, a molecule made from six carbon atoms and six hydrogen atoms. Kekulé realized that a ring formed from six carbon atoms with alternating double bonds between them would preserve the tetravalence of carbon. But there is more to the structure of benzene, as we shall presently see.

ing, Kekulé seemed confined to two-dimensional thought, but then that was the norm for the time. Those that stepped back from the norm and started thinking three dimensionally had to take their share of censure and skepticism.

CHEMISTRY IN THREE DIMENSIONS

Louis Pasteur

The story starts with Louis Pasteur, more famous for discoveries in microbiology and biochemistry (and whose life is examined in that capacity), but whose primary training was as a chemist. In the 1840s Pasteur was starting out in his chemical career and working in the same laboratory as Laurent. Laurent like all chemists before and since was interested in the many beautiful colors and shapes of such crystals as the hexagonal spears of natural quartz, the transparent blue-green of aquamarine, and the delicate snowflakes of crystalline ice. He shared this interest with Pasteur and encouraged him to investigate the intriguing crystals of tartaric acid salts.

Tartaric acid is now termed a *chiral* compound; that is, it has a "handedness." Consider the left hand: It is similar to the right hand, but it is also quite different—the left hand is the mirror image of the right. This is because hands have a top and a bottom; they are three-

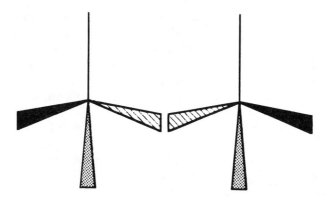

Figure 12.3. A tetrahedron is a three-dimensional object shaped like a three-legged stool with an additional leg on top. If a molecule is formed from one atom at the center of a tetrahedron and a different atom on each leg, then the molecule has a "handedness"; that is, it can be formed in either of two different mirror-image structures, like the right hand and the left hand.

dimensional. Two paper hands cut from a flat piece of paper are indistinguishable, and they can be overlaid; real, three-dimensional hands cannot. So it is with chemical compounds: If they are formed from four different atoms or groups around a central atom and they are flat, then they are indistinguishable. But if the four different groups are arranged around the central atom to form a three-dimensional tetrahedron (a three-legged base with another leg pointing straight up, as shown in Fig. 12.3), then there are ways of arranging them that are mirror images of each other. They are chiral.

Handedness (*chirality*) at this molecular level has some important manifestations on the macroscopic level. For one, a chiral compound affects polarized light. Light is a moving, oscillating electromagnetic field, and normally vibrations go in all different directions. But some orientations of the vibrations can be filtered out (as with Polaroid sunglasses), leaving *plane-polarized* light whose vibrations are confined to a plane. If plane-polarized light is passed through chiral material, the plane of the light exiting chiral material is rotated from the plane of the light that entered. This phenomenon is called *optical activity*.

Some crystals like natural quartz (silicon dioxide) were known at the time to have optical activity, but Jean Baptiste Biot, a crystallographer of this era, was the first to observe optical activity in light passing through liquids or solutions of natural products like turpentine. Tartaric acid is also a natural product found in grapes, and it is a byproduct of wine manufacture. In animals and plants, it was eventually learned, chiral compounds are often produced in only one of their two possible forms, so natural products often have a natural "handedness." But at the time all that was known was that tartaric acid from grapes had optical activity. But another form of tartaric acid, called *racemic* acid, which seemed to be chemically identical to grape tartaric acid, had no effect on polarized light.

Examining crystals of racemic sodium ammonium tartrate under a magnifying glass, Pasteur found that they came in twin forms: One half had a particular crystal face oriented to the left and the other half to the right. He was able to sort them with a magnifying glass and tweezers. (As it turns out sodium ammonium tartrate is one of the few chiral salts that crystallize into mirror-image crystals that can be sorted by hand.)

Pasteur made up a solution of the left-oriented crystals and found that it rotated polarized light in one direction. A solution of right-handed crystals gave the opposite rotation. He made up a mixture of equal amounts of both right-handed and left-handed crystals, and it had no effect on polarized light. (The term racemic mixture has come to indicate a mixture that contains equal portions of the two possible orientations of a chiral compound.) He told Biot, who by now was an elderly man, and Biot did not believe him. Pasteur went to Biot's laboratory, used Biot's chemicals, did the separation in front of him, and let Biot make up the solutions for testing. They had exactly the properties Pasteur claimed. Biot became an enthusiastic supporter of Pasteur's work.

But though this established the existence of the phenomenon, it did not explain it. Pasteur's intuition led him to speculate that it might be connected to spatial properties of matter, but the hypothesis remained to be fully explored through the inspiration of two acquaintances from the same laboratory that had produced Couper: Jacobus Henricus van't Hoff and Joseph Achille Le Bel.

Jacobus Henricus van't Hoff and Joseph Achille Le Bel

Jacobus van't Hoff, born in Holland, started studying chemistry under Kekulé, but he preferred Wurtz's laboratory. After his work there he took a teaching position at a veterinary college, and 2 years later he became a professor in Amsterdam. Joseph Le Bel, born in France, received his education in Paris, and he also eventually went to work for Wurtz. He was acquainted with van't Hoff during this time, but they separately developed the idea that the light-rotating power of compounds (optical activity) was due to their three-dimensional arrangements in space. In 1874 both van't Hoff and Le Bel in separate articles cataloged existing compounds known to have optical activity, and they showed that these compounds all had four different groups around a central carbon atom. Several chemists who had studied optical activity immediately saw the value of van't Hoff's ideas, but the idea was not well received by all. In particular Kolbe, who had been a friend and mentor of Frankland's while he was developing ideas on valence, wrote an almost poetic diatribe:

> A Dr. J. H. van't Hoff, of the Veterinary School at Utrecht, has no liking, it seems, for exact chemical investigation. He has considered it more convenient to mount Pegaus (apparently borrowed from the Veterinary School) and to proclaim in his *La chimie dans l'espace* how the atoms appear to him to be arranged in space . . . .[13]

He concluded

> To criticize this paper in any detail is impossible because the play of the imagination completely forsakes the solid ground . . . and is quite incomprehensible to the sober chemist.[14]

Kolbe was wrong.

The new concept of the three-dimensional organic molecule was the topping to the concepts of types, homologous series, and tetravalent, self-connecting carbon chains. Organic chemistry leapt forward in understanding and synthetic control, and it soon outstripped inorganic chemistry, which had dominated the field for the last 2000 years.

Models that organic chemists now had for their molecules were tactile and intuitive. They could explain chemical properties based on three-dimensional structure as well as elemental composition. They could construct synthetic routes with chalk and chalkboard like generals planning a battle.

The rest of the world was becoming three-dimensional, too. Early in the twentieth century airplanes had gotten off the ground, and submarines were no longer routinely drowning their crews. Europeans had expanded their thinking laterally—to another, far off, country called the United States—that was sending chemists in droves to study at the European laboratories. These chemists returned from Europe to do some fine organic chemistry.

THE UNITED STATES

Because of its newness, organic chemistry, was custom-made for the new breed of chemist, and in the United States materials were readily available from coal and oil reserves. In the 1870s Johns Hopkins, a wealthy Baltimore banker, left seven million dollars to build a hospital and a university, and the university began producing chemists.

The university was located in an old part of the city described by E. Emmet Reid in his autobiography, *My First Hundred Years*, published when he was 100 years old:

> At the time I arrived in 1894, and for about a decade later, Baltimore had no sewage system. Water from the wash basin, bath tub, and the kitchen sink went through a drain which passes under the sidewalk and discharged into the gutter. As you walked along the street you could tell when someone had taken a bath.[15]

But the university was well-equipped for its time, and the chemistry department had the good luck to be headed by Ira Remsen (who also headed the committee to develop a sewage system). Remsen trained in Wöhler's laboratory in Germany, and the two must have gotten along well because they shared a similar sense of humor, as can be gleaned from the following report of an experience Remsen had when he first became interested in chemistry:

While reading a textbook on chemistry, I came upon the statement "nitric acid acts upon copper". I was getting tired of reading such absurd stuff and I determined to see what this meant. Copper was more or less familiar to me, for copper cents were then in use. I had seen a bottle marked "nitric acid" on a table in the doctor's office where I was then "doing time" . . . and [wanted] only to learn what the words "act upon" meant . . .

All was still. In the interest of knowledge I was even willing to sacrifice one of the few copper cents then in my possession. I put one of them on the table; opened the bottle marked "nitric acid"; poured some of the liquid on the copper; and prepared to make an observation.

But what was this wonderful thing which I beheld? The cent was already changed, and it was no small change either. A greenish blue liquid foamed and fumed over the cent and over the table. The air in the neighborhood of the performance became dark red. A great colored cloud arose. This was disagreeable and suffocating—how should I stop this? I tried to get rid of the objectionable mess by picking it up and throwing it out of the window, which I had meanwhile opened. I learned another fact—nitric acid not only acts upon copper but it acts upon fingers. The pain led to another unpremeditated experiment. I drew my fingers across my trousers and another fact was discovered. Nitric acid also acts upon trousers.[16]

Remsen survived his experiments with nitric acid and went on to build such an active department at Johns Hopkins that the *American Journal of Science* complained that if it published all his papers, it would become exclusively a chemistry journal. So Remsen founded the *American Chemical Journal*, which was eventually absorbed into the *Journal of the American Chemical Society*. His department successfully competed with German institutions, keeping potential students in North America, and it trained many of the next generation of U.S. chemists. Remsen also recognized the need for good text books in the United States, so he wrote them. His *Introduction to the Study of Chemistry* went through eight editions, and it was translated into eight foreign languages—including German.

In the Michigan laboratory at the University of Ann Arbor, Moses Gomberg, who had come from Russia with his impoverished father, isolated a trivalent carbon compound that had one of its valencies unfilled. The compound, one of the first examples of a class later to be called *free radicals*, was understandably reactive, but long-lived enough to be identifiable. This work was referred to in Germany as "der wundervollen Arbeit Gomberg's"[17] (Gomberg's wonderful work). These successes on the part of U.S. chemists established their place in the world of chemistry and allowed them to establish their own traditions, too. One of these traditions was to train women in chemistry at universities. Although the start was shaky—M. Carey Thomas, one of the shapers of Bryn Mawr, the women's college modeled after Johns Hopkins, was not allowed to participate in Johns Hopkins graduate school—the United States was soon producing quality chemists of all genders and all heritages as we continue to see.

So although the only organic chemist who ever saw an atom was Kekulé—and then only in his dreams—in the late 1800s organic chemists found they could explain and control their products and their processes by making assumptions about the behavior of atoms and their three-dimensional arrangement in space. During the last half of the 1800s organic chemistry matured as a specialty, and organic chemists defined their art. They not only watched the atoms "gambol before [their] eyes," but now with the development of synthetic organic chemistry, they choreographed the dance. But more grew out of their efforts than synthetic organic chemistry. The observation that elements seemed to have a preferred valence, or combining power, was to point the way to the next major systemization of *all* chemistry—the arrangement of elements into groups or families with similar properties: a Periodic Table. And the effort in this direction came from another upcoming frontier: Russia.

chapter
THIRTEEN

ca. 1848–1914: Inorganic Elements and Ions— New Earths and Airs

A long with the dazzling progress in organic carbon chemistry in the late 1800s came dazzling progress in the understanding of the other 60-odd elements in the Periodic Table—that is, as soon as the Periodic Table came to be. This all-important table is the cornerstone of systematic chemistry as we understand it today, and it is with the construction of the Periodic Table that the story of 1800's inorganic chemistry begins.

THE PERIODIC TABLE

It had been evident for some time that there were families of elements with similar reactivities. Selenium was mistaken for tellurium. Bromine was extracted along with iodine and chlorine from the sea. Gold, copper, and silver were found by the ancients in a naturally metallic state. Davy isolated potassium and sodium by the same method from similar salts, and he watched both burn on exposure to moist air. These few similarities however were not enough to reveal a pattern until organic chemists began categorizing elements by combining power, or valence. When they did, more relationships became evident. Oxygen and sulfur for instance shared an identical valence, as did carbon and silicon.

Tantalized by these hints, several chemists undertook classification schemes based on atomic weights to ferret out the underlying logic. Unfortunately atomic weights varied nearly as much by laboratory as by element, and patterns were obscured. Nevertheless in an attempt to find trends, chemists listed elements in tables, on the surfaces of a cylinders, in spirals, and one even thought he could discern a relationship between the distances of the planets from the sun and atomic weights. John Alexander Reina Newlands made a list based on Cannizarro's atomic weights, and he saw that the properties seemed to repeat after eight elements, reminding him of the octave in music. Newlands made a table, aligning elements with like properties and leaving blank spaces for elements that had not yet been found. He could not get his speculations accepted by mainstream journals, and when he presented his table at a meeting—perhaps hoping to appear less radical—he omitted some blank spaces, so the correlation became much weaker. One critic asked Newlands if he saw patterns when he arranged the elements alphabetically.

Another chemist who make a table with blank spaces for missing elements—and stuck by it—was a German, Lothar Meyer. But Meyer published his table in 1870. Dmitri Mendeleev published a similar table in 1869, and he receives most of the credit as the discoverer of the Periodic Table.

Dmitri Ivanovich Mendeleev

Dmitri Ivanovich Mendeleev, born in Siberia, was the youngest of a family of at least 14 children. His father managed to support the family as a school teacher, but when he became blind, his small pension was not enough. Mendeleev's mother, who had educated herself by repeating her brother's lessons with him when he came home from school, was able to reopen a run-down glass factory that had once belonged to her family. She managed to support her family, and even build a church and school for the factory workers, until the factory burned down in 1848. By this time she was in her late fifties, and her husband had died. Mendeleev was in his early teens and she saw in him a talent. She traveled with him to Moscow, a journey of thousands of miles, so he could receive a better education. He was denied entrance to the University of Moscow because he was a Siberian, so they set out for St. Petersburg, where he was allowed into the Institute of Pedagogy to train as a teacher. His mother died shortly thereafter, and Mendeleev was later to write: "She instructed by example, corrected with love, and in order to devote [me] to science, . . . spen[t] her last resources and strength."[1]

But her instincts had been correct: Mendeleev was worth the effort. He continued at the University of St. Petersburg, writing a dissertation that won him a traveling fellowship. While in Europe on this fellowship, he, like Lothar Meyer, attended the Karlsruhe Conference, the first international chemical conference, which was held to resolve difficulties with definitions of atoms, atomic weights, and nomenclature. The conference was not particularly productive, but on the way out the door, Mendeleev was handed a pamphlet by Cannizzaro, explaining the hypothesis of Avogadro. He read it, and it made sense.

He was already very familiar with the properties of elements, and once back home and faced with the prospect of teaching chemistry, he looked for a logical basis for inorganic chemistry. He wrote out flash cards with the properties of the elements and Cannizzaro's atomic weights. He tried various arrangements of the flash cards, playing a sort of solitaire his friends jokingly named Patience. He pinned these flash cards to the laboratory wall and stood back. He began to see patterns.

He was not sure where to place hydrogen, and that remains debatable today. In various versions of the Periodic Table it is placed top right, top left, or center. Mendeleev solved the problem by leaving it out. In fact Mendeleev's ultimate triumph lay in his omissions: He listed elements by atomic weight but kept families of elements with similar properties together. When this did not work out quite right, he left gaps for elements he suspected were missing. He then went one step further. Using the Sanskrit *eka*, meaning one, he predicted that an ekaboron, eka-aluminum, and ekasilicon would be found to fit in the empty slots, then predicted their properties and the atomic weights they should have. He found some anomalies in his arrangement—a few of the atomic weights decreased instead of increased from one element to the next—but Mendeleev dismissed these as errors in atomic weight (common enough at the time). It turned out that on this point, he was wrong, as we soon see, but in all else, he definitely pointed the way. He summed it all up in the paper he published in 1869.

At first the paper received little notice because it was lengthy and cumbersome to read. Then someone saw that the newly discovered gallium was in fact eka-aluminum. Scandium proved to be ekaboron. A German, Clemens Alexander Winkler, decided to look for Mendeleev's ekasilicon, basing his search on Mendeleev's predicted properties. He found his element and named it germanium. So although unconventional (besides being a Siberian, he cut his hair only once a year in the spring and refused to alter this custom even when presented to the tzar), Mendeleev became a celebrity and a regular guest at scientific gatherings.

But for all his acceptance in Europe, Mendeleev was not approved of in Russia. He was liberal, admitting women to his lectures (although he considered them biologically incapable of the same intellectual achievements as men), and he was divorced and had remarried. Aware first hand from his Siberian youth of the oppressive tactics of the tzarist government, he was outspoken in his politics. The tzar did not like this, so Mendeleev was not elected to the Russian Academy of Sciences. But he lived out his life peacefully, being made director of the Bureau of Weights and Measures, and writing on chemistry, art, education, and spiritualism.

So after the flurry of discoveries in the early 1800s, the collection of elements was finally systematized, and inorganic chemists began working on their chemistry. Ferdinand-Frédéric-Henri Moissan finally isolated fluorine—an element that had long eluded isolation because of its reactivity and toxicity (fluorine, in the form of hydrofluoric acid, dissolves glass, including laboratory glassware)—by electrolysis in a tube made from another element isolated earlier that century: platinum. Other improvements in the chemist's repertoire of equipment—better balances, optical glass, furnaces, and other implements—led to sorting out a bewilderingly similar group of elements huddled together in ores and hiding each other with nearly identical properties—the rare earths.

RARE EARTHS

The group called the rare earth elements (now known as the lanthanides) includes the strip of 14 elements at the bottom of the modern Periodic Table (a rendition of which can be found on the inside back cover of this book). The rare earths include elements 58–71 and a few mavericks from above: lanthanum, yttrium, and scandium. Though in most cases the pure metals were not isolated until the 1900s, chemists of the 1800s were able to distinguish between their various oxides, or earths. Once distinguished however they gave Mendeleev considerable trouble in classification. Though these elements are not so rare as initially assumed, they required rare chemists with rare patience to tackle their separation. One such chemist was Carl Gustaf Mosander, known to Wöhler and Berzelius as Father Moses.

Carl Gustaf Mosander

A protégé of Berzelius, Mosander went so far as to move in with his mentor, wife and all. His wife taught Berzelius the Dutch language, and Berzelius taught Mosander chemistry. In 1839 Mosander showed that ceria, an earth isolated by Berzelius, was really a mixture of two earths. He named the minor component lanthana, from the Greek for *to escape notice*, then showed that *this* was also a mixture with another earth, which he named didymia, from the Greek for *twin*. Another chemist however separated two more components from didymia: praseodymia (from the Greek *prasios* meaning green) and neodymia (which in Greek means *new twin*).

Mosander also separated yttria, another earth thought to be a pure oxide, into three: yttria, erbia, and terbia, all named for the Swedish town of Ytterby. Other workers found other earths including lutetia (named for the ancient name for Paris), holmia (for the Latin name for Stockholm), thulia (from Thule, was the earliest name for Scandinavia), and dysprosia (from the Greek *dysprositos*, which means *hard to get at*). When the Finnish chemist Johan Gadolin isolated yttria, he thought it was pure, but then two other chemists separated yet another component from this earth and named it gadolinia.

By some reports gadolinium was the first element named for a person, but it was probably really named for the mineral that contained gadolinia, and the mineral was named for Johan Gadolin. By the same token samarium was named for the mineral samarskite, which had been named in honor of a Russian mine official, Colonel Samarski. It is difficult therefore to assert that the name gadolinium was meant to immortalize the chemist any more than the name samarium was meant to immortalize military personnel. What *was* unique about samarium however was its discovery using a new analytical technique: spectroscopy.

SPECTROSCOPY

Spectroscopy is the systematic study of the interaction of light with matter, and in the hands of the chemists of the 1800s, spectroscopy became a powerful analytical tool. The observation that light interacts with matter was nothing new—Newton separated white light into component colors with a glass prism, then recombined it into white light by passing it through a second prism; the ancient texts of India report the use of flame color in chemical analysis (though the Indian savants were looking for poisons, not new elements); later chemists used flame colors as their only way of distinguishing sodium and potassium salts. The discovery to be exploited in the 1800s however was that elements in flames have spectra that show characteristic line patterns, and these line patterns can be measured and cataloged. Such was the work of the German chemist Robert Bunsen (whom we met in conjunction with his work in organic chemistry), and the physicist Gustav Robert Kirchhoff.

Robert Bunsen and Gustav Robert Kirchhoff

When Bunsen came to Heidelberg to teach, he was promised a new laboratory. Coal gas had just been introduced for street lighting, so Bunsen had gas brought into his new laboratory. The flame produced by gas in conventional burners was sooty and unsteady. Bunsen needed a steady, essentially colorless flame to analyze flame colors of salts in mineral waters, so he invented the laboratory burner named in his honor, the Bunsen burner. To separate the light of the element from the light of the flame, Bunsen had to pass the light through glass and liquid filters before viewing, and this was cumbersome. Gustav Robert Kirchhoff, a physicist and colleague at Heidelberg, suggested passing the light through a prism to spread out its component parts into a spectrum. Together the two workers assembled the flame, prism, lenses, and viewing tubes on a stand and produced the first spectrometer, and in very short order they used their spectrometer to identify the new elements cesium and rubidium, showing in each case that these new elements produced line spectra that were unique.

Following closely on the heels of this work came the discovery of indium by another German professor, Ferdinand Reich. The colorblind Reich had to leave the spectroscopic analysis of this element with its bright indigo line (hence indium) to his normally sighted assistant, Hieronymus Theodor Richter. In France Boisbaudran performed careful spectral examinations of 35 elements, including his two new rare earth elements, samarium and europium, and he wrote his results in an important reference volume.

The technique of spectral analysis rapidly became widespread, and another who quickly took it up was William Crookes: chemist, physicist, spiritualist, and father of 10. A bit schizophrenic in his transitions between the practical and the ethereal, this chemist, motivated by the financial needs of ten children, planned to do spectroscopic analyses for pay. His other motivation however came from his interest in spiritualism (and in the medium Florence Cook). He regularly published accounts of kinetic, audible, and visual phenomena witnessed at séances, but his interest in the glowing lights—be they from chemicals, cathode rays, or specters—bore more positive results. When a particular unknown gas was sent to him for analysis, he realized that its lines matched lines previously observed only in the spectrum of the sun. This element, helium (named for the Greek *helios* for the sun) therefore has the distinction of being the only element (so far) discovered in space before being discovered on Earth. Helium is one of a group sometimes called the rare gases—now more commonly called the noble gases—which form the rightmost column in the Periodic

Table. These gases led their discoverers on a chase before revealing their nature, but as is now known, not because they are particularly rare but because they are *noble*. Noble materials are highly unreactive, or inert (the noble metals include gold and platinum), and this lack of reactivity means they are hard to isolate and characterize by chemical means. But as always if there is a challenge, there are those who are willing to take it up. In the case of the rare gases the champions were John William Strutt, the Third Baron Rayleigh (Lord Rayleigh) and William Ramsay.

RARE GASES

Lord Rayleigh and William Ramsay

Lord Rayleigh gained his peerage by inheritance but his scientific reputation by sheer personal effort. He carried out experiments in a laboratory attached to the family manor house, using personally purchased equipment. He produced in his career some 430 papers as well as a two-volume work *The Theory of Sound*. He was an established experimentalist when he became interested in the densities of gases.

A conscientious worker, he cross-checked his values by using gases generated by different methods. For hydrogen and oxygen he found good agreement, but when he compared the density of atmospheric nitrogen with nitrogen generated by chemical means, he found a discrepancy. The atmospheric nitrogen was consistently denser than chemically prepared nitrogen, though the difference was small. Puzzled but convinced the effect was real, he published his results and asked for comments. The chemist William Ramsay commented that Rayleigh should try purifying the nitrogen chemically. In close collaboration and consultation, Ramsay and Rayleigh then pursued separate routes. Ramsay reported

> [Rayleigh] thought that the cause of the discrepancy was a light gas in non-atmospheric nitrogen; I thought that the cause was a heavy gas in atmospheric nitrogen. He spent the summer in looking for the light gas. I spent July in hunting for the heavy one. . . . I have succeeded . . .[2]

The mysterious weight factor in atmospheric nitrogen turned out to be argon, the third rare gas in the Periodic Table and now known to constitute about one percent of the air (the rest is nitrogen, oxygen, and small amounts of carbon dioxide, water, and other gases). Rayleigh wrote to Lady Frances Balfour

> The new gas has been leading me a life. I had only about a quarter of a thimbleful. I now have a more decent quantity but it has cost about a thousand times its weight in gold. It has not yet been christened. One pundit suggested 'aeron,' but when I have tried the effect privately, the answer has usually been, "When may we expect Moses?"[3]

The two investigators reported their results in a joint paper in which they named the gas argon, from Greek for *the lazy one*, in reference to its lack of reactivity.

Other scientists were initially skeptical. It would not have been the first time a report of the discovery of a new element turned out to be false, and it would not be the last. Berthelot was sent a sample to subject to an electrical discharge, but he reported null results. He also reported that his assistant hurt himself undoing the iron wire used to clamp down the rubber tubing sealing the sample tube. Years later it was realized that Rayleigh and Ramsay had used copper wire to tie off the tube, so the sample had been opened, presumably by customs officials, and Berthelot's experiment had been on air.

Mendeleev remained unconvinced, preferring to believe that argon was triatomic nitrogen, N_3. But Mendeleev's hesitation can be understood by realizing that there was no position for the noble gases in his Periodic Table. When Ramsay suggested that argon might go after chlorine and before potassium (its present position), it was observed that argon had a atomic weight greater than that of potassium, so it would still not fit Mendeleev's scheme. But the arguments faded as Ramsay filled in the new column. Ramsay found a terrestrial source for helium (the one spectroscopically confirmed by Crookes) in the mineral cleveite less than 2 months after the announcement of argon. Within 4 years Ramsay, working with Morris Travers, isolated krypton (from the Greek for hidden), neon (from the Greek for new), and xenon (from the Greek for stranger) from liquefied air. They also announced the discovery of metargon—but unannounced it when they found it to be a mixture of impurities. Finishing out the group was not accomplished by Ramsay—this honor fell to a Prussian professor, Friedrich Ernest Dorn, who isolated radon, a product of radioactive radium disintegration, to complete the column, though Ramsay was the first to determine radon's atomic weight.[4]

Ramsay probably would have found radon if Dorn had not, because Ramsay was also interested in the new phenomenon of radioactivity, as was the rest of the world. For all the overtones of physics associated with them, the first radioactive elements were just new metals found in ores, and the beginning of the study of radioactivity was a piece of inorganic chemistry.

RADIOCHEMISTRY—THE BEGINNINGS

As Emmett Reid, the centenarian chemist in the last chapter remarked about his early teaching days, "Things were much simpler then, as there were no ions or isotopes . . ."[5] This situation was soon to change. While rummaging around in the nature of matter, chemists applied everything they had to whatever they had, and they eventually got around to applying electricity to "vacuum," or gases under very low pressure. When they did, they found light was produced and an apparent "ray" capable of casting a shadow. In the 1870s the chemist/spectroscopist William Crookes found that these cathode rays (so called because they were emitted from the negative electrode, or cathode) could turn a tiny windmill placed in their path and that they bent in a magnetic field. The English physicist Joseph John Thomson (or J. J. Thomson, as he is better known) carried out experiments that showed these "rays" were composed of negatively charged particles that apparently had about one-thousandth the mass of a hydrogen atom. (He also showed that the particles were not massive enough to have turned the paddle wheel, which really resulted from a heat effect) The same particles were found no what matter the gas in the cathode ray tube. These particles took on the name that had been given to corpuscles of electrical charge, *electrons*, and Thomson and Kelvin modeled the atom as a mass of positive charge interfused with imbedded electrons, promptly christened the plum pudding model for the English Christmas dessert of pudding interfused with imbedded plums.

Then one night in 1895 Wilhelm Conrad Röntgen, experimenting with a cathode ray tube, noticed that a screen on the other side of the room, coated with a barium platinum cyanide compound, was glowing. This took him aback a bit because the room was dark and the cathode ray tube was covered with black paper. He turned the screen away from the cathode ray tube, but it still glowed. He placed his hand between the cathode ray tube and the screen and saw the outline of his bones. He concluded that there were rays coming from the cathode ray tube—rays that passed through cardboard, cloth, and skin. He convinced himself by capturing the image on photographic plates. He wrote a paper on the rays, calling them "X" rays for lack of a better name.

A little later Henri Becquerel, wondering about the connection between X-rays and glowing minerals, decided to see if phosphorescent uranium salts, which glow with a green light when exposed to the sun, emit X-rays. He set up his experiment: uranium salts in contact with photographic plates wrapped in black paper. But this was Paris in the winter. The sun didn't shine, so the salts did not phosphoresce. He put the assembly away in a dark drawer, and then a few weeks later, being a careful scientist, he developed the plates and observed ghostly images of the salts. Becquerel thought the uranium salts were producing Röntgen's X rays, but he soon found that the radiation was even more powerful, and it was produced spontaneously by uranium compounds and the free metal, without any input of energy. The phenomenon fascinated chemists and physicists and caught the attention of the Sorbonne's most promising doctoral student: a Polish woman by the name of Marya Sklodowska, the soon-to-be Marie Curie.

Marie Curie

Poland by the late 1800s had been absorbed into the Russian empire despite fierce nationalistic pride and resistance from Polish nobility. The tzarist government responded with arrests and hangings and by forbidding the use of the Polish language in any official capacity. But resistance remained a political undercurrent, having as its philosophic anchor the Positivism teachings of August Comte, who looked to

Figure 13.1. Marie Curie. (Courtesy Chemical Heritage Foundation.)

empirical science and humanitarianism to bring an age of peace and prosperity. This was the inheritance of Marie Curie.

Her family's fortunes had been ruined by the tzarist regime, and the careers of her parents, both teachers, suffered. Marie excelled in school, using the Russian language, but after her initial schooling she became involved in clandestine, anti-tzarist, student movements.

For the persecuted Polish intelligentsia, Paris was the land of intellectual opportunity. Marie took work as a governess in Poland, so that her sister could study medicine in Paris. When Marie's sister completed her studies and married another doctor, Marie was ready to go to Paris herself to study at the Sorbonne. Riding in a fourth-class car, seated on a camp stool, Marie traveled across Germany by train. Once in Paris she endured the hardships of poverty (reportedly fainting on one occasion from lack of food) to earn the equivalent of a master's degree in physics, at the head of her class, and a master's degree in mathematics, second in her class. A Sorbonne physics professor, Gabriel Lippmann, allowed her to work in his laboratory, but she also needed financial support, so she took on work investigating the magnetic properties of various steels for Le Châtelier and the Society for the Encouragement of National Industry. In an attempt to find laboratory space to conduct experiments, she contacted a professor in the School of Chemistry and Physics, Pierre Curie.

Pierre Curie

Pierre Curie had a rather different beginning from Marie's. He was the child of a comfortable middle-class family, though apparently intellectually inhibited as a child. His parents educated him at home for that reason, but once introduced to mathematics and science, he quickly blossomed. By the age of 19 he was doing original scientific research. He found an experimental law for the effect of temperature on magnetism and studied a strange property of crystals predicted by Gabriel Lippmann—piezoelectricity, the ability of certain crystals to generate an electrical potential under applied pressure. So it was with a established promising career (though restricted by the fact that he had not attended the Polytechnique) that Pierre Curie at the age of 35, fell in love with the Polish scientist of 26.

Marie and Pierre Curie

After some hesitation (Marie had planned to return to Poland, and Pierre at one point suggested moving there with her), Marie decided to stay with Pierre in Paris. After 14 months of marriage their first daughter was born, and Marie began to think about finishing her doctoral work. For her dissertation topic she decided to study the new phenomenon that had been observed by Becquerel: the emanation of rays from minerals.

Marie was given permission to work in a small unused storeroom with no insulation or heat near the School of Physics. One entry in a notebook from this period reads "Temperature in here 6.25°, !!!!!!!!!!"[6]

But Marie had then the dogged determination that would dominate her whole life, so she observed that radiation from uranium did not seem to depend on temperature and continued. Marie decided to investigate all known elements to see if she could find others that were radioactive. She found another: thorium. But then so did others who were investigating this same effect at this same time. Marie's inspiration came when she decided to look at minerals for new sources of radiation. She found that pitchblende, the commonest ore of uranium, was more radioactive than its uranium and thorium content could account for. There was something else in there, and Marie set out to find it.

At this point Pierre, interested in her work all along, decided to suspend his own research temporarily and work with his wife on hers. They isolated a sample that they believed to be concentrated in the new element, and they asked a spectroscopist to examine it, but no new lines were found. Convinced that the element existed, Marie named it *polonium* after her homeland, Poland, and they set about isolating a larger sample. Marie wrote with mild understatement:

> I submitted to a fractionated crystallization two kilograms of purified radium-bearing barium chloride that had been extracted from half a [metric] ton of residues of uranium oxide ore.[7]

The process was straightforward inorganic chemistry—dissolving minerals in acids, precipitating different fractions with reagents such as sulfates and hydroxides, redissolving the precipitates, and crystallizing and recrystallizing the resulting products—but on a huge scale.

They worked in an abandoned wooden shed across from Marie's original storeroom laboratory, with a leaky roof, poor ventilation, and poor heat during the 4 years from 1898–1902. Marie did most of the physical labor involved in handling the carboys and the vats because Pierre suffered from rheumatic-like pain, and he focused on investigating chemical properties of the material. Marie wrote, "We sleep well."[8]

They were plagued by a shortage of funds. Marie took a teaching job at a girl's school. Pierre attempted to secure better posts on several occasions but lost by a hairs' breadths and regretted the time wasted trying. Marie lost a child midterm, writing that she had not slowed down, thinking her body could take it. Their hands, burned by the radiation, became scarred with brown tissue. But instead of one radioactive material, they found two: They confirmed polonium, and the second one, when they managed relatively pure samples, glowed with a blue light. Marie wrote.

> One of our joys was to go into our workroom at night, we then perceived on all sides the feebly luminous silhouettes of the bottles or capsules containing our products. It was really a lovely sight and always new to us. The glowing tubes looked like faint fairy lights.[9]

This time the spectroscopists saw the spectral lines of a new element, too. Marie and Pierre named the second element radium, and finally at the age of 36, Marie submitted her dissertation. Fame was soon to follow, though fortune never did.

Pierre and Marie shared the same political idealism and refused to patent their materials, preferring to make them available to all to wished to use them. Pierre eventually achieved the chair of physics and chemistry, and Marie eventually became his paid assistant. At this time in France women were routinely barred from many professional positions (it would be 40 more years before women won the right to vote in France), so allowing Marie to be Pierre's paid assistant was an acknowledgment of her abilities. Then in 1906 Pierre was struck by a runaway horse-driven vehicle in a crowded Paris street and died instantly. Marie, surprisingly enough, was appointed to replace him. She took the lectern resolutely, spoke in a stilted monotone, and continued the lecture course exactly at the point where Pierre had left off.

Marie also continued the research. She allowed herself to be persuaded to submit her name for membership in the Academy of Sciences (which had never yet admitted a woman member) to achieve a professional status that would help secure funding. A slanderous campaign against her followed in which even her ethnic origins were questioned. On the day of the election, the president opened the meeting by saying to the ushers, "Let everybody come in, women excepted,"[10] and Marie Curie missed being elected by only one vote.

Marie's work continued to unleash the Furies. Other prominent scientists, such as Lord Kelvin, questioned her scientific originality, hinting that her ideas were derived from her husband. She was accused of having an affair with Paul Langevin, a long-time family friend and former student of Pierre. Though the affair was widely reported as fact, there remain the facts that Marie was not named in litigation between Langevin and his wife, and the alleged affair, or rumor of it, did not divert the course of her work.

But then World War I came and with it other things for people to worry about. During the war Marie drove a mobile X-ray unit and after the war, she continued her work on the therapeutic properties of radium, undertaking rigorous international tours to raise money to purchase radium for cancer therapy. Her long exposure to radiation eventually had its effect. Three years before her death in her early sixties, an acquaintance recorded that "she looked frail and pale . . . Her bandaged hands, damaged by radiation burns, twitched nervously."[11] Marie Curie died of leukemia, probably brought on by exposure to radiation. We are left with the impression, however, that had she known her eventual fate beforehand, she would have noted it in her lab book and continued.

On the other side of the ocean, in Canada, the theoretical consequences of radiation were also being sorted out.

STRUCTURE OF THE ATOM

Born on a pioneer farm in New Zealand, Ernest Rutherford reportedly rejected the rugged life and concentrated on winning a scholarship to Cambridge. At Cambridge, J. J. Thomson, the author of the plum-pudding model of the atom, gave Rutherford a sample of uranium to investigate, and Rutherford showed that it gave off at least two types of particles: one that turned out to be the positively charged nuclei of helium atoms, which he named *alpha radiation*, and the other high-speed electrons, which he named *beta radiation*.

Rutherford's successes were enough to win him a position at McGill University in Montreal, Canada, which paid 12 pounds a week (about twice the salary Pierre Curie received). Happy with the chance to build his own laboratory, he wrote home, "They want me to form a research school to knock the shine off the Yankees."[12] At McGill, Rutherford felt the need for a chemist to help him identify disintegration products. He got his chemist in the person of Frederick Soddy, a recently graduated English chemist looking for academic employment. Soddy had applied to Toronto but received no reply, so he bought a boat ticket and went there in person to find that the post had been filled. Because he was already in Canada, he applied for a post at McGill University and was hired in the chemistry department. Rutherford invited him to collaborate, and with Harriet Brooks, a graduate student, they began identifying disintegration products.

Rutherford like most bright young academics was ambitious and upwardly mobile. When the chair of physics become available in Manchester, England, he left Canada for a more central position in the scientific world. In this laboratory he directed a beam of radiation at a thin metal foil, detecting any deflections by observing a screen coated with phosphorescent material. When his collaborators observed deflections, he is said to have danced the *haka* of the native New Zealand Maori. In further experiments with Hans Geiger (of Geiger counter fame), he showed that high-energy alpha particles passed through thin foil, as would be expected if the foil were a plum pudding of soft atoms and electrons—but sometimes the alpha particles, amazingly bounced back. Rutherford said

> It was quite the most incredible event that ever happened to me in my life. It was . . . as if you had fired a 15-inch shell at a piece of tissue paper and it came back and hit you.[13]

Alpha particles bounced back, Rutherford reasoned, because the atom is not a pudding. The atom is mostly empty space with an incredibly small, incredibly dense, positively charged core. Rutherford proposed a new model—the planetary model—in which negatively charged electrons in huge orbits circled a tiny, dense, positively charged nucleus.

As we understand things today, this model is in many ways correct. The atom is mostly empty space. If an electron "orbit" were the size of a garage, the nucleus would be the size of the dot atop an *i*—with nothing between. Though there can be many electrons around one nucleus, electrons are light, so the mass of the atom is mainly in the nucleus, which has a density on the order of a thousand trillion grams per cubic centimeter—this has been compared to the mass of all the cars in the world compacted into a thimble. This picture is certainly not intuitive—but on the atomic level, it is a different world—as Rutherford and his collaborators were just beginning to understand. They also understood that there was a fundamental problem with the picture. Physicists knew that a charged particle emits radiation when accelerated, so an electron circling a nucleus should emit radiation. Emitting radiation, it should lose energy and spiral into the nucleus. But we know from practical experience that all matter does not glow, and the fact that we are here attests to the fact that all atoms do not continuously disintegrate. As we shall see however the paradox was not resolved until later in the 1900s. One of the primary products of Rutherford's laboratory were protégés, and in the early 1900s Henrik David Niels Bohr came for an extended visit and proposed refinements in the model of the atom that would bring it in line with observation. That story however belongs more to the twentieth century than the nineteenth, and so we defer it till then. Here let it suffice to say that dramatic progress continued to be made, albeit on the basis of an acknowledged defective model.

Henry Moseley

One large step forward occurred when Rutherford's student, Henry Gwyn-Jeffreys Moseley showed by analyzing X-ray spectra that the positive charge of each element's nucleus increases by 1 as Mendeleev's Periodic Table is traversed. He called this the *atomic number* and revealed experimentally for the first time the underlying logic of the table: The first element has one positive charge on the nucleus and the second element has two; the third element has three . . . and so on through the hundred plus elements identified today. It will never be known what more the young Moseley could have contributed to science. Like many other patriotic youths, he volunteered for military service when World War I broke out, and he was killed at Gallipoli at the age of 27.

And so with Moseley's work break, it would seem that we have come full cycle. We have now seen a rationalization of the order of the elements in the Periodic Table and an explanation for atomic weights. But finding new elements is not all there is to inorganic chemistry. Each one of these elements once identified has a chemistry and a character all its own that also waits to be discovered. Progress in the investigation of these chemistries hinged on elucidating another special property of materials: the ability to form charged particles, or *ions*. The answer to this problem was in the solution.

IONS

Solutions are formed when one material, the *solute*, is dispersed uniformly throughout another material, the *solvent*. These behave differently from either of their pure components; for example salt water freezes at a lower temperature than pure water. This phenomenon, called *freezing point depression*, is the reason salt is put on icy roads. A solution of ethylene glycol has to be heated to a higher temperature than pure water before it boils; this phenomenon, called *boiling point elevation*, is one reason car radiators are filled with ethylene glycol solutions (antifreeze) in the summer. Another phenomenon associated with solutions is *osmosis*. When a solution comes in contact with a pure solvent through a semiporous membrane—a membrane that allows the solvent to pass but not any material dissolved in it—the solvent flows into the solution in an attempt to dilute it. Osmosis is responsible for the swelling of limp celery placed in clean water: Water flows through cell membranes to dilute the salt solution in the dehydrated cells. Osmosis also explains why pickles shrink in brine: Here the brine is saltier than the solution in the cucumber cells, so water in the cucumber flows out of the cells in an attempt to dilute the brine.

As we understand it now, an important driver for these effects is entropy. Systems tend to evolve to states of maximum disorder—maximum entropy—and a solution with solute particles is more disordered than the pure solvent. Putting energy into a system increases disorder (as shuffling a deck of cards increases disorder). Liquids evaporate when they are heated (energy is put in) because the gas phase is more disordered than the liquid phase. But if there are solute particles in the liquid, evaporation does not change the disorder as much, and the boiling point goes up (more energy is required to put things into the gas phase).

The solid state is more ordered than the liquid state, so liquids have to be cooled to freeze (energy is removed). If there are solute particles in the liquid, freezing changes order even more, and the freezing point goes down (more energy has to be removed). A pure solvent flows into a solution because the mixture is more disordered than the pure solvent, and this drives the osmotic flow.

It is also now understood that these entropy effects depend on the number of particles, and not on what kind of particles they are. So if two solutions are made up—one with salt and the other with an equal number of particles of sugar—the salt solution show these effects to a greater extent because a molecule of sugar remains intact when dissolved, but salt in solution breaks up into *ions*, one positive sodium ion (Na^+) and one negative chlorine ion (Cl^-), creating twice as many particles as would be expected from the formula NaCl. At the time this effect was considered anomalous because it was thought, as Faraday believed, that ions were formed only when electric current passed though a liquid. Then a certain savant came up with the solution.

Svante August Arrhenius

In his dissertation written to complete his studies in chemistry, Arrhenius proposed the existence of permanent ions in solution to explain anomalously high freezing points and low boiling points seen in salt solutions. But his thesis was still rated fourth class and his defense of it third class, which effectively terminated any hopes he might have had for an academic position at the university level. He resorted to sending copies of his dissertation to various chemists in the hopes that he could gain some sympathy, but Clausius, Meyer, and others just filed it. Ostwald received his copy on the same day his wife had a baby and he had a toothache. He later said that it "was too much for one day. The worst was the dissertation, for the others developed quite normally."[14]

Ostwald however was able to repeat some of Arrhenius's experiments, and he eventually championed the idea and the author. Through Ostwald's influence Arrhenius was given a lecturer's position. Van't Hoff also came to support the Arrhenius theory and developed a compatible theoretical description of the behavior of ideal solutions (solutions in which intermolecular forces between all constituents are about the same). With the support of Ostwald and eventually van't Hoff—and because it successfully explained observations—the Arrhenius theory of permanent ions eventually gained acceptance.

Arrhenius remained interested in new approaches and new ideas the rest of his life, and he wrote about one of the more interesting in a book, *Världarnas Utveckling* (*Worlds in the Making*). In this book he proposed a Panspermia theory that suggested life could be spread by spores transported through the universe by radiation pressure. The theory however supposed an interstellar medium instead of the high vacuum we now know it to be.

Arrhenius's theory of ions in solution tremendously advanced the study of the usually brightly colored inorganic compounds called *coordination complexes*. Many dyes and several biologically active materials, such as hemoglobin and chlorophyll, are coordination complexes: stable, isolatable compounds that generally consist of a central, positively charged metal ion surrounded by coordinating groups, or *ligands*, such as ammonia or water. The number of ligands can be anywhere from one to ten, but the most common number is six, with the ligands surrounding the central element in a structure resembling a jack (the metal, six-armed game piece in the game Pick up Jacks) or what amounts to a three-dimensional, six-pointed star.

The fact that coordinating ligands did not always have the opposite charge to the metal was not too disturbing because most chemists by then had resigned themselves to the failure of Berzelius's dualistic theory, which required one positive species and one negative species for a bond. Some had hoped that exceptions to his theory would be confined to organic chemistry, but they were forced to see such seemingly nonelectrical bonding in inorganic compounds, too. What did cause problems was that these complexes threatened to defy the concept of valence, and because valence had worked so well for organic compounds, it was difficult to abandon here.

Actually the originator of the valence theory, Frankland, was able to visualize variable valence—just as a juggler defies the human valence of 2—so he would not have balked at compounds with varying numbers of ligands. In the hands of Kekulé however, the then guru of organic chemistry, the theory of valence had come to mean absolute fixed valence, as characteristic of elements as their atomic weights, and he would not give it up. When faced with the experimental fact of phosphorus *penta*chloride (formed from a phosphorus atom surrounded by five chlorine atoms), Kekulé, who had decided phosphorus should have a valence of 3, said that this was really phosphorus *tri*chloride (formed from a phosphorus atom surrounded by three chlorine atoms) with a chlorine molecule, Cl_2, associated with it. This was something like saying that a cow is a horse with some horns associated with it, and the theory did not prove very satisfactory.

Other workers sometimes leaned toward variable valence in their search for an explanation, but the authority of Kekulé tended to squelch such efforts. One theory that did find some acceptability from the fixed-valence point of view and still accommodated the extra ligands was the chain theory of Christian Wilhelm Blomstrand. In this theory fixed valences were maintained and extra attachments were accommodated in chains that extended from a central group. These chains were visualized as resembling carbon chains found in organic chemistry, an idea that had proved to be very successful in that field. This theory was modified and extended by Sophus Mads Jørgensen at the University of Copenhagen, and with careful experimental work, he found substantial evidence to support it. The apple cart was toppled however by a young organic chemist out to make a name for himself. He succeeded. His name was Alfred Werner, the founder of coordination-complex chemistry.

Alfred Werner

Alfred Werner was born in that unfortunate strip of territory between France and Germany known as Alsace-Lorraine. Fought over, occupied, and reoccupied repeatedly throughout history, the city that Werner was born in, Mulhouse, finally, voluntarily, sought association with France. As a child however Werner experienced the Franco-Prussian War, which resulted in Mulhouse being annexed to the German Reich. Many citizens with religious and ethnic ties to France left the city, but Werner's parents chose to remain, though resistant and disregarding official dictum by using French, not German, in their home.

As an adult Werner had memories of shooting a toy gun at occupying German soldiers, and he saved a shell fragment for a paperweight. He was not a diligent student and often hid cardboard in his pants to ameliorate the corporal punishment that was the consequence of a missed class. But this rebellious spirit is credited by his biographer as the source of his creativity.[15] There are accompanying stories of chemistry experiments performed in the barn behind the house and of a bedroom ruined by an explosion.

Werner grew to be a broad-shouldered boy with light hair and blue eyes, and at the age of 18 he was called on to perform his military duty. On his return from military service he completed work for his doctoral degree (although he failed some courses in mathematics). He qualified to become a *privatdozent*, a lecturer without a salary who could collect fees from students—the first rung on the academic ladder. He then spent the next 2 years working as an unsalaried lecturer and puzzling over the problem of molecular compounds and valence.

Then one morning at two o'clock he woke up and started writing. By five o'clock that afternoon he had finished his most famous paper, *Contribution to the Constitution of Inorganic Compounds*, in which he spelled out the basics of what would eventually become the accepted theory of coordination complexes.

In his treatment Werner softened Kekulé's ideas of a rigid directed valence to a sphere of force coming from the central element. He proposed a *coordination number* that was the preferred number of ligands that a given metal atom sought to acquire. The ideas were controversial but interesting enough to secure him a position at the University of Zurich. He grew a mustache and a beard so that his students would respect him, and he put on enough weight to be considered stocky. He married a Zurich resident, renounced his German citizenship, and settled down to raise a family and prove his theory. The family was complete after 8 years; it took him 20 to prove his theory.

Werner thrived on the work. Through many tedious syntheses he was able to prepare compounds with properties predicted by his proposed structures. For instance the proposed octahedrons had atoms in fixed positions around a center, which meant that they could have a "handedness" and rotate polarized light. If such compounds were found, then their optical activity would be positive proof for Werner's ideas. Werner though did not have the luck of preparing Pasteur's tweezers-sized crystals, so separating right-handed from left-handed proved an enormous task. He succeeded after about 18 years of trying, with the help of a United States doctoral student, Victor King.

But even then critics still surfaced. It had been speculated that optical activity was an exclusive property of carbon, and Werner's optical isomers contained carbon, so it was imagined that this might be the source of optical activity. In 1914 with the help of his doctoral student Sophie Matissen, Werner dispelled even this doubt by isolating an optically active complex that contained nitrogen, hydrogen, and oxygen, but no carbon. The triumph was complete.

To understand anything about Werner, it is necessary to realize that while he was achieving all of the preceding, he was also teaching and training doctoral students, and he was a very popular speaker. The lecture hall was routinely crowded with up to twice as many students as it was designed to hold. They sat in the aisles and crowded around the demonstration table. In the summer students fainted (pre-air-conditioning!), and there were concerns about such disasters as fire or explosion. One of Werner's nicknames was Professor Nunwiegehts, from a favorite phrase, *Nun, wie geht's?* (How's it going?) and thus the joke was, "When does a chemist occupy the minimum volume? In Professor Nunwiegeht's lecture."[16]

Over the course of his career Werner had some 200 doctoral students, and as the University of Zurich is reportedly the first European university to admit women as regular doctoral students, quite a bit of Werner's research was carried out by women. In addition students came from all parts of Europe and the Americas to study under him, though laboratory conditions in the original chemistry building were less than ideal and Werner was a difficult task master.

The original laboratories, unofficially called the Catacombs, were cellars with little ventilation that constantly required artificial light. A humorously contrived air sample was reported to contain "50% evaporated acids, 30% ill-smelling preparation odors ... , 10% cigarette smoke ... 5% alcohol ... , and 5% illuminating gas ... sufficient to send less resistant individuals to the great beyond."[17] Promised a new building, Werner worked tirelessly on the design while still carrying out some of his most productive research. The price he had to pay was dear. His temperamental ways became extreme. He reportedly threw a chair at a student who did poorly in an oral exam on one occasion, and on another occasion, he swept an arm over an untidy workbench and sent solutions crashing to the floor. He became intensely involved with his work and began to stay at the university longer and at home less. Finding little time to relax, he chose alcohol as an expedient and became increasingly dependent on it. He received his new building, but the combination of overwork, overweight, smoking, and alcohol came together in a general breakdown in health. He began suffering from poor circulation and headaches. He was finally forced to resign his position, then died in a Zurich psychiatric hospital 1 month later at the age of 53. Werner's lasting contribution however had been made, and subsequent chemists built on this foundation.

With Werner's work we see how far inorganic chemistry in the early 1900s had come from the mineral chemistry of the rock-hunting chemists in the early 1800s. After the Chemical Revolution the emphasis was on identifying new elements, the majority of which were metals, and their chemistry was studied from the perspective of separation and purification. Some of these new elements possessed a unique physical property—radioactivity, which contributed to the understanding of the structure of the atom. Many of the new elements also showed unique chemical properties, which lead to a better understanding of ionization and to refinements in the notion of valence. In Werner's work we see that this new information now allowed inorganic chemists like organic chemists to do more than observe the chemistry: They were now directing the reactions. Werner showed that synthetic inorganic chemistry was as valuable and realizable a goal as organic synthesis.

The study of new inorganic elements and ions also fueled other fires, including the field of analytical chemistry—the art of discovering the composition and form of materials. We next examine the development of this particular branch of chemistry along with two other rather strange bedfellows, industrial chemistry and biochemistry.

chapter
FOURTEEN

ca. 1848–1914: Analytical, Industrial, and Biochemistry— Creations of Coal

By the turn of the century chemists were industriously employing the new theories and techniques of organic and inorganic chemistry to create new products and new medicines. And they had an ideal laboratory in which to work: Peace, progress, and prosperity were the hallmarks of Europe at this time. France was in a comfortable Third Republic, German was unified under the progressive Kaiser Wilhelm, and the strength of the British navy kept English waters calm. Europe was very much the center for science, but science was steadily being transmitted to the other center of stability—North America—

and notably the United States. Ties between the two centers strengthened with improved transportation and communication: The telephone appeared in the 1870s, and trans-Atlantic radio signals were transmitted in 1901. Secure behind two ocean borders, with its government affirmed by having survived a civil war, the United States was ripe to receive. Electricity was coming on line, and internal combustion engines soon powered automobiles, airplanes, and submarines. But in the mid-1800s to early 1900s, the steam engine was still the primary source of power, and steam was powered by coal. Consequently chemists concentrated their new tools on the chemistry of coal. The information they gained led to a upward spiral of new products, new discoveries, new riches, and a new quality of life.

COAL

Coal is formed from swamp plants that died almost 300 million years ago. Shielded from atmospheric oxygen by water, plants decomposed slowly, giving off oxygen and hydrogen and becoming concentrated in carbon. In addition to being a fuel, coal is an excellent reducing agent for metal ores. A form even more concentrated in carbon, coke, is used in steel and iron manufacture. Coke is made by heating coal in the absence of air, which produces a flammable gas called coal gas and a viscous black liquid called coal tar. Though these materials were first considered waste and were burned off or allowed to run off, uses for them were eventually developed in the 1800s.

The first to be used was coal gas. Composed mostly of hydrogen, carbon monoxide, and hydrocarbons, coal gas can be explosive and toxic, but it contains traces of sulfur compounds (natural or added) that give it the rotten-egg smell that indicates a gas leak. This gas was used for illuminating houses and streets beginning in the early 1800s, and it was the gas Bunsen used to fuel his Bunsen burner for spectroscopic studies. Coal tar found its first use as a wood preservative, but coal tar is a complex mixture of aromatic organic compounds, including benzene, toluene, xylenes, naphthalene, phenols, and anthracene; as more compounds were isolated from its matrix, identified, and studied, coal tar became the basis for an entirely new chemistry.

In 1845 when the Royal College of Chemistry was established in London, the prince consort persuaded August Wilhelm von Hofmann, one of Liebig's students, to head it. Hofmann studied coal tar. Charles Mansfield, one of his students, patented a process for separating some of the hydrocarbons in coal tar, but he died from burns when one of his large stills caught fire. This may account for Hofmann's hostility toward industrial applications of these materials before their chemical properties were fully understood. But the industrial application held financial attractions, and despite his objections, much effort was expended in this direction.

The industrial goal was to develop products that used the abundant coal tar as a starting material. One attractive target was quinine because it was the only truly effective treatment for malaria, and the presence of malaria—or threat of it—has shaped cultures and won or lost wars as far back as history recalls. (Romans built their city on hills to get away from malaria-bearing mosquitoes.) But while quinine existed as a remedy, most could not afford it. It was first used by a European in the 1600s, when the wife of the viceroy of Peru, the fourth count of Cinchon, lay dying of malaria, and the court physician, in an act of daring, tried a native cure. It worked, so the curative was brought back to Europe, but it had to be brought back in the form of shipments of cinchona bark (named after the patient, not after the physician). The tree itself would not grow in Europe.[1]

The active ingredient of the bark was extracted with alcohol, and chemists found four alkaloids in this extract, the most effective of which was quinine. Once this was known, the race was on to synthesize quinine. When another of Hofmann's students, 18-year-old William Henry Perkin, attempted to synthesize quinine, he used coal tar as his starting material, but his product was a synthetic dye.

William Henry Perkin

Working in a home-built laboratory in the 1850s, Perkin obtained a black, sticky mess as the result of his first efforts. This material at the bottom of the reaction vessel, a general phenomenon of organic synthesis, is known in the parlance of today's organic chemist as *gok* (God Only Knows). Perkin was attempting to clean the reaction vessel with alcohol when he obtained a purple solution that dyed silk an attractive new shade. Contrary to Hofmann's advice, Perkin resigned from the Royal College and went into business producing the new synthetic dye with his father and his brother. The color quickly caught on, especially at the court of Napoleon III and with Queen Victoria, who proclaimed it acceptable for mourning clothes (Prince Albert had recently died). Perkin's fortune was made. Named after the color of some plants with a similar name, the color is now known as *mauve*.

So though Perkin began with noble intentions (the synthesis of quinine), he found success in more prosaic pursuits. Perkin did not however forget his original intent, and he always preferred basic research to life as an industrialist. Perkin eventually sold his patents to a German concern, retiring when he was 37 and a millionaire. He became a university professor and made substantial contributions to the field of synthetic and structural organic chemistry, as well as physical chemistry.

It is significant that Perkin sold out to a German. In Europe a newly united Germany had resolved to secure self-sufficiency and its place in European and world economies. And it did so with the marriage of science and industry.[2] In the 1910s the Kaiser Wilhelm Institute for Chemistry and the Kaiser Wilhelm Institute for Physical Chemistry and Electrochemistry were endowed, and other institutes with a pragmatic focus were quickly established thereafter (such as the Kaiser Wilhelm Institute for Coal Research). The kaiser was motivated by the success he had already seen from the cooperation of German industrial and academic chemistry—based almost entirely on the success of textile dyes.

GERMAN DYE INDUSTRY

With the peace and prosperity of the late 1800s came the desire for an improved quality of life—real or perceived. This created an increased demand for consumer goods, including decoratively dyed cloth. Before the late 1800s virtually all dyes were natural products harvested from plants that grew in very specific environments—usually outside Europe. Harvesting such plants provided a good income for the areas in which the plants grew and for the importers. German commercial agents, realizing the economic potential, worked to corner the market on all natural-dye sources. India for instance was the source of indigo dye, found naturally in the indigo plant, and the majority of India's production of natural indigo dye was controlled by a German company BASF (*Badische Anilin und Soda-Fabrik*, Baden Dye and Soda Company) located near the village of Baden. However, there was still a considerable and continuous outlay of funds to India.

Britain had a balance-of-payments problem caused by its import of Chinese tea, which it chose to alleviate by selling opium from India to China, which gave rise to the Opium Wars. Germany avoided the warfare solution to its balance-of-payments problem by coming up with synthetic dyes, and these synthetic dyes gave Germany a virtual world monopoly. German exports of dyes went from none in 1860 to eighty percent of the world supply by 1890; by 1914 Germany provided ninenty percent of the world supply. German dye manufacturers formed a cartel (or *interessen Gemeinschaft*, community of interests) that became known as I.G. Farben (*farben* means color). The laboratories of I.G. Farben proceeded to produce not only dyes but fine chemicals and pharmaceuticals. Of course none of this would have been possible without such chemists as Adolf von Baeyer, who in the 1880s synthesized indigo.

Adolf von Baeyer

Baeyer began performing chemical experiments at home at the age of nine and later reported that he began work on indigo when he was 13. He studied at Bunsen's laboratory, the most famous in Europe, and then after a disagreement with Bunsen, he studied with Kekulé. He submitted a dissertation on cacodylic compounds (the noxious material that had injured Bunsen) to the University of Berlin, but the examining committee gave it a poor rating, and Baeyer went back to work for Kekulé.

He eventually took a position at the Berlin Institute of Technology, where the pay was poor, but laboratory facilities were good. Here he isolated barbituric acid, the starting material for the family of barbiturates, and supposedly named it for his girlfriend at the time, Barbara (though the Oxford English Dictionary discreetly reports the word's origin as obscure). Despite this success he still found it difficult to secure a university position because of his dissertation results. After he married he obtained a position as professor at Strasbourg at the age of 39. He had to oversee construction of laboratory facilities, but when they were completed, he continued his research on indigo, working with a student, Emil Fischer. After only 2 years he moved to Munich, where he had to construct a laboratory again. Because many of his students and assistants came with him, he was able to continue his work on indigo. In 1880 he came up with a synthesis for the dye. Successful syntheses of relatively simple natural products like indigo led later generations of scientists to elevate their sights to the more complex materials of biochemistry.

BIOCHEMISTRY

Although students were not yet declaring themselves biochemists (the term biochemistry did not appear until around 1910),[3] the study of natural products—their formation and function in nature—has always been part of the investigation of materials. In the last decades of the 1800s, chemists made considerable progress in identifying the structures and properties of many important natural products. The leader in this field was Baeyer's student Emil Fischer.

Emil Fischer

A group of simple sugars—glucose, fructose, galactose, and sorbose—had been identified in plants, and each of them had the same molecular formula, $C_6H_{12}O_6$. But their individual structures were not understood, and in fact the general class of compounds to which sugars belong came to be known as *carbohydrates* because it was initially thought that they were hydrated carbon: carbon surrounded by water molecules. Fischer brought the arsenal of chemistry to bear on the problem. He invented a new reagent, phenylhydrazine, to make derivatives of sugars (surviving a bout with phenylhydrazine poisoning in the process) and he deduced structures. Fischer recognized that these sugars were chiral (they had a "handedness," coming in two distinct mirror-image forms) and they were produced in nature with a preferred configuration, or "hand." He used this to study enzymes, protein molecules that are nature's omnipresent catalysts and themselves chiral.

Fischer also figured prominently in the elucidation of structures of purines, a class of compounds that have in common a nitrogen-carbon two-ring structure. One example of this class is caffeine (a good deal of this purine was consumed in the production of these very words), and another is uric acid. Uric acid is a purine found in the excreta of humans, birds, reptiles, and it is one source of the nitrogen richness of guano, or bird droppings, harvested in Chile for fertilizer.

Proteins and their building blocks, the amino acids, were also extensively investigated in this era. The first pure amino acid, cystine, was isolated in the early 1800s from urinary calculi (stones from the urinary tract), and leucine was later isolated from cheese. But breaking down proteins (which are large molecules made up of many amino acids strung together) failed to reveal most of the individual amino acids because methods used at the time—which included heating the substance in concentrated caustic soda or concentrated sulfuric acid—generally chewed up everything: protein, amino acids, and all. Only a few of the more stable amino acids, such as tyrosine and glutamic acid, survived to be identified.

Fischer brought his talents to bear on amino acids, too. He used gentler methods to isolate them from proteins, synthesized many amino acids, and developed ways of joining them together into small models of proteins called peptides. Fisher also isolated and characterized many terpenes: Components of fragrant oils that can be removed from plants by steam distillation. This technology had been known since the 1500s, but these substances were not systematically investigated until the 1800s. Even then they gave chemists conniptions because the oils isolated from plants are mainly mixtures of many nearly identical compounds.

Similarly the class of nitrogen- and oxygen-containing organic bases isolated from plants and known as alkaloids (e.g. morphine, codeine, nicotine, quinine) was recognized in the early part of the century, and a number of pure alkaloids were isolated, but their structures defied elucidation for some time. Many of these natural products, particularly the alkaloids, have powerful physiological effects, and some of them, like quinine and morphine, have a long history of use as medicines. After identifying and classifying many biological compounds, chemists began to pursue the next obvious step, which was to attempt to synthesize the more important natural products. Syntheses that were pursued most vigorously were those of compounds with medicinal activity, and it may be recalled that Perkin was looking for quinine when he found mauve dye. Though the synthesis of quinine eluded chemists for another half-century, another folk remedy known to be a powerful fever reducer succumbed to the chemist's efforts more readily: aspirin.

Aspirin

In the mid-1700s Reverend Edmund Stone of Oxfordshire wrote, "There is a bark of an English tree, which I have found by experience to be a powerful astringent, and very efficacious in curing anguish and intermitting disorders."[4] Actually the remedy had been known to generations of country folk long before Reverend Stone, but it took some time after Reverend Stone's report for French and German pharmacologists to find the active principle. Bitter-tasting yellow crystals were extracted from the bark and named salicin after the Latin name of the willow, *Salix alba*; later a cleaner preparation was called salicylic acid. Then in 1860 Hermann Kolbe and his students synthesized the sodium salt of salicylic acid from phenol, carbon dioxide, and sodium hydroxide. The sodium salt preparation was effective in relieving a wide range of pains, but it was rather hard on the stomach. It proved to be particularity difficult to tolerate for the arthritic parent of a dye chemist at the Bayer division of I.G. Farben. This chemist, Felix Hofmann, found that a less irritating derivative, acetylsalicylic acid (prepared from salicylic acid and the anhydride of acetic acid) was easier on his father's stomach but retained the analgesic effect. Bayer called the new drug *aspirin*. The fact that aspirin is the acetyl derivative of salicylic acid accounts for the vinegary smell of old aspirin. (Aspirin decomposes in moist air to acetic acid, the acid found in vinegar.)

Along with the quest for cures for disease, there was also a search for the cause. Great strides were made in this era by a chemist we met in another capacity: Louis Pasteur.

Louis Pasteur

In the 1670s Antony van Leeuwenhoek invented a microscope like the one used later by Hooke. With it he saw "beasties" and "animalcules"[5] so small "that I judged that even if 100 of these very small animals lay stretched out one against another, they would not reach to the length of a grain of coarse sand."[6] In the late 1800s these animalcules became less of a novelty in the hands of Louis Pasteur. Pasteur had initially made his name studying tartaric acid, a product of fermentation. But Pasteur was as interested in the fermentation process as the fermentation product, and while studying the process he came to his momentous conclusions regarding the origin and control of disease.

It was known that yeast caused fermentation, but the mechanism was not understood. Several scientists observing yeast under microscopes declared it to be a microorganism (one of Leeuwenhoek's animalcules), but Berzelius criticized these reports severely, believing that yeast was an inanimate chemical catalyst. Wöhler and Leibig published an article wherein they lampooned the idea of live yeast by portraying them as tiny animals in the shape of a distilling flask, with a visible stomach, intestinal canal, anus (a rose-colored point), and urinary secretion organs, then described fermentation as taking place when these animals ate sugar and excreted alcohol from the intestinal tract and carbon dioxide from large sex organs.[7] Pasteur however showed that fermentation is accompanied by the growth of yeast, and growth is a property limited to living systems. In the end his views won out. Though it was then proposed that fermentation and decomposition were caused by the spontaneous generation of these microorganisms in organic media (such as milk, wine, or meat), Pasteur showed that microorganisms were present in the air and water. He proved this by sealing out or filtering the air that had access to the media and stopping putrefaction. He also showed that when microorganisms were destroyed, as by heat, the media were preserved—a process now known as *pasteurization* and routinely applied to milk and beer.

Pasteur went on to argue that many diseases characterized by a decomposition of animal fluids and tissues (such as gas gangrene) were caused by microorganisms. But despite the hope they held for the control of many heinous diseases and deaths, Pasteur's theories attracted considerable criticism from the medical establishment. The main criticism was that Pasteur was making too broad a generalization based on too few results. We may have believed that with stakes as high as they were, medical practitioners would have chosen to err on the side of caution, but some steadfastly refused to institute the simplest sterilization techniques, such as washing hands. Pasteur answered their objections with harsh rhetoric, which may have weakened his cause. There were others however, such as Joseph Lister, who actively sought disinfectants to use against microorganisms. When he introduced carbolic acid (phenol) as an antiseptic for sterilizing instruments and (in dilute solution) wounds, he reduced mortality from amputations dramatically from their prior fifty percent.

Following a lead given a century earlier by the physician Edward Jenner who observed that people who milked cows diseased with cowpox seldom acquired smallpox and used this information to devise a smallpox vaccine, Pasteur looked for sources of dead bacteria that could be used to trigger immunity and protect against live bacteria. He used such methods successfully to protect sheep from anthrax, then he deliberately injected the 9-year-old Joseph Meister, who had been bitten by a mad dog, with attenuated fluid from rabid dogs and saved his life.

Pasteur also succeeded in rescuing the French beer, wine, and silk industries from crises due to microbial contamination by identifying microorganisms and finding ways of protecting against them. He always emphasized close ties between chemical research and industry to his students, and as dean of Science Faculty at the University of Lille, he instituted night classes for workers and took students on tours of factories.

For Paul Ehrlich this exchange between the chemical industry and pharmaceuticals was also to bear fruit. Employing his four Gs: *Geduld, Geschick, Geld, und Glück* (patience, skill, money, and luck), Ehrlich took what was known about dye chemistry and turned it into medicine.[8]

Paul Ehrlich

Trained as a physician, Paul Ehrlich used dyes to stain microorganisms so they could be better observed under a microscope. Water soluble dyes did not work, so he turned to the coal-tar-based dyes. When he did he observed that some of the dyes killed the microbes in the process of staining them. He developed a theory about why these dyes attached themselves to the microbe, and not the host, and set about synthesizing compounds to test. In the early 1900s he found a dye effective against sleeping sickness. And when the syphilis-causing spirochete was found, he worked on designing a dye to act against it. His assistant Sahachiro Hata directed the testing of hundreds of organic arsenic compounds until it was determined that compound number 606 had some activity. Marketed as Salvarsan (from the Latin *salvare*, meaning to preserve, and *sanitas*, meaning health), it was the first effective medicine for syphilis since Paracelsus's mercury treatment in the 1500s.

There was however another threat to the heath of people in the new industrial age that had nothing to do with little bugs. This was one brought on by the Industrial Revolution itself. Now that more foods were mass produced instead of made in the home, food purity became a significant problem. In 1820 the German apothecary and chemist Fredrick Accum, then working in England, raised questions in a pointedly titled tract, "Death in the Pot,"[9] in which he announced that heavy metal pigments were being used in foods—lead chromate (chrome yellow) colored yellow sweets, and copper arsenite (Scheele's green) colored green ones—and lead equipment was being used in food processing. He was drummed out of England on the contrived-sounding charge of defacing books in the Royal Institute—a charge that was not contested by the food industry. But the work he did stimulated a public demand for purity in food, and this demand, along with others from industrial and medicinal chemistry, stimulated the growth of another branch of chemistry: analytical chemistry. Analytical chemistry had always existed as the essential adjunct to any chemical investigation, but in the late 1800s it took on a life of its own.

ANALYTICAL CHEMISTRY

Analytical chemistry gained further impetus from the work of Karl Remegius Fresenius, the German chemist who at the age of 30 asked his prosperous attorney father for funds to set a up laboratory. In this laboratory he did analyses for government agencies, police departments, and the chemical industry. He also taught analytical methods and compiled two manuals on quantitative analysis that became standard reference works. It can be added that his father did not have to support his grandchildren because the laboratory ran continuously for four more generations of the Fresenius family. And as what may be taken as the christening of a branch of chemistry, Fresenius started a journal—*The Journal of Analytical Chemistry*. The synergism of analytical chemistry with industrial and biological chemistry is summarized in the work of the U.S. chemist Ellen Swallow Richards.

Ellen Swallow Richards

In the United States of the late 1800s, wild-West shows were popular, buffalo herds free-ranged, and George Armstrong Custer, fighting the Sioux, made his famous last stand. In 1861 Yale granted its first Ph.D.

Figure 14.1. Ellen Swallow Richards—front row, second from the right—with MIT chemistry faculty. (Courtesy of the MIT Museum.)

(in English) and Josiah Willard Gibbs received the first Ph.D. in science from Yale in 1863. In 1842 Ellen was born in Massachusetts to Fanny and Peter Swallow.

Ellen was taught at home until she was 16, at which point the family moved to Westford, Massachusetts, so that their only child could attend the Westford Academy. She taught, tutored, and cleaned houses to earn money until at the age of 25, she was able to enter Vassar College—a newly opened institution of higher learning for women. Here she studied chemistry as well as other sciences.

On graduation she tried to get work with commercial chemists, but she was summarily rejected. One however suggested that she apply to the newly organized Massachusetts Institute of Technology (MIT). She did and was delighted to be accepted—doubly delighted when she found out she would not be charged tuition. She assumed at the time it was because of her financial straits, but she found out later that it was so the president could say with a straight face to the board of trustees that she was not really a student. While at MIT she earned a second bachelor's degree and submitted her work on the chemical analysis of iron ore to Vassar, receiving a master's degree. The staff at MIT however would not award her a doctorate because awarding its first doctorate in chemistry to a woman would have made the institution appear trivial and second rate. The faculty and administration that permitted her to attend MIT were actually quite liberal.

While working as an assistant for professors at MIT, Ellen Swallow married Robert Hallowell Richards, a professor of mining engineering, head of MIT's new metallurgical laboratory. The match worked; they had no children but devoted their energies to science. Robert Richards proved to be supportive of his wife's work, promoting it whenever possible.

Finally financially secure, Richards was now able to devote herself to a favorite cause: the scientific education for women. Through her efforts a Women's Laboratory was established at MIT. Teaching there (without pay) Richards introduced the first course in biology at MIT. Her courses also included chemical analysis, industrial chemistry, mineralogy, and applied biology. When MIT started admitting women as regular students, she shifted to the newly opened MIT laboratory of sanitary chemistry, finally receiving an official appointment as instructor of sanitary chemistry (though this was the highest rank she was ever permitted to achieve). Working with Thomas Drown, she did a major water-analysis project for the Massachusetts Board of Health, sometimes working 14 hours a day, 7 days a week. This work resulted in the first state water standards and the first modern sewage treatment plant.

Richards also maintained a private analytical practice. In one consultation her husband related

> My friend, David Browne of Coppercliff, Ontario, was seeking information about his copper ore from the Coppercliff Mine. He sent samples to a number of assayers, and among others, to Mrs. Richards. All the others returned results in copper, and, I dare say, they did not know that they were to look for anything else. She, on the other hand, gave him a percent of copper in the ore and also reported five percent of nickel. This, I believe, was the beginning of the great nickel industry of which Coppercliff Mine was the center. David Browne always said that Mrs. Richards was the best analyst in the United States.[10]

She was chemical consultant to the Manufacturer's Mutual Fire Insurance Company for 10 years, and her research on the volatility of oils drastically reduced fire insurance costs and saved many lives. She authored more than 15 books, including *Home Sanitation, Cost of Living, Air, Water and Food, Sanitation in Daily Life, Industrial Water Analysis*, and *Conservation by Sanitation*, and she advanced the teaching of home economics as a regular course of university study. Although home economics has received some bad press as the allegedly trivial pursuit of husband-seeking coeds, it has historically offered an acceptable route for women to participate in science when they might otherwise be barred. And home economics is a serious concern: The purity and safe preparation of food was, and is, critical, and the homemaker then, as now, has a life or death responsibility. Richards was also active in the organization that was to become the American Association of University Women, being one of the 18 original members. One of the association's first projects was to conduct a survey of the health of female college graduates to refute claims that university studies undermined women's health. (These authors note that this is *not* true. University studies undermine *everybody's* health.)

OTHER INDUSTRIES OF THE LATE 1800s

As growth in industry promoted the growth of analytical chemistry and biochemistry, the analytical chemist and the biochemists promoted industrial growth. Though the direct beneficiary of Pasteur's work was the fermentation industry, the soap industry benefited indirectly, too: Public awareness of microorganisms increased the demand for soap. The manufacture of soap (and glass) soon required a supply of alkali (potassium and sodium carbonates) that far outstripped supplies from natural sources. Forty-five companies merged at the end of the 1800s into the United Alkali Company, a colossal chemical concern based on the synthetic production of alkali. Much of the story of the synthetic alkali industry however begins in the 1700s with the story of the unfortunate chemist Nicolas Le Blanc.

Nicolas Le Blanc

The need for a cheap and abundant source of sodium carbonate was already felt in prerevolutionary France, where sodium carbonate had to be imported or made from wood ash. The French Academy of Sciences offered a considerable prize to anyone who came up with a method of preparing sodium carbonate from common salt, sodium chloride. France had, of course, oceans of salt.

Le Blanc became interested in chemistry through his study of medicine, and when he was appointed surgeon to the Duke of Orleans (a considerable accomplishment), he found time and funds to work on his favorite project: finding the means of producing soda from sea salt. He came up with a novel synthesis, but because there were plenty of people working on the problem, there were those who questioned the originality of his ideas. But Le Blanc weathered all this, as well as the first part of the French Revolution, and he was granted a patent for his process on behalf of Louis XVI when Louis became the constitutional monarch. His patron, the Duke of Orleans, lent him money to build a plant based on his process, and the future looked good. Two years later however the duke was found guilty of royalism and guillotined. The plant, considered to be the duke's property, was confiscated, as was Le Blanc's patent. Napoleon finally returned the original plant to

Le Blanc but without sufficient compensation to bring it back into production. Le Blanc, realizing this was the final dispensation, shot himself.

Though it was eventually replaced by the Solvay process (Solvay was another industrialist whose name will appear in this history again), the Le Blanc process showed the way to improved industrial production through chemical research. And it was important for another reason: The gaseous hydrogen chloride by-product of the reaction of his process was initially released directly into the air, causing extreme and literal acid rain. This provoked one of the first pieces of environmental legislation, the British Alkali Act, that required soda manufactures to run their gaseous waste through acid-absorbing towers. (The dilute hydrochloric acid product from the towers however was discharged into streams, and by-product calcium sulfide cake was stored in fields where it slowly released toxic and evil-smelling hydrogen sulfide.) In the late 1800s the social impact of the chemical industry was still being sorted out, as exemplified by the life of another major chemical industrialist: Alfred Nobel.

Alfred Nobel

The European explosives industry had changed little from the introduction of gunpowder until the late 1800s. Berthollet and Lavoisier had experimented with potassium chlorate as a substitute for potassium nitrate in gunpowder, but casualties caused by an uncontrolled reaction put an end to this. By 1850 Italy's Ascanio Sobrero had added nitro groups (NO_2) to glycerol (an oxygenated organic compound obtained in soap making) by treating it with sulfuric and nitric acids. This *nitroglycerin* (a misnomer; it should be called glyceryl trinitrate) was an explosive several times more powerful than black gunpowder but seemingly uncontrollably unstable. Workshops that produced the material suffered many disasters with loss of life. Then another of Hofmann's pupils at the Royal College of Chemistry, Frederick Abel, showed that nitrocellulose, produced by treating a plant-fiber material, cellulose, with a mixture of sulfuric and nitric acid, was a stable explosive when pressed to remove residual acid. This explosive, known as smokeless powder, found its use in mining and munitions. Then in the 1860s Immanuel Nobel, a Swedish industrialist—perhaps encouraged by the stability of smokeless powder but definitely encouraged by the commercial potential of nitroglycerine—began working with his four sons on the production of nitroglycerine. His third son, Alfred Nobel, was sickly as a child and could not attend school regularly. He eventually studied at the University of Saint Petersburg, took to industrial chemistry, and worked on nitroglycerine. An explosion killed his youngest brother and four assistants, but Nobel continued the work, shifting his laboratory to a boat anchored off the Swedish coast. Nitroglycerine was used in building the Suez Canal and railroads in the United States, but it was too tricky to work with, and after many mishaps and much loss of life, its use was banned by several governments. Nobel continued his research and finally succeeded in stabilizing nitroglycerine by absorbing it in sawdust or diatomaceous earth. He named the new explosive dynamite (in reference to its dynamic force) and took out patents. Because the new stabilized form now needed a detonator, he also patented one of these based on mercury fulminate. The new explosive became the standard in mining and engineering works, and Nobel became one of the richest people in the world.

He kept up his interest in chemistry, equipping most of his mansions with chemistry laboratories. Reportedly convinced that he was too homely to be loved, he never married, though he did have an 18-year love affair with Sophie Hess, a woman whom he met when she waited on him in a flower shop. Hess more or less attached herself to Nobel, and though she eventually proved to be an embarrassment and significant expense for him, he was too kindhearted to cut her off.

In 1876 Nobel took out an ad in the newspaper that read: "Elderly, cultured, wealthy gentleman requires equally mature lady, linguist, as secretary and supervisor of household in Paris."[11] But when Bertha Kinsky answered the ad, she found

> He was then forty-three years old, rather below medium height, with a dark, full beard, with features neither ugly nor handsome; his expression rather gloomy, softened only by kindly blue eyes; in his voice there was a melancholy alternating with a satirical tone.[12]

Nobel hired Kinsky, but she served as his secretary for only a few days before eloping to marry a young baron. She and Nobel remained friends however and kept up correspondence as well as visits. Kinsky was a pacifist, but Nobel, both atheist and socialist, had a more reserved view of the possibilities for humanity. Then his brother died, and a newspaper mistakenly printed Alfred Nobel's prepared obituary. After reading it Nobel realized people would remember him as a merchant of death, and he challenged Kinsky to "Inform me, convince me; then I will do something great for the [pacifist] movement."[13]

She must have argued well, for the following January he wrote to her:

> I should like to allot part of my fortune to the formation of a prize fund . . . to be awarded to the man or woman who does most to advance the idea of general peace in Europe . . ."[14]

In his will (after some minor remembrances to relatives and others) Nobel directed his estate to be invested in safe securities and the interest to be distributed annually:

> One share to the person who shall have made the most important discovery or invention in the domain of Physics;
> One share to the person who shall have made the most important Chemical discovery or improvement;
> One share to the person who shall have made the most important discovery in the domain of Physiology or Medicine;
> One share to the person who shall have produced in the field of Literature the most distinguished work of an idealistic tendency;
> And, finally, one share to the person who shall have most or best promoted the Fraternity of Nations and the Abolition or Diminution of Standing Armies and the Formation and Increase of Peace Congresses.[15]

Though it was Nobel's desire "to place those whose work showed promise in a position of such complete independence that they would be able to devote their whole energies to their work,"[16] the prize has come to mean much more. It is now considered the ultimate recognition of a body of work. Though there have been over the years some questionable judgments by the awarding committee—caused perhaps by political or personal influence—history has proved the selection sound in the majority of cases, and a review of the record of recipients summarizes both the evolution of chemistry and political situations of the times.

The first Nobel laureate in chemistry (1901) was Jacobus van't Hoff, the veterinarian who visualized three-dimensional carbon. The second Nobel Prize went to Emil Fischer, the guru of natural product identification. The prize was the validation of his work, but not of his life. When three of his four sons were killed in World War I, he committed suicide.

The third laureate was Svante Arrhenius, who barely met qualifications for a doctorate, and the fourth was William Ramsay, the discoverer of helium, neon, argon, krypton, and xenon. The fifth was Adolf von Baeyer for accumulated work in organic chemistry—including his work on indigo dye—and the sixth was Henri Moissan, the conqueror of fluorine. The seventh was Eduard Büchner for his work on cell-free fermentation. Pasteur believed that fermentation could take place only in an oxygen-free atmosphere, but Büchner demonstrated that fermentation could take place outside the cell by grinding cells with sand and passing them through a hydraulic press. The resulting "yeast juice" still caused fermentation—the first test-tube life process. Büchner's employers initially frowned on his efforts as unproductive and a waste of time. What was unproductive and a waste was that he died as the result of wounds incurred as a volunteer officer in the German army in World War I.

The eighth Nobel Prize was awarded to Ernest Rutherford for his work on the composition of the atom. While today this work might be considered more in the realm of physics, this award for chemistry emphasizes the overlap of the two fields. Chemistry could not progress without physics, but we dare say physics would not progress without chemists. Wilhelm Ostwald received the ninth Nobel Prize for his work in the field of physical chemistry, as this study of the border area between physics and chemistry came to be known. Otto Wallach, another product of Kekulé's laboratories, received the tenth Nobel Prize for his work in organic chemistry, mostly on elucidating structures of plant oils, terpenes. The eleventh prize went to Marie Curie for her chemical work on polonium and radium. She had already shared the 1903 Nobel Prize in physics with her husband and Henri Becquerel for their work on radioactive elements. The twelfth prize was awarded to Victor Grignard for work on organometallics and the thirteenth to Paul Sabatier for his contribution to catalytic organic chemistry, particularly the addition of hydrogen to unsaturated compounds, an important synthetic and industrial process. The fourteenth prize went to Alfred Werner for his work on the structure and bonding of inorganic complexes. Then in 1914 came the war.

WORLD WAR I

Ellen Swallow Richards recognized that a society rapidly becoming more urban and industrialized was going to face health problems that could easily be translated into social problems, and she was right—but in a way she perhaps had not anticipated. A benchmark for the progress that coal chemists made was the population boom: At the beginning of the 1800s the population of Europe was roughly 50 million; by 1820 it was about 100 million; and by 1914 it had reached 300 million. Along with increased population came increased production, and along with increased production came the need for new sources of raw materials and new markets. As a consequence the late 1800s were the years of imperialism.

One of the consequences of imperialism was the spread of European technical knowledge. India would soon reenter mainstream chemistry, as would China, Africa, and the Islamic and Arab nations. Japan for instance adopted European medicine and chemistry and promptly improved it. In the late 1800s Japanese workers isolated the microorganism responsible for the plague and one that causes dysentery, two historically important mass killers. In the early 1900s they isolated adrenaline in crystalline form and observed the first experimentally induced (indeed, coal-tar induced) cancer.

Unfortunately, another consequence of imperialism was World War I.

At first Germany may have looked like the easy winner, having a seeming corner on industry and resources. But England and France came up to speed with amazing alacrity. For instance acetone was an essential ingredient in preparing smokeless powder, and when

England's demand for acetone accelerated, the need was met by Chaim Weizmann, an immigrant from the Russian-segregated Jewish community called the Pale of Settlement.

Chaim Weizmann

The period 1848–1914 saw the integration of Jewish citizenry into general European society, and all European governments extended some rights to Jewish citizens—except Russia. For Russian Jews, travel outside the Russian Pale of Settlement and access to universities were highly restricted. At 18, Weizmann took a job on a timber barge to escape Russia (passports were not needed for travel in Europe outside Russia and Turkey before 1914),[17] but he returned to Russia often. For him the cause of Zionism had a powerful appeal. He settled in England where his chemical talents were recognized by Perkin, by then rich and retired. Sponsored by Perkin at Manchester, Weizmann developed a fermentation process that produced acetone, and as a result he became director of chemical research in the British Admiralty. In recognition of Weizmann's contribution, the first lord of the British Admiralty, Arthur Balfour, issued the Balfour Declaration, which promised a national homeland to the Jews in Palestine. The ambiguous wording of "homeland" would later cause Britain and the world serious trouble, but in 1948 Chaim Weizmann became the first president of the State of Israel—a state that if not for his efforts might not have existed.

But as World War I dragged on, all parties began to experience shortages. When the British blockade took effect, Germany came up with substitute products, such as synthetic rubber. Cut off from nitrate deposits in Chile, Germany turned to a new synthetic process for ammonia that snatched nitrogen right out of the air. Before the war the author of this process, Fritz Haber, was an instructor at a technical school, and he was helped by a grant from BASF. He found that he could combine nitrogen and hydrogen under pressure, using iron as a catalyst, to produce ammonia. Working in collaboration with a chemical engineer Carl Bosch, Haber developed the laboratory method into factories synthesizing ammonia. Just a couple of years earlier Wilhelm Ostwald had worked out conditions for reacting air and water with ammonia to make the nitric acid needed for explosives. And so

Germany became independent of imported nitrates just in time for the start of World War I. Haber was rewarded with the directorship of the Kaiser Wilhelm Institute in Berlin. An enthusiastic patriot, this was not be his only contribution to the war effort. He began to study the possibility of using chemical gases in combat.

Chemical Warfare

Haber's suggestion was at first resisted by the traditional military, but as the war progressed, it gained grudging approval. The first chemical attack employed chlorine, and it was launched by the Germans on April 22, 1915, during the battle of Ypres in northern France. Personally witnessed by Haber, canisters of chlorine from the I.G. Farben factories were placed in trenches and covered with bags of peat moss soaked in potash solution to absorb any leaking gas. The attack had been scheduled to take place at an earlier date, but the winds had been wrong. When the winds were right, marksmen fired at the canisters, and a thick, green cloud rolled out from the German line.

The Germans were facing French, British, and Canadian troops, but as with the majority of troops in World War I, these were not professional soldiers but draftees and volunteers and unseasoned ones, at that. One of them reported

> Surprise and curiosity riveted us to the ground. None of us knew what was going on. The smoke cloud grew thicker, which made us believe that the German trenches were on fire . . .[18]

The gas first reached a division of Zouaves, native African troops, from imperial French Africa. A Captain Pollard described the troops

> Running blindly in the gas cloud, and dropping with breasts heaving. . . . Hundreds of them fell and died; others lay helpless . . . powerfully sick with tearing nausea at short intervals. They, too, would die later . . .[19]

German soldiers moved up behind the cloud with bayonets fixed until they met their own gas—and the Canadians. The 1st Canadian Division threw back the Germans, and the Germans received no reinforcements. In the end the battle was not decisive—except of course for those who died—but the precedent had been set.

Chemistry, many times, is the whelp of war. The new focus for many chemists became combat gases—finding better ones or antidotes. Phosgene, a chlorinated compound ($COCl_2$) and severe respiratory irritant that causes lesions and congestion in the lungs, replaced chlorine as the French gas of choice at Verdun in February 1916. In July 1917 the Germans countered with dichlorodiethyl sulfide, or mustard gas, a vesicant (blister agent) that has the advantage of attacking the skin as well as the lungs, so a gas mask is not sufficient defense. First time out however the soldiers thought the gas would not affect the eyes, so they used nose clips and mouthpieces for protection. They found themselves blind hours later with huge blisters under their arms and between their legs. Internal blisters caused swelling, and these were eventually fatal, but it was as effective to cause an injury as a kill: an injury not only disabled the soldier, but also took the people out of commission who were required for transportation and care.

There was a great outcry during and after the war over the use of poisonous gas. Haber's first wife, also a chemist, committed suicide in 1915 to protest the use of poison gas. Haber however cannot be condemned as the originator of chemical warfare. In a sense *all* warfare—beginning with the chucking of rocks—has been chemical (certain rocks were chosen because of their favorable rock-chucking composition), and the development of gunpowder and incendiaries was certainly a chemical enterprise. Haber was also not the first to think of combat gases. The British had proposed using noxious sulfur fumes in the Crimean War, though there is some doubt as to their actual use. In France tear-gas rifle and hand grenades were developed for riot control and used.

In 1919 Haber received the Nobel Prize for chemistry in recognition of his work on ammonia synthesis. The award of the prize was protested, but no war crimes were charged against Haber. But in World War I, as with other wars, war crime is a bit of a redundancy. There were battles that left 150,000 dead, as though whole towns—men, women, children—had marched into the fire. The devastation of the war in Europe was so considerable that it essentially ended the age of European dominance. European nations remained important, but they were in debt, crippled, and demoralized. Under Article 297 of the Versailles Treaty, Germany lost many of its chemical patents (such as aspirin), and this forced technology exchange caused a chemical redistribution of wealth, but there was precious little wealth to redistribute. Of all the belligerents, the one that suffered least was probably the United States. The center of finance shifted from London to New York, and it is in the United States that we begin our next chapter in chemical history.

The next chapter will show a shift in emphasis, too: from the practical success of chemistry to the theoretical questions that remained. Of these theoretical questions the most significant was chemical affinity. Starting with Berzelius, chemical affinity had been explained in terms of simple electrical attraction, but this model worked well only in a limited number of cases. It explained nothing about the bonding between like elements. Valence theories of Kekulé and Werner addressed attractions between electrically neutral elements, but they were empirical observations with no fundamental theoretical basis. But after World War I this changed. The postwar world saw the birth of the bond.

part
THREE

chapter
FIFTEEN

ca. 1914–1950: Quantum Chemistry—The Belly of the Beast

By the early 1900s, chemists had thermodynamics for a theoretical basis, but beyond that they were still flying by the seat of their pants/pantaloons. Progress in industry and medicine had clearly demonstrated the benefit of chemical research, but the research was primarily based on chemical art: knowing the reactivities of the elements both singly and in groups and having a repertoire of techniques to bring about a desired result. While this is not to be underrated (it is still the basis of some of the best chemistry today), chemists needed a better understanding of the underlying principles—the theory of chemistry—to make those intuitive shots even better. After World War I, they got it, and chemistry took a quantum leap.

In the early 1900s one of the more pressing problems in chemical theory was the need for an explanation of the nonpolar bond. Chemists felt at least some glimmer of understanding of polar bonds, that is, bonds between elements that preferred to be electrically positive and elements that preferred to be electrically negative. But the nonpolar bond, such as that between like elements (for example the two hydrogen nuclei in hydrogen gas) or between elements with a similar electrical nature (such as carbon and hydrogen) had no underlying theoretical basis. One of the first steps toward a quantitative theory of bonding came from a professor of chemistry at the University of California, Berkeley—G. N. Lewis.

GILBERT NEWTON LEWIS

Born in the 1870s in New England and brought up in Nebraska, Lewis received his initial education at home from his parents, and he was reportedly precocious, reading by the age of three. In any case he was talented enough to travel to Germany to start his chemical education and to begin work on a doctorate at Harvard while still in his early twenties. While teaching undergraduates (as doctoral students often do), Lewis noticed that elements with a certain number of electrons seemed to have a special stability.

Moseley and Bohr had shown that a neutral atom of each element has one more electron than the preceding element; thus hydrogen has one electron, helium has two, lithium has three, and so on. Lewis noted that the first element, hydrogen, with one electron, is reactive, but the second element, helium, is an unreactive noble gas. After helium it took an addition of eight more electrons to again produce another noble gas (neon has 2 + 8 electrons). After that it required eight more (argon has 2 + 8 + 8 electrons). To Lewis it appeared that once a core of eight electrons has formed around a nucleus, the layer is filled, and a new layer is started. Lewis noticed that various ions with eight electrons also seemed to have a special stability, so he proposed his rule of eight, or octet rule: Ions or atoms with a filled layer of eight electrons have a special stability.

A cube has eight corners, and Lewis envisioned an atom as having eight sites available for electrons, like the corners of a cube. (See

Fig. 15.1). Once the corners of the cube were filled, the next cube would start building. Finally Lewis proposed that an atom tended to form an ion by gaining or losing the number of electrons needed to complete a cube. Thus chlorine, with only seven electrons in its cube, tended to gain another electron and became negatively charged. Magnesium, with only two outer electrons, lost these two and became positively charged.

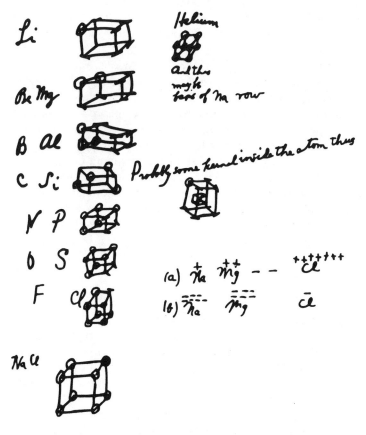

Figure 15.1. A sketch from Lewis's 1902 memorandum showing the possible cubic arrangement of electrons in atoms. (Courtesy of the John F. Kennedy Library, California State University, Los Angeles.)

The model made a good start toward explaining charges on ions and the stability of the noble gases, but it was not appreciated by Lewis's mentor at Harvard. Lewis, without support, did not promote his ideas, and it would be 12 more years before they resurfaced.

Ten years later Lewis was offered the chair of the college of chemistry at the University of California at Berkeley. This sounds a bit more grandiose than it was because at the time the department was run down and Lewis's job was to rebuild it. Given financial support and a free hand, Lewis brought on board a crew of skilled teachers who set about building a department that would train a world-class generation of chemists.

During this time an English graduate student, Alfred Parson, visiting Berkeley for a year, prepared a manuscript in which he suggested that a chemical bond results from two electrons being shared between two atoms. Lewis read this, and his ideas fell into place: Bonding occurred when two electrons formed a shared edge between two complete cubes. In 1916 Lewis published a full-blown theory based on the cubic atom, quoting his unpublished memorandum from 1902.

Lewis no doubt derived some of his ideas from Parson, but then ideas are rarely conceived in a vacuum. A month before Lewis's publication, Walther Kossel of Germany had published a paper that assumed that atoms gained or lost electrons to achieve the same number of electrons as a noble gas atom, but this work was unknown to Lewis while he was preparing his manuscript. And in Lewis's hands the theory did more than explain ions or ionic bond formation. It became a strong rationalization for the nonpolar bond.

Communication was not Lewis's strength (he avoided lecturing because of uneasiness in front of large crowds), and his cubic atom at first appeared to languish. Then there was the interruption of the war, and in 1918 Lewis went to France as a major in the Chemical Warfare Service. In 1919 Irving Langmuir (already famous for reasons we will discover) took up the ideas of the cubic atom and the shared-pair bond and began publishing and speaking about them.

Langmuir *was* a good communicator, so good that the theory became known as Lewis–Langmuir theory, or just the Langmuir theory. Although Langmuir did add to the theory, strengthening the concept of a nonpolar bond and calling it a *covalent* (as opposed to ionic) bond, and he always acknowledged the source of his ideas, Lewis initially was resentful. (Lewis it seems was not a forgiving person; in later years he refused an honorary degree from Harvard, still nursing his resentment of his former mentor). But Lewis eventually acknowledged Langmuir's contribution, and the two remained friends. Lewis was after all a rebel in spirit. He enjoyed upsetting the establishment (he was known to hide his ever-present cigar behind his back when walking past No Smoking signs in laboratories),[1] and if Langmuir carried the rebellion forward, so much the better. At any rate these ideas were soon to become the property of the chemical masses.

The shared-pair bond turned out to be enormously useful. Mechanisms for organic reactions could now be explained in terms of shifting pairs of electrons. The behavior of acids and bases could be understood by the need to fill octets. Inorganic structures could be rationalized and predicted on the basis of satisfied octets and the two-electron bond. A good testament to the usefulness of Lewis's model is that Lewis structures are still taught today as the most intuitive way of envisioning simple structures and bonding, though the corners of the cube have been replaced by eight dots around the symbol for the element.

The initial approach however did have some problems. Nitrogen molecules were known to be formed from two nitrogen atoms (N_2), but each nitrogen has only five electrons in its outer shell, and there was no obvious way that 10 electrons could form two cubes. In addition the Lewis model was a static one, requiring the electrons to be stationary, which did not agree with theories from Europe at the time. Langmuir dropped the theory in the twenties, probably seeing the writing on the wall. Lewis continued to support the static electrons, but the electron did not stay still for long.

OLD QUANTUM THEORY

The theory that killed the static electron—eventually known as *quantum mechanics*—evolved because physicists thought they ought to know more about light. In fact they already knew a lot; for instance they understood that light behaves as a wave. Waves bend around obstacles in their path (a day on the beach observing wave behavior may be in order), and waves interfere with each other: If they are out of phase, they cancel one another to create a calm; if they are in phase they combine to form larger waves. Light also behaves this way. In-phase waves of light projected through parallel pinholes create a pattern of concentric rings of light and dark called an interference pattern.

Waves of *what* was not known until the work of James Clerk Maxwell. Maxwell demonstrated that a moving electric charge generates a magnetic field, and the magnetic field generates an opposing electric field. The opposing electric field generates an opposing magnetic field, and so on, with each field pushing the other out of the way. When Maxwell calculated the speed at which this occurs, he found it to be the known speed of light. He must have said something like "Aha." Light, he had discovered, is an oscillating electromagnetic wave.

There is a story that Maxwell was courting his wife at the time, and on the evening of the day on which he came to this conclusion, they were walking in a garden, gazing at the stars. He supposedly asked her how it felt to be the only other person to know the true nature of starlight. This may be a romanticized version of a true event because we know Maxwell's wife later worked with him on his experiments, so no doubt she discussed this important conclusion with him.

Later (after Maxwell's death at an early age) his conclusion was confirmed experimentally by Heinrich Hertz, who showed that a spark from two charged spheres sets up an electromagnetic wave that can trigger a spark in similar charged spheres a distance away (which eventually became the basis for radio technology). But then Hertz took his experiment one step further. He showed that the spark came easier when ultraviolet light was shining on the spheres. He did not have an explanation for this observation at the time, but we soon will have.

Having found an explanation for one problem, physicists sought the answer to another: This time it was the intensity of radiation from a strange construct called a *blackbody*.

A blackbody is so called because it is an ideal light sponge: It absorbs all wavelengths of light and therefore appears black. A blackbody is a hypothetical construct, but it can be well approximated by a cavity with a pinhole aperture. Any light falling on the aperture bounces around the cavity with a low probability of escaping from the hole again. Because the blackbody absorbs all light, it should also emit all wavelengths of light when heated (as a heated stove burner glows). Experimentalists simulated blackbodies with cavities (the material of the blackbody made no difference), heated them, and measured the output of light. They found that the total energy output at all wavelengths, the *energy density*, increased as cavities were heated, but that most of the light emitted was in the visible or infrared region—and not much in the short-wavelength ultraviolet region. The problem was that the best physical models of the time predicted that the energy density should continue to increase at shorter wavelengths—theoretically to infinity.

To explain why physical models predicted that energy density should continue to increase, first consider pushing a child on a swing. If the swing is pushed down exactly at the time the swing moves down (*in phase* with the swing), then energy is added to the swing. If the swing is given a downward push on the upswing (*out of phase*), the result is a jarring crash, and the energy of the swing is diminished. Now consider a whip. When you crack a whip, you create a wave that travels down the whip and dissipates as a snap at the end. If the end of the whip is fastened to a wall, the energy is reflected when it comes to the wall. If the whip is at the lowest point of its movement when it arrives at the wall (in phase with the wave on the whip), the energy is smoothly reflected into the motion of the whip and added to the energy of the whip. If however the whip is at some intermediate point in its wave when it arrives at the wall, the reflected energy is out of phase and fights the motion of the whip, thereby diminishing the energy of the wave—which self-destructs.

So it is with light in a container. Light that is of a wavelength that fits perfectly in the container is reflected smoothly at all walls and sets up what is called a *standing wave*. This happens when there is exactly one-half wavelength between walls or one complete wavelength or one-and-a-half or two: any multiple of a half-wavelength, as long as the waves are complete when they hit the wall. Now consider a blackbody cavity as a flat circle that supports only horizontal waves (not very real physically but certainly a simple model). If one-half wavelength is the same length as the diameter of the sphere, then only one position in the sphere supports a standing wave: directly through the diameter. A wave that is a little off to one side will not survive. If it takes one complete wavelength to cross the diameter, there are three positions that support standing waves: directly through the diameter and at positions to the top or bottom that are a distance of one-half wavelength. If it takes two complete wavelengths to cross the diameter, there are seven possible standing waves, and so forth (See Fig. 15.2). We see that as the wavelength gets shorter, the number of waves completed before hitting the wall goes up. So as the wavelength gets shorter, the number of possible standing waves goes up. Therefore if heating the cavity produces standing waves, and more standing waves at shorter wavelength, then the total energy density should contain more short-wavelength radiation than long. But this is not experimentally observed: In the ultraviolet, short-wavelength region, the amount of radiation starts to fall off. The dilemma was called the *ultraviolet catastrophe*.

Many of the leading physicists of the day tried to solve this blackbody problem, but in 1900 the solution came from an unassuming German physics professor in his mid-forties, Max Planck.

Max Planck

Initially Planck did not have a spectacular career. When he submitted his doctoral dissertation on thermodynamics, one professor did not read it, another disapproved of it, and the third could not be reached.

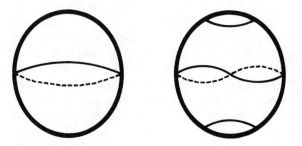

Figure 15.2. Horizontal standing waves in a circular cavity. If the circle is one half wavelength wide, there can be one standing wave. If the circle is one wavelength wide, there can be three standing waves. If the circle is two wavelengths wide there can be seven standing waves, etc.

He finally managed to complete his requirements, though he did not expect his showing to earn him much of a position. It was a surprise all around when he obtained a position at the University of Berlin. In fact Boltzmann, of statistical mechanics fame, had just turned the position down.

Planck had a firm belief that all physical phenomena must eventually succumb to analysis, and he chose to work on the blackbody problem. He tackled the problem from the standpoint of classical thermodynamics because he was familiar with the area and classical thermodynamics does not assume the existence of atoms (and atoms were still controversial among physicists at the time). Planck knew that if heat were added to a blackbody cavity, then there had to be an entropy change, so he tried to calculate an expression for entropy that would match experimental observations. But his efforts resulted in failure until he did what he characterized as "an act of desperation."[2] He assumed that the energy of the light was not continuous but came in discrete packets, called *quanta*, and that the size of these packets became larger at shorter wavelengths. He did some creative curve fitting (technically known as *interpolation*), and his formulation fit.

We can now explain why Planck's quanta worked. For the same total amount of energy, there are more waves excited at long wavelengths because the size of the permitted energy packets is small and there can be more of them. At short wavelengths the size of the permitted energy packets is large, so fewer are excited for the same total energy. There is a story that Planck, walking with his son, said he had a thought that day worthy of Newton. However even if this story is true, Planck did not have a physical interpretation for his formulation at the time, and when one was offered, Planck was a long time in accepting it. The lack of a firm explanation did not however prevent Planck from presenting his result to the German Physical Society, which met on October 19, 1900. Following the presentation attendees at the meeting worked through the night to check Planck's law against experimental results. The next morning they reported agreement.

Planck tried to interpret the physical significance of his lucky "guess."[3] He believed that energy packets were a temporary adaptation and that the size of his packets could become vanishingly small until he would be left with comfortable, continuous physics. But this was not to be the case. Not only did the quanta not disappear, but people started finding these pesky packets elsewhere. The first person to do so was someone whose word would eventually be difficult to doubt. But at the time Albert Einstein was only a Technical Expert, third class, in a Swiss patent office.

Albert Einstein

Einstein was not much of a rising star. Continually in conflict with his teachers, he did not complete secondary school as a result. He found a Swiss technical school where a high school diploma was not required and managed to finish the doctoral requirements, though he flunked a few entrance exams first.

Newly married and in his position as patent inspector, he apparently had the opportunity to think. In 1905 he published the results of three quite distinct lines of thought, any one of which would have established any scientist's reputation. One of the things he thought of was the best evidence to date for the existence of atoms. Minute particles suspended in liquid are in constant motion, and this *Brownian motion* can be seen under a microscope. Einstein extrapolated the motion to the molecular level to show that it arises from buffeting by molecules. Another thing he thought about was his theory of relativity. The third thing he thought about was an explanation for the photoelectric effect based on Planck's quanta.

The Photoelectric Effect

When light of the proper frequency shines on certain metals, it can cause an electron flow from the metal surface, and this is called the photoelectric effect. This is how electric eyes on some automatic doors work: A light beam creates a current, and when people step into the beam, they interrupt the current. This was also the effect seen by Hertz: His ultraviolet light caused the spark to jump more easily. The odd thing is that for electrons to be ejected from the metal, the light has to have greater than a particular frequency. If electron flow were simply caused by light transferring its energy to and heating the material, then any light source would produce electrons, provided there was enough of it. But experimentally it is found that packets of at least a threshold frequency are required: quanta of a minimum energy. Therefore though he did not know it at the time, Hertz had shown light behaving as a particle.

This idea of the dual nature of light, wave and particle, can be dizzying at first (in a remark attributed to the physicist Eddington, light should be treated as a wave on Monday, Wednesday, and Friday, and as a particle the rest of the days). But the idea is not so odd as it seems. It is a matter of models. Light is sometimes modeled as a particle, sometimes as a wave. Similarly we can model a dinosaur as a bird when we want to think about its metabolism, or as a giraffe when we want to learn about the physics of its locomotion. In truth the dinosaur was neither, but because we have no dinosaurs around to experiment on, we develop our models. Similarly our models of light have to relate to our macroscopic world because we cannot shrink down to experience things on the molecular level. Therefore the best models we have are waves and particles, and depending on our purpose, one model or the other may work.

Einstein was not the only one to find Planck's concept useful. In the laboratory of J. J. Thomson, Niels Bohr was in conflict with his mentor over the proper model for an atom. Thomson adhered to his plum pudding model, and Bohr preferred the planetary model of Rutherford. Finally Thomson suggested that Bohr work with Rutherford (who by this time had relocated in Manchester), and Bohr obliged.

Niels Bohr

Bohr felt instinctively that Planck's quantized energies were related to the discrete lines of elemental spectra—and to the planetary model of the atom—but he could not find the connection. Thirty years earlier Johann Jakob Balmer, a teacher at a girls' secondary school, part-time lecturer at the University of Basel (where, we may note, Paracelsus burned the works of Galen), and mathematics hobbyist had found a numerical relationship between frequencies of the lines in the hydrogen spectrum. The relationship was not obvious because it depended on the reciprocal squares of integers, and this was the very feature that caught Bohr's attention. He later said, "As soon as I saw Balmer's formula, the whole thing was immediately clear to me."[4]

Bohr postulated that there can be only certain discrete orbits for the electron around a nucleus—called *stationary states*—and that to go from one state to another, an atom must absorb or emit a packet of just the right amount of energy—a quantum. He then proceeded to predict the position of the lines in the hydrogen spectrum based on Balmer's formula, Planck's energy packets, the mass and charge on an electron, and his quantized orbits.

In Bohr's theory, electron orbits, or energy levels, are given numbers—called *quantum numbers*—and these numbers serve as addresses for electrons the way street names serve as addresses for people. The first quantum number is called the *principal quantum number*, which can be 1, 2, 3 . . . theoretically any number greater than zero. These principal quantum numbers define the *shell* in which the electron is located: An electron with principal quantum number 2 is in the second shell, just as your house might be located on Second street. Bohr also found that orbits behaved as though they had discrete amounts of angular momentum (the force that holds a spinning top upright) and angular momentum for the orbit was described by a second quantum number. Letters were used for these quantum numbers so that they would not be confused with the principal quantum number. Orbits with zero, one, two, and three units of angular momentum became known as the *s, p, d,* and *f* orbitals by analogy with spectral lines that early spectroscopists described as *sharp, principal, diffuse,* and *fine*. Then just as you give your house address as a street and a street number, 343 Second street, the electron can be specified as in the 2*p* orbit. Orbits can have zero angular momentum (a perfectly spherical orbit), but for those orbits that did have a net angular momentum, this angular momentum should interact with a magnetic field similar to the way a spinning top interacts with the gravitational field, slowly circling its center when angled to the side. In the 1920s it was confirmed experimentally that atoms with angular momentum behaved differently in a magnetic field depending on the orientation of their angular momentum with respect to the magnetic field, which provided a third quantum number. Now just as apartment dwellers may give their address as 343 First street, number 4, an electron can be described as in the $2p_z$ orbit. Unfortunately though the same type of experiment showed that for some atoms there were additional interactions with the magnetic field that were not explained by the Bohr model. These interactions came to be called the *anomalous Zeeman effect* after the physicist who described it. The story is related about one quantum theoretician who was met by a colleague when walking down the street and told, "You look very unhappy." The theoretician answered "How can one look happy when he is thinking about the anomalous Zeeman effect?"[5]

Fortunately an explanation was offered quickly enough by two physicists in their twenties, George E. Uhlenbeck and Sam A. Goudsmit, though they tried to recall their paper after it was submitted, because they were afraid they had been wrong. Their explanation was that electrons could be thought of as spinning on an axis, thereby providing another magnetic moment. Although it has been pointed out that for this to be *literally* true, the electron would have to spin at ten times the speed of light, the *model* worked well, and the fourth and last quantum number was born.

Before this model Wolfgang Pauli, the same unhappy theoretician who made the comment about the Zeeman effect, had postulated the existence of a fourth quantum number and enunciated his Exclusion Principle: "There can never be two equivalent electrons in an atom for which . . . the values of all the quantum numbers are the same."[6] The Pauli Exclusion Principle essentially says that no two electrons can have exactly the same coordinates any more than you can park your car in exactly the same space that your neighbors have parked theirs. The beauty of this observation was that it explained the periodic build up of elements. The question had been asked: *Why do not all electrons in an atom go to the lowest energy level* (as ball bearings roll to the bottom of a bowl)? Now there was an explanation. The bottom of the Periodic Table bowl is compartmentalized. An electron can assume one of two orientations, spinning with angular momentum pointed up or down (as certain tops can be made to spin upright or upside down), but once a compartment holds two electrons—one with spin up and one with spin down—it is filled; the next electron has to go to the next compartment. Hydrogen's one electron goes into the first orbit. Helium's two electrons also go into this orbit—but with opposite spin directions—and fill the orbit. Lithium, with three electrons, has to put its last electron in a new orbit, beginning the second row of the Periodic

Table. This idea was given the name *Aufbau* (build-up) principle by Bohr: Electrons go into the lowest unfilled energy level.

The second level though holds eight—not two—electrons. An explanation for this was provided by a short, squat, mustachioed, fencing-scarred German physics professor, Arnold Sommerfeld. Sommerfeld proposed subshells within the principal energy level. Spherical orbits with no angular momentum form the first type of subshell, and these subshells hold two electrons. Orbits with one unit of angular momentum form the second type of subshell. Because these orbits are not spherical (the design of a discus is flat, so that when it is thrown it develops angular momentum for lift), Bohr proposed three possible orientations for orbits: parallel, antiparallel, or perpendicular to a defining magnetic field. Therefore subshells with one unit of angular momentum have three orbits and can hold six electrons (two in each orbit).

Here at long last was an explanation for why selenium was mistaken for tellurium; why Mendeleev categorized sulfur with oxygen; and why copper, gold, and silver were the only three metals gathered 100,000 years ago in a natural metallic state: Groups of elements with similar reactivities have the same *electron configuration*; that is, they have the same number of electrons in the same types of *outer* orbits and subshells. Selenium has its last 4 electrons in a $4p$ subshell; tellurium has its last 4 electrons in a $5p$ subshell. Copper has its last 10 electrons in the $3d$ subshell; gold has its last 10 electrons in a $5d$ subshell; silver has its last 10 electron in a $4d$ subshell.

There were however serious problems with Bohr's neat picture, the most catastrophic of which was that the model was exact only for hydrogen. Helium, with two electrons, was already too complicated for the model to handle exactly. But a new approach would soon be born and born out of war. During World War I, Louis de Broglie, a young French aristocrat who had been interested in history, was assigned to a radio communications unit, and he became interested in radio waves.

NEW QUANTUM THEORY

Louis Victor Pierre Raymond de Broglie

The second influence on de Broglie was his brother. Orphaned at an early age, he had been raised by his much older brother, and when his brother served as secretary of a Solvay conference (physics conferences sponsored by the industrialist Solvay who invented the soda process that displaced that of Le Blanc), the younger de Broglie wanted to go. However attendance was by invitation only and de Broglie could not get an invitation. Annoyed, he vowed he would be asked to the next conference—and asked because of his discoveries. In fact this did happen but not until a later conference in 1927.

De Broglie submitted a doctoral thesis to the Sorbonne in which he proposed a radical idea: If light behaved as a wave or a particle, then why could not a particle, like an electron, behave like a wave? Using Planck's expression for the energy of a quantum of light (based on its wavelength) and Einstein's expression for the energy/mass equivalence (the famous $E = mc^2$), he came up with a wavelength for a particle the size of an electron. In the early 1920s the wave properties of the electron had been observed (accidentally) in the Bell Telephone laboratories, but at the time the Sorbonne faculty did not know quite what to do with de Broglie's dissertation. They suggested that he try to demonstrate the idea experimentally, so he tried to interest one of his brother's acquaintances. But the acquaintance was too involved in work he was doing on a new communications device: television. The examining committee eventually concluded that De Broglie should be awarded the degree based on his "effort which had to be attempted in order to overcome the difficulties besetting the physicists,"[7] but praise from other quarters was stronger. Langevin sent the paper to Einstein, and Einstein thought it was a most important work. Langevin also circulated de Broglie's dissertation to another physicist, Erwin Schrödinger, at the University of Zurich, who at first pronounced it "rubbish,"[8] but then, convinced by Langevin, he decided to give it another look.

WAVE MECHANICS
Erwin Schrödinger

Schrödinger worked through the consequences of treating the electron as a wave and came to a pleasing conclusion: The fact that only certain energy states were allowed—which Bohr had postulated as a device to explain atomic spectra—came as a natural consequence of wave mechanics. This can be understood by visualizing a wave traveling around a circle. If the distance around the circle is exactly an integral number of wavelengths, then the tail of the wave meets the origin of the wave in exactly the same phase of motion, and the wave is a standing wave. If the circumference is not an integral number of wavelengths, the wave is out of phase, and it will self-destruct. Because the electron has a given wavelength for a given energy, there can be only certain orbits that support standing waves: orbits with circumferences equal to an integral number of wavelengths.

The astute reader by now has noticed that we announced we were going to talk about a new quantum theory but there has been nothing up to this point all that radically new. We are still talking in terms of macroscopic, mechanical models and aside from a bit of fuzziness here and there, a fairly intuitive picture. But the next advance, though fruitful, is based on a much less intuitive concept, and it marks the beginning of the new quantum theory.

Werner Heisenberg

The true philosophic differences between the old quantum theory and the new quantum theory have their origins in the work of another German physicist, Werner Heisenberg. Heisenberg was really a first-generation quantum physicist, having been brought up in part under the tutelage of Bohr. As a youth Heisenberg was a street brawler and mountain climber, and he brought much of this approach to his physics. (One biographer had attributed Heisenberg's combative spirit to an upbringing in which sibling rivalry was encouraged in an effort to foster combativeness. As a consequence he and his brother engaged in physical battles, once armed with chairs, and as adults they were largely estranged.[9]) Heisenberg was dissatisfied with pseudophysical models, and he wished for a purer quantum theory based solely on such observables as the energies of lines in a spectrum. He devised his system using tables of these values, or matrices, and employed matrix algebra to arrive at his conclusions. He published his approach a few months before Schrödinger published his, but Schrödinger soon showed that the two approaches were equivalent. Heisenberg stoutly defended his approach and said, "What Schrödinger writes about the visualizability of his theory . . . is [manure]."[10] However after seeing the matrix algebra of Heisenberg, most scientists were pleased to return to the intuitive picture of waves.

One additional result that came out of Heisenberg's work however was that his matrices did not allow the computation of the position and the momentum of the electron at the same instant in time. This interesting result, called the Heisenberg uncertainty principle, has a physical interpretation. Atomic properties are measured by the interaction of light with matter. In our everyday world we are constantly interacting with light, but we are mechanically unaffected by its properties; light does not beat us up unless we sunbathe too long. On the atomic level this is not so. The energy of ultraviolet light is sufficient to knock the tiny electron around (which is what ultraviolet light from the sun does to electrons in your skin while you lie in the sun). Therefore if we want to measure the position of an electron, we can do so by bouncing light off it; but when we do, we hit it with enough energy to change its momentum, so now the measurement of momentum is uncertain. If we measure the momentum with light, then we change the electron's trajectory, so its position is uncertain. This problem of not being able to know the position and momentum of an electron at the same time casts a shadow on all the nice intuitive models with macroscopic analogs. Now it had to be said that an electron "behaves" as a wave, but never again could it be said that an electron "is" a wave—or even that it is in an "orbit."

When Heisenberg originally assembled his tables of values, he did not realize he would have to use matrix algebra to work with them. It was his mentor at the time, Max Born, who recognized the form and with another student, Pascual Jordan, the three developed the system. Then Born dropped the next bomb. He knew that Schrödinger's waves had to be retained, but in light of Heisenberg's uncertainty principle, they had to be reinterpreted. Born decided that they were *probability* waves, which was greeted with a great hue and cry, and in fact the debate remains lively today.[11]

Since the days of Newton and Descartes, scientists had been riding high on the idea that they were dealing with a clockwork universe and that the motions of all things were ultimately predictable if one could just get the equations right. Boltzmann and Gibbs had shaken this notion by using statistical probabilities when it was difficult to calculate each particle's exact position and momentum, but now Heisenberg and Born were saying that for the atom's electron, calculating the position and momentum was not just difficult—it was impossible. The best that could ever be known was the probability of position and momentum.

If not the most outspoken, then certainly the most eloquent objections came from Einstein. Heisenberg, he said, "has laid a large quantum egg." God, said Einstein, "does not play dice with the universe." In addition to philosophical objections, there were aesthetic objections. The neat, pleasing orbits of Bohr were replaced by obscure, dissatisfying "regions of probability"—and probability "waves" at that—which would soon be called "orbitals." The atom was no longer a neat, visualizable, little solar system.

THE CHEMICAL BOND

Disregarding Einstein's opinions about the gaming habits of the gods and the need for visualizable pictures, some physicists found the probability wave concept useful and built on it. Notably important for chemists was the work of the physicists Walter Heitler and Fritz London. In the late 1920s using the Lewis shared-electron-pair bond as their starting point, they considered the equation for the bond orbital to be a product of equations for atomic orbits and applied quantum theory to develop the first quantitative model of a chemical bond. Though equations for the forces involved in two electrons interacting with two atomic nuclei were too complicated to solve exactly, they used a simplifying device developed by Lord Rayleigh for the theory of sound. This device assumed that the approximate answer that yielded the lowest energy was nearest the truth. Their method gave them a reasonably good approximation to the experimental energy holding together the two hydrogen nuclei in a hydrogen molecule. In addition they found that two electrons sharing a space between two nuclei were attracted by both nuclei, and this lowered the energy of the system as a whole. Like the ball bearing in the bowl seeking the state of lowest energy, two hydrogen nuclei formed a bond because a bond minimized electrostatic interactions: The two positive nuclei were drawn together by attracting each other's negative electron cloud but only until this attraction was balanced by the repulsion of like charges. Berzelius was vindicated. The attraction was electric in origin after all but in a manner Berzelius could never have imagined.

Given a physical model for the chemical bond, chemists needed no further invitation.

QUANTUM CHEMISTRY
Linus Pauling

One of the first champions of quantum chemistry was from the United States and from the Wild West at that. Linus Pauling was born in Portland, Oregon. Cowherders, Native Americans, and his father's drugstore formed the backdrop for his early years. He enjoyed playing with drugstore chemicals (like sulfuric acid), reading his father's old chemistry books, and working in the laboratory he set up in in his family's basement. But when his father died when Linus was eight, the family faced financial hardship. He was able to study chemical engineering at Oregon State Agricultural College, but he had to drop out periodically to take jobs to help his family.

He did graduate work at California Institute of Technology (CalTech), and afterward he went to Europe where he studied with Bohr, Sommerfeld, and Schrödinger. In England, Pauling met Lawrence Bragg, a figure important in the development of the X-ray diffraction technique for determining structures of crystals. Back at CalTech, which was quickly becoming a premiere research institution (Pauling turned down a Rhodes scholarship to Oxford to study at CalTech), Pauling used X-ray diffraction to study bonding in crystals.

Pauling learned that many previous assumptions about bond lengths were in error because they underestimated the strength of the two-electron bond and overestimated covalent bond length (a covalent bond length is the distance between covalently bonded nuclei in a molecule). Pauling investigated as many crystals as he could get his hands on, and he developed an impressive working knowledge of bond distances and strengths. Based on this knowledge and the success of the Heitler–London model, Pauling developed a model for bonding called the *valence-bond* approach, which assumed that orbitals were formed in molecules by the overlapping of atomic orbitals. This recaptured much of the intuitive picture of bonding because now atomic orbitals could be sketched and bonds drawn where these orbitals overlapped. The device offered an explanation for bonding in many molecules, and where it failed—as in the tetrahedral bonding of carbon—Pauling showed that a *hybrid* atomic orbital system sufficed: Just as two ripples on a pond can come together to form a differently shaped wave, one s and three p orbitals can mix to form four hybrid orbitals extended tetrahedrally in four directions in space.

Pauling also used the notion of *resonance*: If a molecule can be described by two valence-bond structures, then its real electronic structure is a *resonance bond*, a blend of the two. (A German chemist F. G. Arndt also proposed this idea in the early 1920s but without the advantage of quantum mechanics.) The benzene molecule can be described in terms of resonance bonds. Its structure can be written as a six-carbon ring with three double bonds: one double bond between every other member of the ring. But because there are six bonds, there are two ways of placing the double bonds. So according to Pauling the real molecule is a resonance blend of these two structures. But the idea of resonance created problems for Pauling, some of which may have been caused by the choice of names. *Resonance* to many implied an oscillation or something going back and forth, and some thought that Pauling meant to imply that several valence-bond structures actually existed and that the molecule literally flipped back and forth between various possible structures rather than a blend. But in the late 1930s when Pauling wrote *The Nature of the Chemical Bond and the Structure of Molecules and Crystals: An Introduction to Modern Structural Chemistry*, in which he presented his valence-bond ideas and described how they could be used to understand—and predict—chemical bonds and molecular shapes, it quickly became one of the most important textbooks of the century.

Pauling's valence-bond method however is a conceptual device for understanding molecular bonding, and there are other possible devices. One of these, called the *molecular-orbital* approach, eventually challenged Pauling's sovereignty in the field.

Robert S. Mulliken

The champion of the molecular-orbital system was Robert Mulliken, whose father was a professor of chemistry at Massachusetts Institute of Technology (MIT). While still young Mulliken proofread a four-volume chemical treatise written by his father and helped him in the laboratory. He had little thought for any career other than chemistry.

Mulliken stayed at MIT to study, where he reportedly "loved molecules in general, some molecules in particular,"[12] but his undergraduate work was not outstanding. He graduated in the middle of World War I, so he spent some time working on war gases before going into industry. He read the theories of Lewis and Langmuir and learned the old quantum theory, then traveled to Europe to study with Rayleigh, Sommerfeld, Bohr, and Born.

In Europe he also met the physicist Frederick Hund. In the period between the mid-1920s to the mid-1930s, he developed with Hund through correspondence a bonding model based on orbitals that extended over the entire molecule. These molecular orbitals were numbered, as were atomic orbitals, and electrons in them had various quantized amounts of angular momentum, magnetic moment, and spin. Mulliken and Hund developed a notation for the molecular orbitals for diatomics (molecules with just two nuclei) using the Greek letters σ (sigma), π (pi), δ (delta), and φ (phi) to be analogous with the atomic s, p, d, and f orbitals.

In the molecular-orbital model, energies of molecular orbitals are calculated, then their electron "occupation" is determined by filling the orbitals with electrons (two per orbital, vis-à-vis Pauli), starting with the lowest energy orbital first. Whether or not the molecule is stable depends on how many electrons are in bonding versus antibonding orbitals. Bonding orbitals have a high probability of electrons being between nuclei; antibonding orbitals have a high probability of electrons being outside the internuclear region. Reactivity is based on the shapes of occupied molecular orbitals.

Because this proved to be a powerful method for predicting reactivity and sometimes offered information where Pauling's valence-bond methods did not, there arose two camps: those who regarded the valence-bond approach as *the* method, and those who regarded the molecular-orbital approach as *the* method. Instead of acknowledging the value of the atomic-orbital picture, Mulliken criticized Pauling for oversimplifying. Mulliken said this was popular with chemists but actually inhibited their understanding of the true complexity of electronic structure. Pauling responded by becoming more one-sided himself. When he could have acknowledged the utility of the molecular-orbital approach in treating large molecules, he stuck doggedly to his hybridization of atomic orbitals, invoking 560 different resonance structures to explain the bonding in one inorganic molecule in his last edition of *The Nature of the Chemical Bond*.[13]

The irony is that the argument between proponents of the valence-bond and molecular-orbital models was another argument over dinosaurs: Both approaches are models, not exact representations—both approaches sacrifice some mathematical rigor for a conceptual picture, and both have areas in which they are particularly useful.

A better analogy for the models in this case might be maps: A road map and a topological map can describe the same territory, but they offer different kinds of information—and neither one would ever be confused for the terrain itself. Pauling's valence-bond map can be used to visualize the structure and reactivity of literally thousands of compounds but Mulliken's molecular-orbital map is used to understand the structure and reactivity of others, particularly those in which nonbonding and antibonding orbitals play a significant role. Pauling's valence-bond road maps may not include all the contours and detail of Mulliken's topographical maps—but they clearly show how to go from point A to point B. Mulliken's topographical maps are able to offer more information—Mulliken's molecular-orbital approach, for instance, predicts oxygen's observed magnetic behavior whereas Pauling's approach does not—but their interpretation requires a more highly trained eye.[14]

Pauling received the Nobel Prize in 1954 "for his research into the nature of the chemical bond and its application to the elucidation of the structure of complex substances."[15] Mulliken received the Nobel Prize in 1966 for "fundamental work concerning chemical bonds and the electronic structure of molecules."[16] Hund curiously was not mentioned by the Nobel committee, though in the early stages of development, the molecular orbital theory was often referred to as the

Hund–Mulliken theory. Mulliken said he would have been "glad to share"[17] the prize with Hund.

Nowadays both models are still very much in use. Pauling's method is used to teach bonding and to visualize three-dimensional structures. An extensive library of calculated molecular orbitals has accumulated and is helpful in explaining some aspects of another particular type of reactivity: interactions of molecules with light, or spectroscopy. The most immediate triumph of quantum chemistry has been the explanation it offers for molecular spectroscopy.

SPECTROSCOPY

When we talked about spectroscopy before, we talked about the absorption or emission of visible light. But as we have seen in conjunction with our discussion of X-rays, there is more than one kind of light. (Our eyes are sensitive only to visible light because this is the light that reaches the surface of the Earth from our sun—along with some ultraviolet, which is what sunscreen is for, and some infrared, which warms things up. If we had evolved under a different star, our eyes might be sensitive to other frequencies of light). In fact the electromagnetic spectrum of light extends theoretically over an infinite range of wavelengths. For practical purposes however it extends from high-energy gamma rays, to X rays, to ultraviolet, to visible, to infrared, to microwave, and then to low-energy radio waves.

All light can interact with matter in some fashion, and the various interactions of light with matter comprise the field of spectroscopy. Shortly after it was realized that electrons move between quantized energy levels when the right frequency of light is absorbed, it was realized that vibrating molecules also have quantized vibrational energy levels and rotating molecules have quantized rotational energy levels. It takes light in the infrared range to excite changes in vibrational levels and light in the microwave region to excite changes in rotational level. Quantum theory relates the energy of vibrational and rotational energy-level transitions of molecules to such fundamental quantities as molecular mass (if two masses are connected by a spring, the lighter the masses, the more rapid the vibration for the same amount of energy) and bond distances (if an ice skater is spinning, outstretched arms cause the skater to spin slower, while drawn-in arms cause the skater to spin faster). The quantization of frequency changes of scattered light was described by Chandrasekhara Venkata Raman, born and educated in India. Efficiency of this process—called Raman scattering—goes up with the frequency of light, which explains why the sky is blue. Light from the sun covers the visible range, going up in frequency from red to orange, to yellow, to green, to blue, to violet. Blue and violet are scattered more efficiently, but our eyes are less sensitive to violet (which is close to ultraviolet), so the sky appears blue. The sun appears yellow because the blue color has been subtracted by being scattered.

G. N. Lewis Revisited

Lewis also contributed to the field of molecular spectroscopy. One important contribution was his creation of the word *photon* to denote a quantum, or packet of light, and his research, begun in his sixties, into photoluminescence, the process in which absorbed light causes one electron in a molecule to go to a higher energy level, then the light is re-emitted at the same or slightly shifted frequencies as the electron returns to the ground state. Lewis found that some re-emissions appeared to be immediate but some were delayed. Lewis at first tried to avoid a quantum mechanical interpretation, but by the early 1940s he attributed the difference to the two possible states for the excited electron: spin parallel to the spin of the remaining ground-state electron or spin antiparallel to the ground-state electron. These two types of photoluminescence—called phosphorescence and fluorescence, respectively—formed the starting point for the field of photochemistry, and these were the subject of Lewis's last papers. He died in 1946 while performing an experiment in photoluminescence.

So quantum theory has proved to be enormously useful in chemistry. Organic, inorganic, analytical, and biochemists now use quantum theory to calculate thermodynamic properties, interpret spectra, and determine such molecular properties as bond lengths and bond angles. We should not leave the impression however that quantum theory is a complete and polished theory, so that feeding the proper numbers into a computer produces all that needs to be known about a molecular system. There is in fact a great deal of work to be done and much room for improvement. After all the hydrogen atom is still the only system that can be solved exactly. Paul Dirac, one of the founders of quantum theory, stated

> The underlying physical laws necessary for . . . a large part of physics and the whole of chemistry are thus completely known . . . the difficulty is only that the . . . equations [are] much too complicated to be soluble . . .[18]

With all due respect to Dirac, this "only" difficulty is a large one. His statement is akin to saying that a sieve is a perfectly good bucket, the only difficulty being the holes in the bottom. Einstein was a bit more reserved in his praise. He said:

> Quantum mechanics is certainly imposing. But an inner voice tells me it is not yet the real thing. The theory says a lot, but does not really bring us closer to the secret of "the Old One."[19]

So there remains work to be done, and in fact we have to wonder how much more work would have been done if it had not been for the disruption of World War II. But the disruption came.

WORLD WAR II

World War I had done little to diminish militarism or nationalism. The peace negotiated with Germany had been punitive, and the Italians, who had fought and sacrificed for the side of the Allied powers felt that they had been shorted in the division of the spoils. It was in Italy that fascism—with its totalitarianism, nationalism, authoritarianism, and militarism—started, but in Germany, struggling to deal with defeat, humiliation, inflation, and shortages, fascism was seen as a ray of hope.

And at first there was reason to hope. Using a curious but well-proven method for focusing national energy, a new leader, Adolph Hitler, identified some enemies—the Socialists and the Jews—and began a systematic military buildup. Because of this rearmament, in the space of a few years unemployment was down and business was back up.

Other nations followed the German example. Militarists in Japan used a military buildup to revive their economy and seized Manchuria to secure a source of iron and coal. Germany likewise looked to expansion to continue its recovery and growth. On September 1, 1939 a column of German tanks crossed the Polish border, and World War II began.

War is disruptive by definition, but World War II was particularly virulent. In his *A Short History of World War II*, James Stokesbury characterized the period from 1939–1945 as "the most profound and concentrated upheaval of humanity since the Black Death."[20] Among those affected was Haber, the German hero of World War I. But hero status did not protect Haber the Jew, who was forced to flee. Born likewise left Germany in 1933. (He returned to Germany after the war, though prudently remaining a British subject.) Schrödinger left in 1933 because of his opposition to Nazi policies and went back to his native Austria. He was forced out again by the German invasion of Austria.

In the 1930s Nazis established anti-Semitic laws, one of which prohibited faculty members with Jewish wives from teaching at universities. Hertzberg and his wife decided not to bother the regime with their presence at all and moved to the University of Saskatchewan in Canada. Here however as a German national, Hertzberg's work was restricted.

Not all scientists opposed the Nazi regime. Johannes Stark, discoverer of the Stark effect—the splitting of electron energy levels in an electric field—was energetically pro-Nazi and anti-Semitic. Heisenberg, though not actively pro-Nazi, stayed in Germany and tried to preserve what was left of German science. He ultimately headed the German atomic bomb project.

Einstein would not even attend conferences held in Mussolini's Italy as early as 1927, and he eventually emigrated to the United States. Fritz London, forced from Germany by Nazi persecution, became a professor of theoretical chemistry at Duke University in the United States. Bohr in Copenhagen actively helped to relocate scientists leaving Germany. When the Nazis occupied Denmark, Bohr, about to be arrested, left the country in a fishing boat that made it to Sweden. He was smuggled from there to England in a small plane that had to fly at a high altitude. Bohr had a large head (he is easy to identify in pictures for this feature alone), and his oxygen mask did not fit well. He lost consciousness in the plane, and at one point the crew though he was dead.

Planck, originator of the quantum concept, remained in Germany, though he was forced from his university position. He had already lost two daughters, who died in childbirth, and one son in World War I. Another son was executed after taking part in an attempt to assassinate Hitler. Planck lost his home in a bombing, but when the allies advanced into Germany the U.S. forces sent a car to take him to safety. He died in 1947, a few months before his ninetieth birthday.

World War II put a decisive end to any corner Europeans may have had on chemistry and physics. European scientists, like seeds, were scattered to the four winds. It was not be long before German chemistry students were making pilgrimages to learning centers in the United States.

In the development of quantum chemistry, physicists and chemists came together from all over the world to arrive at a theory for the chemical bond. This mass amalgamation was promoted by global wars—wars that also represented a philosophic turning point for many: Humanism replaced nationalism as the ultimate loyalty. The nation was replaced by the world.

Besides being global, the new wars were unique for another reason: technology. Technology, which had been changing by the century, was now changing by the year. One of the most influential of the new technologies for chemistry was the automobile. As we see in the next two chapters, this invention inspired as much chemistry as the search for gold: theoretical chemistry, synthetic chemistry, and, interestingly, biochemistry. The contribution of the automobile to biochemistry began with the need for rubber for its tires.

chapter
SIXTEEN

ca. 1914–1950: Polymers and Proteins—Links in the Chain

When summarized neatly as a list of accomplishments, the work of the chemist may seem a continuous flow of productive pursuits and rewarding discoveries. But in truth chemists spend thousands of hours at the laboratory bench for every hour spent at awards banquets—if any—and it is more than instant gratification that keeps them at their labors. One motivation has always been and always will be pure love for the beauty of the system. Seeing and understanding the marvelous way in which elements interact is motivation enough, for most. But beyond that is the knowledge that from time to time discoveries can be made that contribute greatly to the quality of life and even bring us closer to understanding the essence of life itself.

In any case chemists in the era of the world wars had no lack of motivation from any quarter whatsoever. There were enough theoretical advances to please the most rigorous purist, enough new materials to permanently alter the standard of living, and enough new information on the structure of biologically important materials—the stuff of life—even to make hypotheses about their origin—the origin, that is, of life. Interestingly this avalanche of understanding came from the antithesis of life: war. Wars caused a shortage of rubber, and rubber was needed for tires for the Kaiser's car.

Rubber originally came from rubber trees. Joseph Priestley used it to rub out pencil marks (hence *rubber*), Charles Macintosh used it to waterproof clothing (hence *mackintosh*), and Charles Goodyear found that rubber mixed with sulfur and heated (hence *vulcanization*) kept rubber rubbery at high and low temperatures. He patented the process, and rubber was used to make buggy wheels. When it was determined that vulcanized rubber made equally adequate tires for the new automobile, the world consumption of rubber doubled. Britain in an effort to protect Malayan rubber prices restricted exports. Other countries in an effort to protect their economies tried to make synthetic rubber. With the world wars efforts redoubled. The German Fritz Hofmann had used a material produced cheaply from acetone to make synthetic rubber by World War I, and by World War II many countries including Russia and the United States were producing a synthetic rubber, too. In the process of sorting out the synthesis of rubber, chemists learned a lot about natural products, polymers, and how biological molecules are put together.

POLYMERS

Rubber, like many other biological materials, is a *polymer*: made up of chains of thousands of individual groups called *monomers*. In a polymer the monomer units can all be identical, X-X-X-X- (isoprene, a branched five-carbon hydrocarbon, is the monomer of rubber), or it can be made of different types of units, alternating positions in the chain, X-Y-X-Y-X-Y-. It is also possible to join two different molecules with identical reactive heads, X-X and Y-Y, to form X-X-Y-Y-X-X-Y-Y-

or branched chains or any multitude of combinations. Polymer properties depend on the monomers that make up the chains, the length of the chain, and the branching of the chain. The success of synthetic rubber inspired a search for other polymeric materials, and the search was rewarded.[1]

In the 1920s Joseph Patrick invented a nasty-smelling sulfur-containing polymer called *Thiokol rubber* that was resistant to solvents. It was used to seal gas tanks in airplanes in World War II. The material now called *polyvinylchloride* (PVC) had been known as a brittle solid since the mid-1800s, but in the mid-1920s Waldo Lonsbury Semon, a U.S. chemist, looking for a better glue, accidentally found that adding high-boiling solvents to PVC made the polymer more pliable. British chemists at Imperial Chemicals Industries found that at high pressures ethylene and benzaldehyde combined to produce a white waxy solid but ethylene by itself exploded. When they rebuilt their apparatus they unknowingly introduced a leak and let in oxygen, which turned out to be necessary to catalyze the reaction. In their next experiment ethylene alone gave a tough, useful polymer, *polyethylene*. Polyethylene made an excellent insulation for World War II radar cables. Polystyrene plastics and foam (*Styrofoam*) were introduced in the 1930s, as were *Lucite* and *Plexiglas* (organic glass) polymers. Plexiglas has excellent optical properties, and during World War II it was used as a lightweight material for airplane windows. The largest immediate postwar use was in jukeboxes. By the end of the 1920s, the U.S.-based company Du Pont hired chemist Wallace Hume Carothers to head its fundamental polymer research. Du Pont's most immediate goal was to replace silk supplies from the Orient, which were dwindling because of Japanese militancy.

Wallace Carothers

Though Carothers's first academic degree was in accountancy and secretarial administration, he eventually received a doctorate in chemistry from the University of Illinois. His first goal at Du Pont was to understand the composition of such natural polymers as rubber, cellulose, and silk, and then to imitate them. Carothers used established organic reactions but applied them to molecules with two reactive centers—

one on either end—to form links in his chains. Carothers and his group learned many interesting things about polymers but produced no silk until someone decided to have a little fun.

One day one of Carothers's assistants, Julian Hill, playing with a gooey mess of polyester in the bottom of a beaker, noted that when he picked up a glob of the goo on a glass stirring rod, it trailed strings that got silky as they stretched out. On a later date when the boss was downtown, Hill and some other technicians decided to see how far they could stretch the goo by running down the hall. They ended up with *very* silky strands, resulting from the orientation of the extended polymer. But they knew that the melting point of these first polyesters was too low and they were too soluble in water to make a good clothing material (this would later be improved). So they went back and tried their running-down-the-hall method with a tougher polyamide they had on the shelf. It worked.

When the boss came back, they told him, and *nylon* was developed—over a period of 10 years—into a silk-like material. Advertised at the 1939 New York World's Fair as synthetic silk made from coal, air, and water, its application in hosiery for women was demonstrated by a model in a giant test tube.[2] The material was an instant success. Reportedly four million pairs of nylon stockings sold in a few hours in the first New York City sale, but almost immediately sales were restricted and the material commandeered for parachutes. William Carothers did not live to see the success. He died by his own hand in a Philadelphia hotel in 1937.

Two other synthetic materials quickly followed. *Dacron*, a useful high-melting and insoluble polyester, was developed by the British chemists Rex Whinfield and James T. Dickson and marketed in Britain by Imperial Chemical Industries as Terylene. In Germany chemists of the I. G. Farben developed polyurethane foams, and in the United States Roy J. Plunkett, a Du Pont chemist, 2 years beyond his Ph.D., discovered *Teflon* when he opened a gas cylinder supposed to contain a fluorinated hydrocarbon but nothing came out. He became curious, and after inserting a wire through the valve to make sure the tank was indeed open, he cut the tank in half and found a waxy white powder in the bottom. He tested the properties of the material, found it was inert to acids, bases, heat, and solvents, and very slippery.

Nothing seemed to stick to it. The material was expensive to produce and probably would have remained a laboratory curiosity if the atomic bomb project had not needed a gasket material that was inert to corrosive uranium hexafluoride.

Plunkett's Teflon remained a military secret until after the war, when it became a modern miracle. It has been used to coat frying pans and seal plumbing joints. Teflon is one of the few artificial materials that the body tolerates, and it has been used for artificial corneas, bones, joints, tracheas, heart valves, tendons, bile ducts, and dentures. The United States is even said to have produced a Teflon President.

The success of polymeric materials had made the study of polymer synthesis, structure, and properties a science unto itself: polymer chemistry. To those of us sitting in plastic chairs in cozy houses insulated with plastic foam, reading books through plastic lenses, benefits of the science are obvious. But there have been more subtle benefits, too. Through an understanding of natural and synthetic polymers, insight has been gained as to the structure and functionality of biologically important macromolecules and polymers, such as ribonucleic acid (RNA), and the process by which cells replicate and produce life. But the road from rubber to RNA was rather rough and owes its navigation to chemists who were willing to work up to their elbows in "grease."

Hermann Staudinger

The major figure in early polymer chemistry was the German chemist Hermann Staudinger. Though originally interested in botany, his father convinced him that a background in chemistry would strengthen his understanding of botany, and he stayed with the field. He, like many chemists of his day (ca. 1920), was encouraged to study the synthesis of polymers, but his interests extended more to the nature of the materials than their synthesis. He began work in the general field of high-molecular-weight compounds: *macromolecules*. Though these materials were appreciated enough by industry at the time, actual research on macromolecules was considered "grease chemistry" by purists in the academic establishment.

A *dalton* is a unit of atomic mass, roughly equal to the mass of a hydrogen atom (and named of course in honor of one of our earlier heroes, John Dalton). When Staudinger determined molecular weights for some macromolecules, he found they were on the order of 100,000 or 300,000 daltons, compared to a molecular weight of 18 daltons for water or perhaps 500 or so for an organometallic. The weights were considered absurd. The prevalent opinion was that macromolecules were really just smaller units somehow clumped together, and if they were sufficiently purified, smaller units would emerge. Staudinger (described as a large, soft-spoken man) was refused research funds on several occasions, but he stayed with his opinion, one time reportedly defending his views with Martin Luther's words, "Here I stand and can do no other".[3]

Steadily Staudinger and his collaborators produced experimental evidence that macromolecules really were covalently bonded, very large molecules. By 1935 Staudinger's ideas were accepted, and in 1953 he received a Nobel Prize for his work. Experimental work bolstering Staudinger's ideas came from his collaborators, including Magda Woit Staudinger, his wife and a plant physiologist, who coauthored a number of publications with her husband. The Staudingers were also supported by the X-ray crystallography work conducted by Herman Mark.

X-RAY CRYSTALLOGRAPHY

Herman Mark

In the early 1900s Max von Laue had predicted that X rays would be diffracted by the atomic nuclei in a crystal. The father and son team of William Henry Bragg and William Lawrence Bragg developed equipment and equations, respectively, for extracting information about the structure of molecules from the X-ray diffraction pattern in a process that has been described as a three-dimensional jigsaw puzzle.

Herman Mark however solved the puzzle for polymers and used their X-ray diffraction patterns to show that they were indeed macromolecules. He had to leave Germany because of ethnic persecution, and he went to the United States. In the United States ideas about polymers were not so entrenched as they were in Germany, and Mark's conclusions were more readily accepted. However he also concluded that macromolcules have loose, easily rotated chains, and this did not sit well with Staudinger, who had decided that macromolecules had a rigid structure. Staudinger became rigid in his belief, but with the new generation of polymer chemists, free rotation of units in the polymer became established.

Polymer chemistry by now had also became respectable and well established, and the principles discovered by the polymer chemists were applied to important classes of natural macromolecules critical for all life processes: the stuff of life.

THE STUFF OF LIFE

Proteins are polymers of amino acids. Nucleic acids are polymers of nucleotides. Carbohydrates can be polymers of simple sugars, and fats, oils, and waxes—though not polymers—are certainly large molecules. The chemist tackled these materials in the first half of the 1900s with an aim to analyzing, understanding, and imitating. Though the assault was on all fronts, leading inroads were made in understanding the ubiquitous body substance: protein.

Proteins (from the Greek meaning *first*) are the main constituents of hair, skin, nails, muscles, tendons, and blood vessels, and they perform many biological functions. Enzymes, catalysts for biological reactions, are also proteins (with a single exception discussed shortly). Proteins are polymers of amino acids: an organic acid (COOH) and an amine (NH_2), joined through a carbon center. The joining carbon also has another group attached, called the *side chain*, and differences in side chains make amino acids different (see Fig. 16.1). Theoretically this side chain can be almost any organic group, but actually only about 20 different amino acids are found in biological systems. Proteins are made up of anywhere from 50 to about 1000 of these different amino acids, bonded through the acid end of one to the amine end of another in a link called a *peptide bond*. Amino acids are bonded in a very specific order, and the

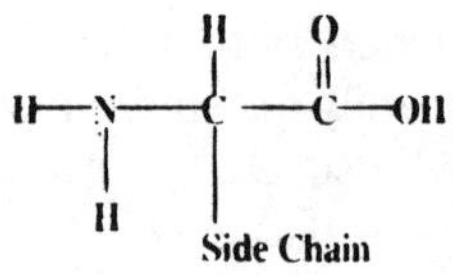

$$H_2N-CH(\text{Side Chain})-C(=O)-OH$$

Figure 16.1. A generic amino acid. Different amino acids have different side chains. For instance alanine has a CH_3 side chain, glycine has an H side chain, and cysteine has a CH_2SH side chain.

misplacement of just one amino acid can totally alter the function of a protein. Sickle-cell anemia is caused by such a change in the structure of protein hemoglobin.

Initially amino acids were hard to identify because reaction conditions used by chemists to break up the protein usually destroyed amino acids in the process. But as techniques improved, these units were identified by such chemists as Emil Fischer, Sidney W. Cole, and Frederick Gowland Hopkins. The difficulty was then to decide how amino acids were joined to make up different proteins. In the mid-1940s the first complete analysis of the sequence of amino acids that makes up a protein was achieved. More quickly followed, such as the work by Frederick Sanger in which he unraveled the amino acid sequence for insulin.

Frederick Sanger

When Sanger began his work the protein insulin had just become available in a pure, crystalline form. His research director at Cambridge, Charles Chibnall, suggested he determine its amino acid sequence. Using a fluorine compound newly synthesized in conjunction with gas warfare research and a combination of acids and enzymes, he was able to chop off pieces of varying lengths. He used this grid of information to determine the sequences of the two chains that make up insulin and where the two chains were linked.

But the sequence of amino acids in a protein could be deduced only from purified proteins removed from their original cellular environment, and for some chemists this raised the question whether structures were the same in the cell as in purified proteins. And they had a point.

Living materials, including the human body, are made of cells or a single cell (for example a bacterium). The cell serves as a minuscule beaker for carrying out biological chemistry, providing a controlled chemical environment for these complicated and exacting processes. Proteins can behave very differently outside the controlled environment of the cell because their functionality depends in part on the three-dimensional shape that the long chain assumes inside the cell (just as the functionality of a chair would be seriously impaired if it were a stack of sticks instead of a seat, legs, and a back). This shape is very specific, and it is determined by attractions within the chain and between chains. Though these attractions (called intermolecular forces) are weaker than covalent bonding forces, there are so many interactions that they are enough to dictate three-dimensional structure. Elucidating the three-dimensional structure then required understanding these forces, and it was in this era of the world wars that this work was done, too.

INTERMOLECULAR FORCES

Although the previous discussion of quantum chemistry lent some insight into chemical bonding, there is more to the story than this. Molecular bonds, as *intramolecular* forces, are certainly important in the structure of matter, but equally important and not to be neglected are *inter*molecular forces. Without molecular bonds there would be no compounds, but without intermolecular forces, all compounds would be gases. Intermolecular forces are the sticky tape that holds materials together, allowing liquids to condense and solids to freeze. These forces are tens of times weaker than bonding forces, but without them materials would come apart at the seams.

One tremendously important intermolecular force is the so-called *hydrogen bond*. Hydrogen bonding is the attraction of the electropositive hydrogen portion of a molecule for the strongly electronegative oxygen, nitrogen, or fluorine portion of either another molecule or such a group within the same molecule. The best example of hydrogen bonding is the attraction of the hydrogen of one water molecule (H_2O) for the oxygen of another. Hydrogen bonding is responsible for many of the properties of water, including its tendency to expand on freezing. Most other liquids *contract* on solidifying, but water has to align its hydrogen bonds as well as its molecules in crystalline ice, and the net result is expansion. This increased volume (decreased density) explains why ice floats on water and wet sidewalks crack in winter. Hydrogen bonding also accounts for some of the solvent properties of water. Through hydrogen bonds water molecules can form a *cage* around solvent particles and increase their solubility. Hydrogen bonding also occurs between protein chains.

Once out of the delicately balanced natural environment however, heat and other perturbations can "denature" or disturb the three-dimensional structure of the protein held in place by such intermolecular forces as hydrogen bonds. Alcohol is a disinfectant because it denatures the protein of bacteria, and cooking visibly alters the form of the protein in eggs or meat. The importance of a three-dimensional structure to the functionality of biological molecules is well illustrated by enzymes.

ENZYMES

While it was previously believed that all enzymes were composed of protein, it appears that this view is currently undergoing some alteration, as we will see. But it can certainly be said that the vast majority of enzymes are proteins (there are over 2000 known), and each has its own specific three-dimensional structure that is the key to its functionality. In the late 1800s Emil Fischer expressed this as the "lock and key" model: An enzyme has a particular shape so that reagent(s) for the reactions it will catalyze fit into it and are held there for reaction—as a key fits into a lock (see Fig. 16.2). John Cornforth, an Australian chemist, used this model to explain why natural molecules are formed in only one of two possible mirror images—a mystery since Pasteur's work with tartaric acid and tweezers. Cornforth saw that the enzyme acted as a three-dimensional template and only one shape would come

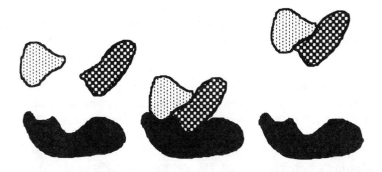

Figure 16.2. The lock-and-key model for enzyme activity. The enzyme is shaped so that reactants fit on its surface. Once they are on the surface, they are held together so that they can react. The product does not fit as well on the enzyme, so it will separate and leave.

off the mold. Cornforth, totally deaf since his midtwenties, had as one principal collaborator another Australian chemist, his wife Rita. Further elucidation of the structure and functionality of this important class of biochemicals came from the work of a U.S. chemist, James Sumner, and a German chemist, Richard Martin Willstätter.

James Sumner and Richard Martin Willstätter

Willstätter studied with Baeyer, and Baeyer, confident of Willstätter's talent, advised him to be baptized a Christian to further his career. Willstätter refused, but as future events showed, it would not have mattered: Any advantage it would have given him would have been short-lived. He became a professor, and he was given responsibility for teaching dyestuff chemistry. As a related area of research, he began work on chlorophyll, the plant pigment involved in photosynthesis.

His work on chlorophyll was hampered during the first World War because many assistants were drafted and because of his own war efforts, which included developing gas masks effective against chlorine and phosgene. For this work he was awarded the Iron Cross Second Class. In 1915 he was awarded the Nobel Prize for his work on plant pigments, but he was not able to receive the award until after the war.

In the period between wars Willstätter did further work on substances of biological interest, including the synthesis of the alkaloid cocaine, and he also investigated enzymes. As a result of his work on enzymes, he came to the conclusion that enzymes were not composed solely of protein, though they might contain a protein carrier. This produced quite a debate with James Sumner, a U.S. chemist.

James Sumner, who was left-handed, had his left arm amputated following a hunting accident. He was told he ought to consider a career other than chemistry because he would have difficulty doing laboratory work. He decided otherwise, stayed with chemistry, and was able to crystallize an enzyme for the first time and show that the known enzymes were proteins.

When Sumner reported in the 1920s (after a 9-year investigation) that he had a pure enzyme and it was a protein, Willstätter was considered at the time the leading enzyme chemist, and his opinion carried great weight. Sumner was told by some critics to go to Germany to take lessons from chemists there. Sumner steadfastly defended his results, and in the 1930s other chemists began duplicating his findings, which vindicated Sumner.

With World War II impending Willstätter found his position increasingly uncomfortable in Germany, and he finally retired in protest over anti-Semitic policies at universities. He remained involved with research, communicating with his assistant, Dr. Margarete Rohdewald, over the telephone. He was finally required to turn in his passport, and at one point the gestapo went to his home to arrest him. He was in his garden at the time, however and the gestapo did not think to look for him there. He eventually managed to gain permission to go to Switzerland but only after most of his possessions—including the gold Nobel medal—had been confiscated.

As the functioning of enzymes was sorted out, it became apparent that not all enzymes worked on their own. Sometimes important accompanying reagents were necessary—coenzymes—one well-known class of which is vitamins. Another group of highly specialized protein structures are hormones, chemical messengers of the body.

HORMONES

Produced by various glands and carried to target organs by the blood, hormones stimulate and regulate a variety of functions, including growth, digestion, and reproduction. The Japanese chemist Jokichi Takamine, working in the United States in the early years of the 1900s, isolated the first hormone, adrenaline. Insulin was isolated in 1921 by Canada's Frederick G. Banting and Charles H. Best, and its structure was determined, as we already learned, in 1954 by Frederick Sanger.

The Croatian chemist Leopold Ruzicka converted cholesterol into androsterone, a male sex hormone and a steroid (a derivative of a specific ring structure) in 1934, and he received the Nobel Prize for his work on ring structures. Adolf Butenandt was also named by the Nobel Prize committee for his work on steroid hormones, but as a German under Hitler, he had to wait until after the war to acknowledge the honor, though he declined the prize. (Hitler was angered that Carl von Ossietzky, a writer and concentration camp prisoner who had been tortured nearly to death, was awarded the Nobel Peace Prize in 1936. He would not allow any German to accept a Nobel Prize.) After the war Lewis H. Sarrett developed a 37-step process for synthesizing cortisone, a steroidal hormone useful in the treatment of inflammatory conditions. Percy Julian, an African-American biochemist and holder of 130 patents, improved on this method. But no steroidal hormone has had a greater impact on society than the one that became known simply as "The Pill."

THE PILL

Russell Marker

Russell Marker had an unusual career. He completed a dissertation in organic chemistry for a Ph.D. degree but did not receive the degree because he refused to take some required courses in physical chemistry. Though warned he would end up as a "urine analyst,"[4] he progressed through several industrial positions until in the 1930s, he received a research fellowship at Pennsylvania State University, supported by the pharmaceuticals company Parke-Davis, to do research on steroids. While at Penn State, Marker found a way of making progesterone, a female sex hormone, from a steroid that found in a certain type of wild Mexican yam. He was unable to drum up any support for his work, so he quit his job, went to Mexico, and collected the yams himself. He joined with a small laboratory in Mexico City to manufacture progesterone, naming the joint venture Syntex. Marker could not get along with his partners, left Syntex, and shortly thereafter abandoned chemical research. Syntex hired another chemist, George Rosenkranz, who continued Marker's work and also synthesized the male sex hormone testosterone from the same yam source. The venture became financially successful, and another chemist, Carl Djerassi, was hired to produce cortisone and estradiol, another female hormone. Though Searle Pharmaceuticals had already demonstrated a market for chemical contraceptives, this was not the original goal of the Syntex research. However in trying to make estradiol, Syntex chemists created an analog of progesterone, 19-norprogesterone, which was off one carbon unit from progesterone. One of the functions of progesterone is to inhibit ovulation (as during pregnancy), and it was found that 19-norprogesterone did this better than its natural counterpart. At first it could be administered only by injection, but chemists at Syntex found an oral version in short order, and "The Pill" was born.

With all this new information on cellular and body function came a new understanding of body dysfunction and ways of treating it. In this era medicines progressed from antiseptics, which act on microorganisms on the body surface, to antibiotics, which seek out microorganisms within the body. Initial progress on these materials was made by the same dye manufacturers who had provided aspirin and antisyphilis drugs.

Observing that sulfonamide derivatives (containing the SO_2NH group) were good dyes for silk—which meant that they had an affinity for protein molecules (silk is a protein)—the German dye chemist Gerhard Domagk found that one of these dyes, prontosil, showed strong activity against bacterial infections, and rats and rabbits seemed to be able to tolerate large doses of it. Here the stories diverge. Some have it that Domagk used it to treat his dying daughter, while others suggest that it was used to treat a dying 10-month-old baby boy. In any case the sulfonamide drugs, or sulfa drugs, thus discovered saved many subsequent daughters and sons. Domagk was also prohibited by Hitler from accepting his Nobel Prize.

In 1928 the British bacteriologist Alexander Fleming observed the antibiotic effect of the mold *Penicillium notatum*, and the active agent *penicillin* was isolated by the Australian pathologist Howard Florey and the German refugee biochemist Ernst Chain. The need to treat massive numbers of infected wounds in World War II led to a major development effort to produce penicillin in quantity. In 1964 the chemist Dorothy Hodgkin received a Nobel Prize for her X-ray determination of the structures of penicillin, as well as vitamin B_{12} and insulin.

Dorothy Mary Crowfoot Hodgkin

Hodgkin was born in Egypt, where her British archeologist father and botanist mother found themselves in 1910. She grew crystals in school at the age of 10, and she used a soil chemist's test kit, given to her by a friend of her father, to set up a home laboratory in the attic. In 1925 when she was 15, her mother gave her Bragg's book, *Concerning the Nature of Things*, and she read for the first time about the X-ray crystallographic technique. At 18 she rejoined her parents on an archeology dig in Palestine and almost chose archeology as a career, but she returned to Oxford and to chemistry and physics.

She received her doctorate, married (her husband was an historian), and was appointed university lecturer and demonstrator at Oxford. She worked on the X-ray determination of three-dimensional structures of biological molecules and in 1949 published a structure for penicillin. The structure of vitamin B_{12} required collecting data from several researchers over a 6-year period and the then-novel use of punch card computers. The structure of insulin was determined after 30 years of research.

Structures of virus crystals were also determined eventually by the X-ray diffraction technique, but at the beginning of this era no one really knew what a virus was. One of the first to tackle the question was Wendell Stanley, though he allowed the project to choose him and not the other way around.

Wendell Stanley

Stanley majored in chemistry and mathematics in college, but he also excelled in football and planned to become a coach. After visiting a university with another student and seeing the equipment and the possibilities, he decided on a career as a chemist. When in the 1930s Stanley and his wife were expecting their first child, Stanley became the chemist on a plant-virus project so that his family could move to a rural setting. His job was to isolate the virus, but no one could tell him if the thing he was to isolate was an organism, a macromolecule, or a chemical mixture. After two and a half years and processing a ton of tobacco leaves, he isolated a teaspoon of crystals of tobacco mosaic virus. He reported that this virus was a giant protein molecule that came to life when inserted in a living cell. The media picked this up as the missing link between animate and inanimate matter, and Stanley became a media hero. Then British scientists showed that Stanley's crystals were not so pure as first thought: There were nucleic acids as well as proteins in the crystals, which made the virus a living material because nucleic acids are the *essential* stuff of life.

NUCLEIC ACIDS

If life is the ability for self-replication, then it can be seen why nucleic acids are considered the essential stuff of life. Deoxyribonucleic acid (DNA) and ribonucleic acid (RNA) are the molecules in the cell that store and transfer genetic information during self-replication. When the cell divides, DNA first replicates, or copies, itself. Then in the new cell as in the old, DNA serves as the template for protein construction—a job that depends on the other nucleic acid, RNA.

Both DNA and RNA are polymers; their monomer units are called *nucleotides* and consist of three parts: a phosphate group, a pentose (5-carbon sugar) and a nitrogen-containing base. The DNA molecules vary in size, but they can have as many as 30,000 base pairs, and they are the largest of all natural molecules. We now know that the three-dimensional structure of DNA is a double helix: a form resembling a spiral staircase. The outside handrails are the sugar-phosphate backbones, and the inside steps are formed from pairs of bases joined by hydrogen bonds. The key to the unique three-dimensional structure of DNA is that only certain bases pair up, and this "base pairing" also helps explain the replicating capabilities of DNA.

In the early part of the 1900s however, the only thing known for sure about DNA was that it was a polymer. Pauling, whom we first met in conjunction with quantum mechanics, was interested in the three-dimensional structure of DNA, and he started working on the problem before World War II. But like everyone else he was distracted by war efforts, and he did not return to the problem until the late 1940s. At this point he took a piece of paper on which he had drawn a chain of linked amino acids and began playing with it, twisting it around to see what he could see. He saw that a coil-like configuration (later called the *alpha helix*) brought amino acids into a favorable position for hydrogen bonding.

Pauling then extended his ideas to DNA. In February 1953 he proposed a structure that contained three strands twisted around each other. This structure was not correct, but at the time Pauling was rather outspoken in his criticism of the U.S. government's position on nuclear weapons and their testing, and having recurrent passport problems, he was not able to attend all the international scientific meetings he might have otherwise attended. Specifically Pauling had not seen some interesting X-ray work on DNA that was being done in England by Rosalind Franklin and Maurice Wilkins. However James D. Watson, a postdoctoral student from the United States, had.

James Watson and Francis Crick

James D. Watson had received his Ph.D. from Indiana University in 1950, and he was studying at Cambridge University in England when he met and had profitable discussions with Francis Crick, a biophysicist. These discussions resulted in the April 1953 publication of a correct structure for DNA. And the story might end there if it were not for two statements: one of which denies the use of Rosalind Franklin's X-ray diffraction data—and a second that contradicts the first.

The first statement appears in the one-page communication announcing the Watson–Crick discovery:

> So far as we can tell, it is roughly compatible with the experimental data, but it must be regarded as unproved until it has been checked against more exact results. Some of these are given in the following communications. We were not aware of the details of the results presented there when we devised our structure, which rests mainly though not entirely on published experimental data and stereochemical arguments.[5]

One of the "following communications" contained the X-ray data of Rosalind Franklin.[6]

The second statement, from Maurice Wilkins, was recorded during an interview with Rosalind Franklin's biographer. Wilkins said

> Perhaps I should have asked Rosalind's permission [to show Watson the X-ray diffraction pattern], and I didn't. Things were very difficult. Some people have said that I was entirely wrong to do this without her permission, without consulting her, at least, and perhaps I was . . . If there had been anything like a normal situation here, I'd have asked her permission . . . I had this photograph, and there was a helix right on the picture, you couldn't miss it. I showed it to Jim, and I said, "Look, there's the helix, and that damned woman just won't see it."[7]

Rosalind Franklin

Franklin came from a wealthy, scholarly, and politically active family. She attended St. Paul's Girl's School in London and received an excellent education in chemistry and physics. She entered Cambridge in 1938 where she stayed to complete her graduate work on the physical chemistry of coal in 1945. She then secured work in Paris in a laboratory engaged in X-ray diffraction, a technique which at that time was only about 30 years old. Because of her facility with the technique, she was offered a 3-year research fellowship at King's College, London, in the newly formed biophysics department. She was charged with building up the X-ray facilities needed for DNA studies.

King's College did not have the same congenial atmosphere as Paris. There was formal segregation of the sexes in dining facilities and social areas, though not in the laboratories. There were no formal prohibitions against members of Franklin's Jewish faith, but there were few representatives. There also existed a well-documented and fully acknowledged personality conflict between Franklin and her coworker, Maurice Wilkins. Wilkins was quoted as saying that Franklin "took a very superior attitude from the beginning,"[8] but this same characteristic of Franklin's was described by the graduate student, Raymond Gosling, who did get along with her well, as "She didn't suffer fools gladly . . ."[9]

Wilkins and Gosling had already taken some X-ray data on DNA, but with Franklin's improved equipment and technique, she obtained improved data. In addition she found the conditions for converting the DNA crystal into another, hydrated, form suitable for X-ray analysis, which provided her with additional information. In November 1951 she presented her preliminary results in an informal talk. Watson attended the talk, but he thought that it was inconclusive. In her notes for this talk, despite Wilkins's claim that she downplayed a helical model, she wrote, "Conclusion: Big helix in several chains, phosphates on outside . . ."[10]

The irony seems to be that Franklin had publicly presented her results, but Watson (by his own admission)[11] had not paid attention. It was only when Wilkins surreptitiously showed him Franklin's X-ray data that he became convinced about the helix formation and "presumed that Rosy had hit it right in wanting the bases in the center and the backbone outside."[12]

Equally unfortunate was Watson's, Crick's, and Wilkins's reluctance to acknowledge Franklin's contribution. In their combined Nobel lectures there were 98 references, but not one direct reference to any of Franklin's papers. There is only one textual reference, and this by Wilkins. There is no doubt as to the originality of their insight and the uniqueness of the model they constructed. It would not have distracted at all from their efforts to have given credit to Franklin.

Watson wrote a book, *The Double Helix*, in which he may have revealed the underlying problem. His unflattering descriptions of Franklin grudgingly acknowledge her abilities as a chemist, but he also suggests that she ought to do something about her hair, and he concludes that "the best home for a feminist is in another person's lab."[13]

The biochemist Joseph S. Fruton commented

> On my first reading of *The Double Helix,* its most striking feature was, for me, its amorality. The unabashed account of Watson's striving for success during the 1950s would have been, in my opinion, less deplorable if, in his maturity, he had acknowledged more generously that what Crick and he had achieved owed much to the published work of other scientists.[14]

André Lwoff wrote

> His portrait of Rosalind Franklin is cruel. His remarks concerning the way she dresses and her lack of charm are quite unacceptable. At the very least the fact that all the work of Crick and Watson starts with Rosalind Franklin's X-ray pictures and that Jim has exploited Rosalind's results should have inclined him to indulgence.[15]

However it may be noted that such rivalries are not uncommon in competitive pursuits, and Franklin was perhaps already suffering from the symptoms of her final illness—malignancy of lymphoid tissue possibly brought on by X-ray exposure—whose onset is known to be prolonged and insidious and to have no specific symptoms other than a general feeling of not being entirely well.[16] Such a condition may have made her difficult to work with. She did refuse a formal collaboration, and she could have been playing her cards close to the chest.[17] But for whatever reasons—competition-based, gender-based, ethnic-based, or simply personality-based—the question of the extent of Franklin's contribution has been raised. But by the time *The Double Helix* was written and the Nobel Prize in physiology or medicine handed to Watson, Crick, and Wilkins, Franklin was dead. The Nobel Prize is not awarded posthumously, so it is not known how the Nobel committee would have weighed the case.

In 1953 Franklin continued her work, but she shifted her location to Birkbeck College in London. Aaron Klug, a young South African, collaborated with her for the next 5 years, and they produced some 17 publications on virus structure. Klug received a Nobel Prize in 1982 for this work. In 1956 Franklin took up the study of the polio virus. She was warned that this was dangerous because the virus was so highly infectious, but by then she knew she was dying of cancer and did not fear polio infection.

Nucleic acid research continued. The structure and function of RNA were sorted out and the "code" of nucleic acid sequences, which dictate the synthesis of each protein, deduced. While some scientists were watching the amazing machinery in the cell chug along, others were wondering why such a sequence started in the first place. In 1871 Charles Darwin had written in a letter to a friend, "But if (and oh what a big if) we could conceive [of] some warm little pond, with all sorts of ammonia and phosphoric salts. . . ."[18] In the 1950s a U.S. biochemist, Stanley Miller, not only conceived of such a pond, he actually made one.

ORIGINS OF LIFE

Speculations on the origins of life goes back at least to the earliest writing, and they are found in the religious tradition of many cultures. Darwin made an early scientific attempt at an explanation, and some of his fundamental assumptions were explored by the Russian scientist Aleksandr Oparin in the mid-1920s. In the 1950s Stanley Miller, a graduate student under Harold Urey (a chemist whose name we encounter again) passed an electric spark (simulating lightning) through a mixture of water, methane, hydrogen, and ammonia (simulating the atmosphere of the Earth billions of years ago) and made amino acids. Leslie Orgel froze a dilute water solution of hydrogen cyanide and ammonia—gases probably present in the atmosphere of primeval Earth. After several days the mixture produced adenine, one of the four bases in DNA. Another U.S. biochemist, Sidney Fox, produced a proteinlike polymer by subjecting a dry mixture of amino acids to moderate heat. Under certain conditions he found that these polymers formed microspheres in water solution, which in time would bud. But while these were proteinlike, they were not known proteins, and for all the speculations on the origin of life, these chemists were finding that the reproduction of life molecules was not going to be so simple as passing current through a soup. But it is a problem that promises to inspire and motivate research effort for many years to come.

So from Henry Ford's inspiration to mount an internal-combustion engine on four wheels came the inspiration for synthetic rubber, and from the synthesis of rubber came techniques for sorting out the stuff of life. But rubber was not the only need inspired by the automobile. Petroleum is required to make the engine run, and in the era of the world wars, many new methods were developed for locating, refining, and producing petroleum. By World War II, the Age of Oil, with its abundance of materials, provided grease for the gears of organic and inorganic chemistry, too. Progress in these fields is where our story takes us next.

chapter
SEVENTEEN

ca. 1914–1950: New Materials and Methods—Organic and Inorganic Chemistry Grow

Chemical synthesis is the art of taking known materials and manipulating them to produce new materials. When we last left our organic chemists, they were making impressive progress toward understanding structure and bonding in hydrocarbons, and they were using this understanding to develop methods for organic synthesis. When we last left our inorganic chemists, they were making impressive progress toward understanding structure and bonding in inorganic complexes, and they were using this understanding to develop methods for inorganic synthesis. This blossoming came in the wake of the advances in theory in the late 1800s: three-dimensional structure and valence. In the wake of the new bonding theories in the 1900s—quantum chemistry—the synthetic fields bloomed.

The boon of new analytical instrumentation before and during World War II—improved methods of X-ray crystallography, improved spectrometers and meters for measuring acid strength,[1] development of chromatography (a method for separating gas- or liquid-phase compounds based on solubility), and development of mass spectroscopy (a method for separating ions by mass), sometimes referred to as the Instrument Revolution—also contributed to the growth of these fields. Growth of the synthetic fields was also stimulated by a new availability of starting materials from an old friend—petroleum.

Petroleum deposits are considerably rarer than coal, but in the early 1900s there looked like enough to go around. The British Navy decided to switch its fleet from coal to petroleum, confident that petroleum supplies (mostly from the Middle East) would stay available. The immediate chemical beneficiaries from this bath of oil were the organic and inorganic chemists. The petroleum industry provided stimulus for long-range research as well as research for immediate industrial application. This research produced information and theory that benefited industrial production, which in turn promoted more long-range research. So just as rubber for the kaiser's car spawned new chemistry, so did fuel for the British Navy.

Germans, lacking petroleum supplies (the collapse of the Ottoman Empire cut off Germany's eastern oil supply), sought methods of adding hydrogen to coal to turn this solid carbon source into liquid gasoline. The German chemist Friedrich Bergius, who was familiar with the work of Ostwald and Haber, invented a process in the early 1910s for hydrogenating finely divided coal particles. The process involved high pressures, but Carl Bosch, the chemist who brought the high-pressure Haber ammonia process on line, got this one running, too. The Nobel committee usually recognizes fundamental rather than applied research, but they awarded the Nobel Prize in 1931 to Bosch and Friedrich Bergius for the development of industrial high-pressure processes. When Germany lost its industrial patents as a result of the Treaty of Versailles and its war factories were ordered destroyed, Bosch convinced the French to leave his ammonia factory alone in exchange for his help in building a similar plant in France. In 1935 Hitler decided to ignore the Treaty of Versailles and put Hermann Goering in charge of national rearmament. Aiming for rubber and oil self-sufficiency, he enlisted the cooperation of I. G. Farben and Bosch, now at the head of I. G. Farben. Bosch had difficulty with Hitler's anti-Semitic policies, in particular Haber's dismissal, but he, like others, hoped to weather the storm and keep intact the country and company to which he had devoted his life. He died in 1940, not having seen the worst. Under the National Socialists, I. G. Farben worked closely with Hitler's government and used forced labor from concentration camps.[2]

The Germans Franz Fischer and Hans Tropsch, who were coal chemists for the Kaiser Institute, made gasoline in the 1920s with a carbon monoxide and hydrogen mixture called *water gas*, generated by passing air, then steam, over red-hot coal. They used an iron oxide catalyst at elevated temperatures and moderate pressures. South Africa, a country remote from petroleum sources, currently makes gasoline by a modified Fischer–Tropsch process. With Fischer's work German gasoline supplies were secure, though by the end of World War II, Fischer was not allowed to own a car. His daughter and son-in-law were interned at Dachau concentration camp, and his home was bombed.[3]

The United States however enjoyed a steady supply of natural petroleum. Natural petroleum is a mixture of hydrocarbons, the majority of which are straight-chain and singly bonded (saturated), but by the 1920s a Chicago chemist, George Curme, developed electric-arc and thermal methods for cracking petroleum to produce ethylene, propylene, and other doubly bonded (unsaturated) hydrocarbons. These reactive, doubly bonded hydrocarbons were then used to produce ethylene oxide, ethylene glycol, dichloroethane, ethyl alcohol, and other materials for use in the syntheses of drugs, dyes, and polymers, and for a multitude of other industrial applications. Armed with this plenitude of starting material, chemists went on to see what else they could produce.

The art of synthetic chemistry has been compared to a game of chess. The synthetic chemist must know the reactivities of all the compounds involved and be familiar with the synthetic routes used by others just as a good chess player knows the moves of each piece and is familiar with past games. The best synthetic chemists of course are those who come up with gambits of their own. This game is played by organic and inorganic chemist alike, but we focus on the organic chemists first.

1914–1950: ORGANIC CHEMISTRY

By the mid-1930s the American Chemical Society's *Journal of Organic Chemistry* had been founded, and it was routinely filled with reports of new reaction pathways or novel applications of ones already known. But for all the accumulation of knowledge on synthetic alternatives, there was still a need for underlying theory to explain the observed reactivities. Before one could be found, work had to be done to elucidate structures—three-dimensional and electronic—of the reacting molecules.

Conformation

A large step forward was made when Le Bel and van't Hoff convinced chemists that there was a real three-dimensional structure to molecules and this three-dimensional structure could be used to explain a physical behavior—the rotation of plane-polarized light. But pointing out that there was a three-dimensional structure was not the same as coming up with the three-dimensional structure, and there were a lot of molecules out there whose structures had not been determined. In the process of studying three-dimensional structures, chemists came up with another pearl: Not only does three-dimensional structure influence *physical properties*—the properties that a substance has on its own sitting on the shelf—but three-dimensional structures also influence *chemical properties*—properties that a substance displays when you take it down from the shelf and allow it to react with something else. It was this observation that ultimately shed light on the three-dimensional structures of singly bonded carbon rings.

Baeyer, a leading German organic chemist whom we met in conjunction with his synthesis of indigo, addressed the problem in the late 1800s. Baeyer convinced himself with models that five- or six-membered hydrocarbon rings—with bonds directed to the four corners of a tetrahedron—would be in their most natural conformation when flat. Based on Baeyer's authority, it was then assumed that all rings had to be flat. But smaller or larger rings would have very strained bond angles if forced flat, so it was assumed that these rings would be unstable. Baeyer's *strain theory*, as it was called, did have some validity for small, three- or four-membered rings, but it ran into problems when applied to *all* rings. One of the first twists in the theory came from the work of Leopold Ruzicka.

Leopold Ruzicka

Born in Croatia, Ruzicka originally planned on becoming a priest, but then he became interested in science and math and traveled to Germany to receive an education in these areas. He came to work for Staudinger, but he lost his apprenticeship when he asked to work on some of his own ideas. This problem with the influential Staudinger hampered his career somewhat, but he eventually obtained a position as lecturer at the University of Zurich. Poorly paid, he was glad for the chance to work for a Geneva perfume factory on the side. There he did research on perfume essences and bases extracted from animal and plant sources: civetone, muscone, irone, and jasmone. With this work he eventually won the Nobel Prize (though World War II interfered with his trip to Stockholm to receive it), and he also cast doubt on the ring theory of Baeyer. Baeyer said that any ring larger than six carbons would experience strain and have a fleeting existence at best, but Ruzicka's compounds had rings of 15 and 17 carbons, and they were stable.

Ruzicka explained that the rings presented no anomaly because they could be thought of as two parallel carbon chains joined at either end. But other chemists said that even smaller rings could avoid ring strain if the requirement for flatness were discarded. In particular Herman Sachse in the late 1910s proposed that a six-membered ring (cyclohexane) could assume a boat shape in which opposite ends of the ring point up or a chair shape in which one end points up and the other points down (see Fig. 17.1). These different arrangements are called ring *conformations*, and the question remained which if either of these conformations cyclohexane assumed. This question was addressed by the Norwegian chemist Odd Hassel.

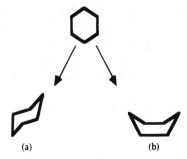

Figure 17.1. A six-member carbon ring (cyclohexane) in (a) chair conformation and (b) boat conformation.

Odd Hassel

Hassel received his doctorate from the University of Berlin but returned to Norway to teach. There he headed a physical chemistry program and developed an X-ray crystallography facility on a shoestring budget. He also assembled equipment for studying the diffraction of electrons by gas-phase molecules, a technique complementary to X-ray diffraction on solids, which gives information about molecular shapes. He used this technique over the next several years to show experimentally that the chair conformation was preferred by six-membered carbon rings. When Germany occupied Norway in World War II, he stopped publishing in German journals, though he did continue to publish in small, little-known Norwegian journals. The Germans eventually closed his university and interned Hassel and several other faculty members, though he was released by 1944. Shortly thereafter an early-bird British chemist introduced Hassel's work, with some wrinkles of his own, to the world.

Derek Harold Richard Barton

Derek Barton habitually gets up at 3:00 A.M. to read, keeping up with some 15 journals. His work day extends from 7:00 A.M. to 7:00 P.M., though he does break for lunch. His son, William Barton, runs a small business making custom tow bars in his garage, a profession reportedly chosen to avoid a life such as his father's. The father also rebelled as a youth, shunning the family trade of carpentry because "after two years of doing my share of manual labor in the wood business [I] felt there must be something more interesting in life."[4]

Barton obtained his doctorate during World War II, then in 1949, serving as a sabbatical replacement at Harvard for a year, he happened to hear a talk by Louis Fieser.[5] Fieser had written a book with his wife, Mary, on steroid chemistry, and during the lecture Fieser listed several unsolved problems in steroid reactivity.

Steroids constitute an important class of naturally occurring compounds that fill many important biological roles: Cholesterol is a steroid, and sex hormones are steroids. People were—and still are—interested in the synthesis of steroids for medical use (the introduction of the culture-altering birth control pill hinged on a steroid synthesis). At this time however they were being confounded by some unexplained chemical behavior of steroids. During Fieser's lecture Barton, aware of Hassel's work because of his reading habits, realized he could explain the anomalous reactivity of steroids.

Steroids are built from hydrocarbon rings joined at the edges. If as had been previously assumed, rings were flat, then all positions on the ring should be equal and no one position should have a special reactivity. If on the other hand rings were in a chair or boat conformation, some positions would be more accessible than others and some angles of attack more successful. Barton showed that if steroids were assumed to be in the chair conformation, as Hassel had found for cyclohexane, their preferred reactivity could be explained. He wrote up his views in a short paper, and 20 years later his work resulted in a Nobel Prize that he shared with Hassel.

After he left Harvard, Barton went to Birkbeck College, the only night school at the University of London. His reasons, he explained, were that "One could carry out research all day and teach from 6:00 to 9:00 P.M. This system was excellent for research, but it was not appreciated very much by wives!"[6] (Rosalind Franklin also moved to Birkbeck College because "Birkbeck College has only part-time evening students, and consequently they really want to learn and to work. And they seem to collect a large proportion of foreigners on the staff which is a good sign. King's [College] has neither foreigners nor Jews."⁷) Barton now teaches at Texas A&M University in the United States and continues to make advances in chemistry. He is currently involved in phosphorus chemistry, having come full cycle from the six-membered ring.

Another carbon ring compound to have its three-dimensional structure confirmed in this era was benzene: a compound always present in small but aromatic amounts in petroleum—a fact appreciated by anyone who has filled up a car with gas and then ended up smelling like gasoline. Though the ground work had been laid down by Kekulé and his visions of hoop snakes, benzene was still not completely understood in the early 1900s. While it was generally agreed that benzene was flat, there was no direct proof until the work of Kathleen Lonsdale, one of the first two women to be elected to the Royal Society and a peace activist who spent a month in prison for refusing civilian wartime duty. Lonsdale solved the structure of hexamethylbenzene by X-ray crystallography and showed that it was both flat—and symmetrical. Kekulé's structure for benzene predicted that it would be flat—so no surprises there—but when Lonsdale showed that the ring was completely symmetrical, too, she showed that there were not three distinct double bonds in benzene, but bonding electrons were evenly distributed over the ring (see Fig. 17.2). Christopher Kelk Ingold, an important organic chemist, said of Lonsdale's work: "one paper like this brings more certainty into organic chemistry than generations of activity by us professionals."[8]

ELECTRON THEORY OF ORGANIC MECHANISM

The problem that professionals had been working on was the reactivity of benzene derivatives. There was a wealth of experimental data that showed different groups tended to substitute at specific positions on benzene's six-membered ring depending on what other substituents were on the ring, but there was very little solid rationale for this behavior. A theory had been advanced that carbon positions around the benzene ring alternated in polarity, but there was little justification for this assumption, and benzene did not display any other polar behav-

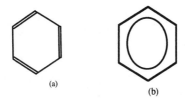

Figure 17.2. (a) Benzene as a Kekulé visualized it: a six-member ring with three double bonds; (b) benzene as Lonsdale's X-ray diffraction showed it to be: a six-member ring with its electron density spread evenly over the ring.

ior. Then in the 1920s the shared-electron-pair theory of G. N. Lewis was used to explain reactivity much more successfully. Instrumental in the application of the Lewis theory were Robert Robinson and Christopher Ingold. They were very similar people in that they were both English, both theoretical organic chemists, and both collaborated with their wives. They were also complete and bitter opponents.

Robert Robinson and Christopher Ingold

In a benzene ring there are enough electrons for six single bonds between the six carbon centers and three fixed double bonds, but Lonsdale's work had shown that the electrons were evenly distributed around the ring. Now Robinson suggested that these electrons could also have mobility on the ring and be able to go wherever needed. Robinson (following ideas of Arthur Lapworth and other workers) used arrows to indicate the movement of electron pairs the way football coaches diagram plays on a chalkboard or generals plan the movement of troops (see Fig. 17.3). This device proved to be as valuable for understanding organic reactions as Lewis dot structures were for understanding molecular bonding. With this device Robinson explained observed reactions of benzene by appropriate movements of electrons around the ring.

Initially Ingold argued against the theory and even reported some experimental results that apparently conflicted with it. Robinson and

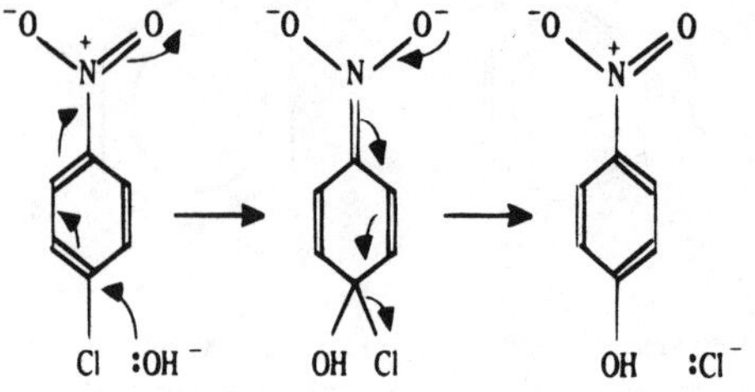

Figure 17.3. Robinson showed how benzene reactivity could be understood in terms of mobile electrons moving around the benzene ring. He used arrows to show electron movements.

others showed Ingold's experimental results to be in error, and in the process they clarified their own theory. Not seeing that a graceful retreat was in order, Ingold persisted, finally coming up with his own theory for benzene reactivity. Ingold's theory also invoked an electron shift and looked too familiar to Robinson. Robinson accused Ingold of plagiarism, and in fact it was later assumed by many that the ideas had originated with Ingold, and Robinson's contribution was overlooked. Robinson was awarded the Nobel Prize in 1947, but it was for his work in natural product synthesis, not the electron theory of organic reactions. Ingold was at one point nominated for the prize, but he did not receive it. The quarrel between them was never resolved.

Robinson made other major contributions to the systemization of organic synthesis that may have arisen from his original attraction to mathematics, "I wanted therefore to be a mathematician, but my father's wish was clearly expressed, and I decided to accept the inevitable and become a chemist."[9] (His father operated a bleaching works and needed a chemist more than a mathematician.) However Robinson brought his passion for mathematical logic to bear on complex organic syntheses (he included in this a passion for Gertrude Maude Walsh, a research student of Chaim Weizmann at Manchester, who became his wife and lifelong collaborator), and he became "fascinated by the beauty of the organic chemical system."[10] This systemization had such appeal that other chemists sought to expand it by coming up with explanations for reactivity that predicted chemical reactions. An attempt in this direction was made by Pauling and George Wheland: They said that the relative reactivity of different molecules could be explained by the stability of the reaction complex formed on the way to product. Other workers tried other approaches, and by the early 1950s the proliferation of reactivity scales threatened to get out of hand. Then in 1952 Kenichi Fukui of the University of Kyoto, building on the work of Mulliken and Hund, published his own reactivity index that said that the reactivity of the molecule as a whole was dictated by the shape of the highest energy-occupied molecular orbital: the *frontier orbital theory of reactions*.

MOLECULAR ORBITAL THEORY

Kenichi Fukui

Molecular orbitals can be pictured as "clouds" of electron density that have a defined shape and extent. Nested like layers of an onion (though like an onion they sometimes overlap or are convoluted), the shape of these molecular orbitals can be calculated and pictures of them can be drawn.[11] Fukui's insight was that orbitals that extend furthest from the nuclei—the top layer of the onion, as it were—should be the most important in reactions. This frontier orbital theory of reactions had intuitive appeal and predictive power: The highest energy occupied molecular orbital of one reagent could be drawn on a blackboard, and the overlap of this orbital with the lowest unoccupied reagent orbitals could be pictured. When R. B. Woodward and Roald Hoffmann used the symmetry of the highest occupied molecular orbital to explain a mystifying collection of experimentally observed reactions that had heretofore gone unexplained, the frontier orbital theory opened new frontiers.

Robert Burns Woodward and Roald Hoffmann

In 1965 Woodward received the Nobel Prize, and a review of his work leaves little doubt why. Needing something to talk about at his Nobel lecture, he stepped up work on the synthesis of a cephalosporin antibiotic and finished it on time. He was a product of the Boston school system and a single-parent family (his father died when he was two); after skipping several grades on his way through school, he enrolled at MIT when he was 16. He eschewed classes, preferring the laboratory and the library, and showed up only for exams. Course work however was required, so he was asked to leave after a year and a half. When he was allowed back, he took 2 years' worth of courses in 1 year and graduated at 19.

Woodward stayed at MIT for his Ph.D. work and received it after a single year. He worked for a summer at the University of Illinois, then took a position at Harvard, where he became full professor at the age of 33. He was allowed to avoid undergraduate teaching and take only research students, and he remained at Harvard, doing research, for the rest of his life. A bit ostentatious, he always wore a dark blue suit with a light blue tie, drove a blue Mercedes Benz sedan and had his office painted blue. (His graduate students painted his parking space blue and made him a blue sedan chair on his sixty-first birthday.) He delivered four-hour and five-hour lectures, drawing precise chemical structures on a chalkboard, and he always ended the lecture just as he reached the last available inch of space. He sometimes worked 16 hours a day, not caring for vacations.

Woodward had an amazing knowledge of the literature, and he used his repertoire of synthetic tools to synthesize (among others) quinine, cholesterol, cortisone, lysergic acid (the core of the notorious drug LSD), strychnine, reserpine, chlorophyll, tetracyclines, and vitamin B_{12}. His Nobel Prize was for "contributions to the *art* of organic synthesis"[12] (emphasis ours).

In the course of one particular synthesis, Woodward noticed that a thermally induced reaction produced only one of two possible products. He discussed the problem with Roald Hoffmann, a Jewish refugee from Poland currently at Harvard completing some work in theoretical chemistry (he gave up experimental chemistry after one experiment sprayed purple dye on the walls of a new laboratory). Together they came up with the Woodward–Hoffmann rules, which use the symmetry of frontier molecular orbitals to predict the outcome of certain photochemical and thermal reactions. Not only did the Woodward–Hoffmann rules explain a backlog of previously unexplained results, they suggested new reaction routes. This work contributed to Hoffmann's winning of the Nobel Prize, and it probably would have earned Woodward a second if he had not died at the age of 62.

So we see in the work of Hoffmann and Woodward the extent to which organic chemists could now choreograph reactions—and to cap our discussion of synthetic organic chemistry in this period, we now discuss a crown. One of the most interesting manifestations of all the progress that was made in organic chemistry during this period was the discovery by Charles J. Pedersen of the crown ethers.

Crown Ethers

Crown ethers are so named because they are ethers—compounds containing a carbon–oxygen–carbon bond—and formed in a zigzag circle like a crown, with carbon on the zigs and pearls of oxygen on the zags. The discoverer of these materials was a pearl in his own right. In his own words:

> Imagine this sequence of events, ca. 1900: An engineer in Norway decides to go halfway around the world to Korea, where he works in a gold mine. A Japanese family, having suffered some financial reverses in Japan, decides to move to Korea, where markets are opening up. The brother starts a business close to the mine. The sister meets the young Norwegian; they marry. Some years later, their son travels to the United States for his education. He becomes a chemist and wins the Nobel Prize.[13]

Charles J. Pedersen

Maybe it was because of the unconventionality of his beginnings that Pedersen did not allow himself to be restricted by convention. He did not earn a Ph.D., but stopped at a master's degree because he did not want to further burden his family financially. He became an industrial chemist for DuPont but stayed in research, which was unusual for someone with no doctorate. He published his Nobel Prize winning research when he was 63, then retired 2 years later. He waited 18 years to receive the Nobel Prize when he was 83.

The novelty and utility of his crown ethers are that they can form complexes with metal ions: The metal ion sits in the hole at the center of the crown, and to some extent the crown ethers can be customized to bind different metals selectively. Because of their structural relationship to biological molecules that bind metal ions, crown ethers are being used to model these biological systems. Crown ethers have also been used to carry metal ions into organic solvents—where they would normally not dissolve—opening up new routes to syntheses that would otherwise have been difficult or impossible. These ethers can also be used to carry metals *out* of organic solvents, and in fact that is how they were found. We said at the beginning of this chapter that petroleum research was the stimulus for synthetic advances: Pedersen found crown ethers when he was doing research on ways of cleaning metals from—no surprise—gasoline.

But as Pedersen's work on crown ethers points out, there is more to petroleum chemistry than organic chemistry. Products from the soil and products from living or used-to-be living material contain associated metals. Metals are the bailiwick of the inorganic chemist, and petroleum chemistry provided a stimulus in this area, too.

1914–1950: INORGANIC CHEMISTRY

One of the first ties of inorganic chemistry to the petroleum industry was the discovery by Thomas Midgley that tetraethyllead aided the smooth combustion of gasoline and helped prevent engine knock caused by premature combustion. Midgley also discovered that dichlorodifluoromethane, later called Freon, made an efficient nontoxic refrigerant, making him a two-time industrial hero or a two-time environmental villain, depending on your point of view. The success of new bonding theories in explaining structures and spectra of inorganic complexes led to a resurgence of interest in their formation and reactions. There had been some lag in activity after Werner received his Nobel Prize in 1913, a gap of some 60 years before another Nobel Prize in the inorganic field was awarded. When it was, it went to Geoffrey Wilkinson and Ernst Otto Fischer for their work on ferrocene, the first "sandwich" compound—a metal atom between two parallel hydrocarbon rings.

Ferrocene

Wilkinson was an English chemist who had worked on the Canadian atomic bomb project during World War II. After World War II, in the 1950s, he was an assistant professor at Harvard, teaching inorganic chemistry. One afternoon while catching up on his reading in the chemical literature, he came across an article that described a new and very robust inorganic compound that had iron as its central metal and two five-membered hydrocarbon rings attached as ligands. The structure was given as linear, but Wilkinson saw that this structure did not account for the compound's reported stability. He redrew the structure with the iron nucleus sandwiched between the two rings, and he saw that this allowed an overlap between double-bond orbitals on the ring and the d orbitals of the iron. There had been speculation about this type of bonding, but there had been no examples until now. It turned out that Wilkinson's friend (and ours), Woodward, had also seen the paper and had come to the same conclusion, so they agreed to collaborate on the research. The name ferrocene was proposed by a graduate student, Mark Whiting, and it was meant to indicate the special stability of these compounds, similar to the special stability of benzene.

Sandwich compounds are of fundamental interest to chemists because of their unique bonding: Metal ion orbitals bond with hydrocarbon orbitals spread over the hydrocarbon ring. One interesting result of this type of bonding is a rapid intramolecular rearrangement first described as "ring whizzing."[14] This type of behavior is now known as *fluxional*, and it was first observed in ferrocene. Subsequently a tungsten compound with four tungsten nuclei in a ring joined by double and single bonds has been shown to undergo a rapid shifting in the double and single bond. This fluxional motion has come to be known as the Bloomington shuffle, because it was first described in Bloomington, Indiana.[15]

This tungsten molecule however contains no carbon, which points up that not all inorganic chemistry is organometallic chemistry; that is, it does not necessarily involve carbon. In fact there is lots of inorganic chemistry that does not even involve metals. There is a whole group of nonmetals that form the block of elements on the right of the Periodic Table—called the *main group* elements—that are all nonmetals (except for those in and to the left of the stair step formed by aluminum, germanium, antimony, and polonium). They, too, have an interesting chemistry that is still being explored.

For instance boron, having the same orbitals available as carbon but one less electron to contribute to bonds, has been shown to form molecules in which two boron centers are joined by a hydrogen bridge. This violates the single-bond behavior that hydrogen shows in carbon compounds (and can come as a bit of a shock to students who have had this rule drummed into their heads). But these interesting compounds perform all kinds of tricks, including the formation of inorganic polymers, which have also been found with boron–phosphorus, aluminum–nitrogen, boron–nitrogen, silicon–phosphorus, and silicon–oxygen linkages. Inorganic chemists have also made inroads in the chemistry of the noble gas elements. These elements were once believed to be completely inert, but in the early 1960s Neil Bartlett, studying the properties of platinum hexafluoride, an extremely powerful oxidizing agent, was able to oxidize the noble gas xenon to form a yellow, solid platinum fluoride derivative of xenon. Subsequently krypton and possibly radon have been shown to have a halide chemistry, too. But while it cannot be denied that many of these nonmetal materials have potential practical and theoretical interest, the big winners over all are still the metal complexes. One reason for this is their role in biological materials. Metal ligand systems are found in vitamins and other coenzymes, hemoglobin, and such important chemotherapeutic agents as cisplatin, an anticancer drug. Henry Taube was awarded the Nobel Prize in 1983 for his research on reaction mechanisms in inorganic reactions, leading what has been called a revival in physical inorganic chemistry.

So it seems the field of inorganic chemistry has built up a bit since tetraethyllead was invented as an antiknock agent for gasoline. In fact whole economies, industries, and even cultures have been built around the use of petroleum. This process continued smoothly after the Second World War until the first shock waves began hitting in the 1970s. An oil cartel was established among the Middle Eastern oil-producing countries, and the price shot up from 3 dollars to 40 dollars a barrel. Along with rapidly rising cost came the realization that the supply of oil on Earth is not infinite and oil reserves could run out in the foreseeable future.

But luckily oil is not the only organic feedstock, and chemists are already considering alternatives. Luckily, too, some of the ground work in this area has already been done. It was done in the early 1900s by George Washington Carver.

BIOMASS

George Washington Carver

Born of slave parents in Missouri in 1864, Carver and his mother were kidnapped by night riders before he was 2 months old. He was ransomed by his original owner for a horse, but all trace of his mother was lost. He left the farm where he was born when he was about 10 years old and went to Kansas where he was able to work his way through school. He graduated at the age of 30 from Iowa State College of Agriculture and Mechanical Arts. He joined the faculty there and continued his research in systematic botany. He became director of the Department of Agricultural Research at Tuskegee Normal and Industrial Institute (now Tuskegee Institute), where he developed nearly 300 products from peanuts, including ink, cosmetics, dyes, soaps, oils, and substitutes for flour, butter, cheese, and coffee. Peanut farming in the southern United States became so profitable that a peanut farm later financed the career of a president of the United States. Carver also developed shoe blacking, library paste, vinegar, starch, candy, and more than a hundred other useful substances from the sweet potato. He formed synthetic marble from wood shavings and extracted dyes from tomato vines, beans, dandelions, onions, trees, and clays. Carver's materials were what is know as *biomass*, a source of organic material. When it becomes economically practical to do so—when petroleum becomes sufficiently scarce—we may find ourselves digging out old Carver processes and giving them a second look.

So petroleum continues to serve as an inspiration to synthetic chemistry—down to the synthesis of its own replacement—and this fits with the pattern we have seen throughout: Much of the progress in organic and inorganic chemistry during this period found its impetus in the petroleum industry, but the desire to understand the fundamental nature and reactivities of materials—the theory of chemistry—also fed into the loop of improved experimental results, improved product, back to improved theory. There is in fact a whole breed of chemists who choose to focus on the theory of chemistry—the physics of chemistry—and members of this bidisciplinary breed are aptly named physical chemists. Areas that fall within the bailiwick of the physical chemist include some we have already encountered—thermodynamics, quantum mechanics, and statistical mechanics. But while thermodynamics predicts whether a reaction will occur and quantum mechanics and statistical mechanics help explain why, nothing said so far has addressed *how fast* the reaction will occur. The importance of the speed of a reaction—the reaction rate—can be understood with a single example: Graphite is a more thermodynamically stable form of carbon than diamond, so diamonds should eventually turn into graphite. But if diamonds routinely turned into graphite (at a reasonable rate), then several national economies would be altered (and no doubt several engagements to be married would be called off). The importance of reaction rate is reaffirmed by a second example: Under controlled conditions, such as in an electrochemical cell, the hydrogen–oxygen reaction to form water takes place at a controlled rate, and its energy can be harnessed; by contrast the explosion of the space shuttle *Challenger* was an uncontrolled hydrogen–oxygen reaction. The study of reaction rates is called *chemical kinetics*, and we examine the development of this area of physical chemistry next.

chapter
EIGHTEEN

ca. 1914–1950: Chemical Kinetics— Boom or Bust

Although the study of chemical kinetics has the bulk of its lifetime in the 1900s, its history goes back much further. In the fourth edition of his classic text, *Physical Chemistry*, Walter J. Moore opened his discussion of reaction rates with a 1660 quotation from the chemist Daniel Sennert:

> the Elements, unless they be altered, cannot constitute mixt bodies; . . . [but] they [cannot] be altered unless they act and suffer one from another . . . they [cannot] act and suffer unless they touch one another, [therefore] we must first speak a little concerning contact or mutual touching, Action, Passion and Reaction.[1]

The quotation is appropriate. By the turn of the century a lot was understood about the tendency for chemical reactions to occur, the types of chemical bonds that formed when they occurred, and the properties of the materials created, but not much was known about Sennert's passion—about *how* these chemical reactions occurred—at the moment of conception, as it were. The difficulty was, and is, that we cannot witness a single chemical reaction at the molecular level. The best we can do is see the results of *many* chemical reactions (the *bulk*) and from this bulk behavior attempt to divine what happened on the molecule-to-molecule level. As it turns out, one of the best handles we have on this problem is our ability to measure rates of chemical reactions, that is, the speed at which they take place—the subject of chemical kinetics.

The logistics of how a reaction occurs—how molecules come together, form and break bonds, and move apart—is called the *mechanism* of the reaction, and this is the mating that is so difficult to see. The time it takes for the reaction to occur however provides some clues. For instance if a reaction requires only two bodies to come together, it has one rate. If it requires three bodies to come together (reaction by committee), it has another rate—usually a slower one. In kinetics, like kissing, approach is as important as speed: If reacting bodies have to come together with a specific orientation (head-to-head or side-on), that alters the rate. If the reactants must come together with some minimum amount of energy to effect the desired reaction (such as wine glasses clinked for a toast must have a minimum energy to shatter), then the energy level affects the rate, too. So measuring the bulk rate can impart a lot of information about what is happening on the molecular level, but it cannot tell all. It must be remembered that the bulk rate helps suggest *possibilities*—and perhaps eliminate a few—but it never *proves* a specific mechanism. For that we would have to shrink to the molecular level and ride along.

But the possibilities have always been enough to intrigue chemists, and Thenard in the early 1800s was already attempting to measure the rate of a chemical reaction. The measurements made by Ludwig Wilhelmy around 1850 were more quantitative: The quantity that he measured was the ability of sugar to rotate polarized light.

LUDWIG FERDINAND WILHELMY

It was known that a solution of cane sugar in the presence of acids slowly turns into a mixture of glucose and fructose. Wilhelmy, born in what is now Poland, found that the initial rate of the reactions was proportional to concentrations of both the sugar and the acid.

Looking at the *initial* rate was an important simplification. A chemical reaction normally begins with all reactants—no products—so it takes off at a good rate. But like a ball rolling down hill, it gradually slows as the hill becomes less steep. In a chemical reaction the buildup of products slows down the forward reaction because the presence of products means the presence of the reverse reaction (recall Berthollet's observation on the shores of Lake Natron: If A and B form C and D, then C and D can form A and B). But because the initial concentration of reagents is so high, the initial rate is insensitive to the fact that a small amount of reagent is used up. And because the initial concentration of products is so low, the initial rate is insensitive to the small buildup of products. So the initial rate of the reaction is the best measure of the forward reaction rate.

Wilhelmy found a mathematical expression for the rate, then he retired from chemical research. There his work sat for some 30 years until it was noticed and expounded by Ostwald. But there were others interested in the rate of chemical reactions, and significant progress was made around 1865 by the team of Harcourt and Esson.

Augustus George Vernon Harcourt and William Esson

Harcourt, an English chemist, was Reader in Chemistry at Oxford University. He was a very competent experimentalist but knew little mathematics. Because rate equations are differential equations—slope equations—and a topic for treatment with calculus, he enlisted the aid of the mathematician Esson. Together they came up with methods for interpreting reaction rates that are essentially the same as those used today.

The results of Harcourt and Esson were complemented by the work of van't Hoff, who wrote an important treatise on chemical kinetics in the 1880s. In addition to discussing experimental rates, van't Hoff proposed several equations showing that these rates depend on temperature. One of these equations, adopted and expanded by Arrhenius, is now commonly known as the Arrhenius equation. In this equation the rate goes up exponentially with temperature, and this is indeed the behavior almost universally seen in chemical reactions. Intuitively this can be understood by considering our wine glasses again. If shattering represents a successful reaction, and temperature increase represents an increase in energy available to those offering the toast, then the more energy with which the wine glasses collide, the more likely the reaction. (Also with more energy, celebrants are more likely to toast more often, thereby causing more successful reactions per unit time.)

With the increased clarity of treatment and approach, the study of kinetics became more attractive. One of the first to dedicate his efforts exclusively to the study of kinetics was the German chemist Max Bodenstein, who had a passion for mountain climbing as well as chemistry.

Max Bodenstein

A workshop in the basement of his father's brewery provided Bodenstein's first exposure to chemistry. His first exposure to climbing was on the roof of the brewery. His first exposure to gas-phase chemical kinetics was in his doctoral work with Victor Meyer, but he found kinetics to be an acquired taste. Believing the study rather dull at first, he gradually became caught up in the construction of experimental equipment—an outlet for his considerable technical skills. He became more interested when his results were pleasingly explainable in terms of the kinetic theory of gases. Bodenstein continued his independent work in this area, eventually becoming director of the Institute for Physical Chemistry in Berlin and succeeding Nernst. There he was reportedly a rugged taskmaster. He believed that students should learn to assemble their own equipment, so when he found one student with a glass manifold made by a technician, he smashed it.

Bodenstein concentrated his research on gas-phase reactions for good reason: In the gas phase, particles are far apart, so there are fewer interactions to account for; in solutions the situation becomes much murkier. Although some of his interpretations were later shown to be incomplete, no one has challenged his experimental work. One of the first reactions he investigated was between hydrogen gas and iodine gas, both diatomic molecules. This reaction appeared to follow simple, single-collision kinetics. He then moved on to hydrogen and bromine, an almost identical system, probably expecting to see the same thing. He obtained instead a complicated rate law that was not interpreted for several more years. When it was explained, the explanation also covered the similar complex reaction between hydrogen and chlorine. The problem (or solution) was the formation of *radicals*: species with an odd number of electrons, an unstable and highly reactive breed. A radical, being a highly reactive species, can react to form more radicals in a chain reaction, thereby accelerating the rate way beyond what would be expected for simple, single-collision kinetics.

However Bodenstein's results, pleasing as they were, did not resolve a big question in chemical kinetics of the day: unimolecular reactions. In a unimolecular reaction a molecule decomposes or rearranges seemingly spontaneously and seemingly without another reagent. This behavior puzzled researchers for some time, giving rise to what could be considered the physical chemist's phlogiston—a theory that *seems* to explain the observations and *seems* to have an experimental basis but is not quite right. For the kineticists this was the *radiation hypothesis.*

The radiation hypothesis stated that the excess energy in dissociating molecules comes from absorbed infrared radiation. Though authorship of the theory is unclear, it had avid proponents. William Cudmore McCullach Lewis, one of the first to apply quantum mechanics and statistical mechanics to reaction rates (and certainly nobody's fool) believed that the distribution of energies in a gas depended on the absorption of infrared radiation. He developed an expression for the rate of reaction between hydrogen and iodine, based on the kinetic theory of gases, that agreed well with Bodenstein's results. Jean Baptiste Perrin, one of the first to interpret Brownian motion correctly, adhered to the radiation hypothesis. In the 1920s Richard Chase Tolman did experimental work to test the radiation hypothesis. Although he did not find conclusive evidence for the hypothesis, he continued to support it. He may have had some justification: Infrared radiation can heat samples—the amount of heating depends on the intensity of the source—and heat *does* increase reaction rates. In the early 1920s however another explanation was offered, and this one was verified experimentally. Frederick Alexander Lindemann suggested that unimolecular decompositions were collisionally induced; that is, a molecule gained excess energy from collisions, even with its own species, and therefore the rate of its reaction appeared not to depend on the concentration of a second reagent.

The investigation of reaction rates and the factors that influence them has its practical application in the design of industrial syntheses, and they also hold interest for the theoretician trying to comprehend the reaction on the molecular level. Because reaction rates are the average result of billions of encounters, it can be imagined that information about individual encounters could be extracted if they could be deconvoluted. Deconvolution did not however turn out to be simple. For although chemists were able to measure the final, bulk result, they soon found that underlying this were many fleeting conformations and species—reaction intermediates—that were not easily pinned down and measured. However progress in understanding reaction intermediates was made in the 1930s by such workers as Henry Eyring.

Henry Eyring

Born in Colonia Juarez, Chihuahua, Mexico (he referred to himself as the little Mexican), Eyring was one of 15 children, and as a youth worked as a cattle herder on his father's ranch.[2] Eyring's first professional vocation was as a mining engineer for the Inspiration Copper Company, but he found the constant danger to the workers for whom he was responsible too troubling. He went back to school to finish a Ph.D. in physical chemistry, which involved additional study in Germany. He worked with Michael Polanyi and together they produced a map of the *potential energy surface* for the reaction between a hydrogen atom and molecular (diatomic) hydrogen. A potential energy surface is an important visual aid for understanding chemical reactions, which like many other processes, tend to follow the path of least resistance. A potential energy surface plots the energy as a function of inter- and intra-atomic distances, showing that potential energy changes as reactants move closer and farther apart. If enough is known about the system, these energies can be plotted as a function of these distances, and the path of minimum energy appears as a valley between energy mountains. Although potential energy surfaces can be calculated for only very simple systems, when they correlate with experimental results, they encourage us to believe we are nearing understanding.

For all his substantial contributions to chemical kinetics, Eyring was not one-dimensional. He was an active member of the Mormon church, choosing the University of Utah, near the center of the Mormon religion, over Princeton. He enjoyed his family and obviously familiar with the important neurotransmitter acetyl-choline, he named one daughter Colleen and promised to name a second one acetyl-Colleen. He could also apparently jump flat-footed from the floor to a table and would cheerfully demonstrate the skill.

Eyring's theoretical work gained much experimental support in the mid-1950s when the *molecular dynamics* approach to kinetics evolved. In this experimental technique, beams of molecules—all traveling in the same direction and all with very close to the same energy—are used to study reactions, which is about as close as we can come to mounting the molecule and riding along. Confined to gas-phase interactions, the technique has offered insight into intermolecular forces and reaction mechanisms. Molecular beam chemists have described *stripping* reactions (where products continue to travel in the general direction of the reactant beam); *rebound* reactions (where the product's path veers from the reactant beam); and *harpoon* mechanisms (where an electron is transferred across a large distance, then the ionic attraction reels the reagent in).

In 1986, the Nobel Prize for molecular beam work was awarded jointly to two U.S. chemists, Dudley R. Herschbach and Yuan Tseh Lee. Born in Taiwan, Lee received his early education during the Japanese occupation of the island and his family had to flee to the mountains to avoid bombing in World War II. He learned to speak Chinese, Japanese, English, Russian and German and became interested in science by reading a biography of Marie Curie.

In addition to the gas-phase reactions of Bodenstein and solution-phase reactions of Wilhelmy, much attention has always been paid to reaction rates at surfaces. There are several reasons for this, but a compelling one is that there is always a surface present in a reaction, even if it is only that of the reaction vessel. Others may also exist: the surface of the stirring utensil; surfaces of any colloids or powders present; surfaces of bubbles; or in the human body, surfaces of enzymes. Reactions at surfaces become important when there is a great deal of surface or when the surface—such as an enzyme—serves as a catalyst, accelerating the reaction.

Chemists have always been interested in catalysis from both an intellectual and a practical aspect. Humphry Davy found that a platinum wire catalyzed the reaction of hydrogen and oxygen to form water; Germany would probably not have survived the World War I British blockade without Haber's catalytic nitrogen-fixing process used to produce nitrates for gunpowder; and without enzyme catalysis, life would not exist. In the early 1900s systematic studies of catalysis were carried out in Germany by Wilhelm Ostwald, in France by Paul Sabatier, and in the United States by Irving Langmuir.

Irving Langmuir

Born in Brooklyn, New York, in the late 1800s, Langmuir received his first degree in metallurgical engineering at Columbia University, then he obtained a Ph.D. in chemistry in Germany. He taught for a short while, then joined the General Electric Company, where he stayed until he retired in the 1950s.

At General Electric, Langmuir's first assignment was to extend the life of tungsten-filament electric light bulbs. These bulbs gave a very nice light, but slowly the inside of the glass bulb blackened, cutting down the light. The bulbs were evacuated to reduce the oxidation and burnout of the tungsten filament, but Langmuir solved the problem, to the delight of his management, by filling the bulbs with an inert gas to reduce the tungsten evaporation that was causing the blackening of the bulbs. This led him to study surface films and reactions on surfaces, which he continued through the 1920s. He proposed that surface catalysis took place on the catalyst's surface rather than in thick, absorbed layers, as had been previously assumed. He used this idea to interpret a number of surface reactions. He also developed a description of adsorption behavior, called the Langmuir isotherm, which was characteristic of surface coverage by a layer one molecule thick. His work in surface-catalyzed reaction kinetics helped disprove the radiation hypothesis, but he found the controversy more personal than chemical, so he abandoned this line of research. He achieved his greatest fame for his work on molecular films and surface chemistry, for which he received the 1932 Nobel Prize; the American Chemical Society journal of surface and colloid chemistry is named *Langmuir* in his honor.

As part of his work on surfaces, Langmuir studied monolayer films on water surfaces, with techniques pioneered by a chemist who started her studies in greasy dishwater: Agnes Pockels.

Agnes Pockels

Born on Valentine's Day and raised in the German states in the mid-1800s, Pockels was a member of a chronically ill family, which nearly as much as the social mores of the day, kept her confined to her home. She was able to attend the Municipal High School for Girls, where she acquired an "enthusiastic interest in the natural sciences, especially physics,"[3] but institutes of higher education did not yet accept women. Later in her life when they did, her parents would not allow her to enroll. She read many of her brother's textbooks when he studied for his doctorate in physics, but her primary occupation was the care of her parents and her home.

Undamped curiosity, will however, have its way, and as her sister-in-law wrote

> This is really true and no joke or poetic license: what millions of women see every day without pleasure and are anxious to clean away, i.e., the greasy washing-up water, encouraged this girl to make observations and eventually to . . . scientific investigation.[4]

By the age of 20 Pockels had invented a surface film balance and was conducting studies of surface films and monomolecular layers. The surface film balance consists of a fixed barrier and a movable barrier on a water surface. A film of some insoluble material like a soap or detergent is spread on the water between the two barriers, and the movable barrier is then used to compress the surface film. The amount of pressure necessary to squeeze the film increases fairly smoothly until it suddenly changes. This is the point (the Pockels Point) at which molecules have been squeezed to their points of closest contact. From the area at this point, and the number of molecules in the film, the surface area per molecule can be calculated. Pockels ensured clean surfaces by sweeping them with the movable barrier, and she deposited some layers by dissolving them in a volatile solvent first. These two techniques are still routinely used to study surface films on liquids. Her brother, aware of her experiments and their significance, told her to write to Lord Rayleigh, who was engaged in similar studies. A translation of part of Pockels' letter reads

> My Lord,
> Having heard of the fruitful researches carried on by you last year on the hitherto little understood properties of water surfaces, I though it might interest you to know of my own observations on the subject. For various reasons I am not in a position to publish them in scientific periodicals, and I therefore adopt this means of communicating to you the most important of them . . .[5]

Lord Rayleigh however *was* in a position to publish her work in scientific journals, and he did so. Her work was published in *Nature* the following year. In his introductory note Rayleigh wrote

> I shall be obliged if you can find space for the accompanying translation of an interesting letter which I have received from a German lady, who with very homely appliances has arrived at valuable results respecting the behavior of contaminated water surfaces.[6]

After this Pockels was for a time able to attend meetings relevant to her work and to contribute, but as she wrote

> Since my time was much in demand for home nursing, I was only rarely able to conduct experiments after 1902. . . . When my brother died in 1913, the alarums of the war and post-war period engulfed me . . . I was no longer in a position to obtain relevant literature in my field, the deterioration in my eyesight and in my health altogether being a contributory factor.[7]

However Ostwald published a tribute to her work in 1932 on the occasion of her seventieth birthday, and the same year the Carolina-Wilhemina University of Brunswick awarded her an honorary doctorate.

In the 1930s building on Pockels' work, Langmuir investigated a number of organic monolayers on water. He was assisted by Katharine Blodgett, who became prominent in the field in her own right, perfecting a technique for transferring successive monomolecular layers from a water surface to a solid and creating what are called Langmuir–Blodgett films. Blodgett had her own adventuresome scientific career. She was the first woman to earn a Ph.D. in physics from Cambridge University after Langmuir used his influence to have her accepted into the program. A mere 40 years earlier women were allowed to attend kinetics lectures at Oxford "but only by special permission in each case and with the accompaniment of some elderly person."[8] The function of the elderly person remains undisclosed, but if it were to serve as censor, it seems difficult to believe that a nonchemist, elderly or otherwise, would remain fully alert through lectures on chemical kinetics. At any rate Blodgett managed to remain alert and secure her position in surface science. And surface science, as it has become known, continues to be a fruitful field.

The general field of chemical kinetics continues to be fruitful, too. In the 1950s novel types of behavior of chemical reactions in solution started to be noticed, including an aberrant behavior in which reactant concentrations do not decrease smoothly from the beginning to the end of the reaction but instead oscillate. These oscillations, which can be temporal (pulses in time) or spatial (waves of material), are fascinating to watch. When such chemists as B. P. Belousov and A. M. Zhabotinskii first observed these reactions, they had a great deal of difficulty getting their observations published. It was firmly believed that a reaction proceeded smoothly from beginning to end. But more oscillating reactions are now being found—and in some interesting places—including reactions that regulate the heartbeat and reactions involved in metabolism and tissue formation and differentiation.

Why a reaction oscillates can be understood by looking at a simplified predator–prey model.[9] If the population of rabbits is taken as the concentration of one reactant, and the population of wolves taken for the other, then if the rabbits' food supply is held constant, the rabbit population will increase, causing a wolf population increase, because it feeds off the rabbits. This in turn will cause a rabbit population decrease (the wolves are feeding well), which in turn will cause a wolf population decrease as rabbits disappear and wolves starve. Once the wolf population reaches a sufficiently low level, rabbits will start to increase again, and so on. This type of feedback loop (the wolf population depends on the rabbit population, which depends on the wolf population) seems to be a necessary condition for an oscillating reaction. Several examples of chemical reactions with a similar feedback loop have been found, and these reactions can be made to oscillate.[10] Another apparently necessary condition can be gleaned from the predator–prey model: The supply of rabbit food has to be held constant. If the rabbits' food supply were used up, the reaction would halt—the system would come to equilibrium. So the second necessary condition for oscillating reactions is that the reaction has to be far from equilibrium.

NONEQUILIBRIUM THERMODYNAMICS

One worker to tackle the thorny problem of extending Gibb's equilibrium thermodynamics to nonequilibrium systems was the 1968 Nobel Laureate chemist, Lars Onsager, but Onsager was an unlikely hero. According to an anecdote told of the then 68-year-old Onsager:

> On first setting eyes on his new postdoc, [Onsager] embraced him in Russian style and took him to his office to show him a reprint. There was chaos on every surface, including the floor. Suddenly Lars disappeared, and [the postdoc] found him underneath the desk, where he had located the reprint (which turned out to be a 400-page thesis) and a two-month-old paycheck. Observing Onsager's contortions, [the postdoc] thought to himself: "Here's a fellow who scratches his left ear by reaching round the back of his head with his right hand. I wonder how he ties his shoes!"[11]

Lars Onsager

But the scope of Onsager's contribution was enormous. In the mid-1920s when he was in his early twenties and before he had completed his doctorate, he traveled from his native Norway to Zurich to tell a premier researcher in solution chemistry, Peter Debye, how his theory of the electrical conductivity of solutions was flawed. Debye listened, then hired Onsager as a research assistant. Still without a doctorate, he came to the United States and got a job teaching first-year chemistry. But he did not communicate well with first-year students, so he moved on to become an associate in chemistry at Brown. There he developed the work in thermodynamics that was later to win him the Nobel Prize, and he submitted it to his former university in Norway as doctoral research. It was not accepted, so he did not receive his degree. Thirty years later however the university gave him an honorary doctorate.

The work that was rejected was a development of the thermodynamics of nonequilibrium systems. Onsager made the assumption that on the molecular level, at equilibrium, the forward reaction and the reverse reaction take place at the same rate (the principle of microscopic reversibility). If there are small displacements from equilibrium, the flow of material or heat will be proportional to a thermodynamic force. One can then calculate the behavior of the system on its return to equilibrium. In the Great Depression of the 1930s Onsager lost his job at Brown. He was offered the Gibbs postdoctoral research fellowship at Yale until an embarrassed administration realized he had no doctorate degree. He was asked to submit a body of work for evaluation by Yale, but he did not want to submit work that had already been rejected, so he submitted other work that neither the chemists nor the physicists were willing to evaluate. The mathematicians however endorsed the work, and he obtained the degree and the job. He stayed at Yale for the next 39 years, working in a small, windowless office in a chemistry laboratory. Onsager was not allowed to teach first-year Yale students (though he did enjoy eating lunch with them), but even graduate students found his courses challenging. His statistical mechanics courses were referred as Advanced Norwegian I and II.[12]

In the 1940s and 1950s, based on foundations laid down by Onsager, Ilya Prigogine (a Nobel Prize winner in 1977) developed an extension of nonequilibrium thermodynamics for systems far from equilibrium. Born in Moscow just before the Bolshevik Revolution, his family eventually left Russia to settle in Belgium. Though he originally planned to be a classical pianist, his family wanted him to study law. While reading about criminal psychology, he got caught up in the chemistry of the brain and was lost to the judicial world.

It is interesting that Prigogine's introduction to chemistry was through the chemistry of a biological system, because his further work—though seemingly far removed from nature—may have its most intriguing implications in the biosphere. Prigogine's treatment of systems far from equilibrium surprisingly predicted oscillating reactions and self-organizing systems. A high degree of order in the flow of matter and energy is the essence of life, and there is no doubt that this work will be vigorously pursued in the future.[13]

Physical chemistry as a separate pursuit has its roots in the nineteenth century, but it grew, matured, and blossomed in the twentieth. The two areas in particular that belong to the 1900s—quantum chemistry and chemical kinetics—have had tantalizing success with simple systems, but they are far from complete theories. Likewise kinetics and thermodynamics, which have enormous predictive and explanatory powers for equilibrium systems, have yet to be completely developed for nonideal and nonequilibrium systems. But the work that has been done shows promise of remarkable revelations just ahead.

Radiochemistry, the study of the chemistry of radioactive materials, is another chemistry born on the cusp of the centuries. But this, too, is really a twentieth-century subject. After its beginnings in the work of Becquerel and the Curies, it showed continued growth—and twists. This is the subject we examine next.

chapter
NINETEEN

ca. 1914–1950: Radiochemistry— Dalton Dissected

No doubt some readers will be surprised by the brevity of this chapter because they are aware that the history of radiochemistry makes a fascinating account: a Nobel Prize awarded for work partly in error, spies and cloak-and-dagger code words, security guards in raccoon coats. We hope that our bibliography and references will aid the intrigued. Other readers may be surprised by the existence of this chapter, considering radioactivity to be solely a subject for the history of physics. But radiochemistry is basically inorganic chemistry, and no study of radioactivity is possible without chemical separation, identification, purification, and manipulation—all the ancient arts. So the history of radiochemistry is part of the history of chemistry—although, we admit, there were a few physicists involved.

After the discovery of radioactivity at the close of the 1800s, it became a hot area of research, pursued by groups throughout the world, notably in Germany, England, Denmark, France, Italy, the United States, Canada, Russia, and Japan. As a common starting point, they knew that radioactive nuclei gave off several forms of radiation, including high-energy electromagnetic radiation (light) called gamma rays; high-energy, positively charged helium nuclei (alpha radiation); and high-energy, negatively charged electrons (beta radiation). A bit more information was added directly following World War I when Rutherford in England, showed that nitrogen atoms bombarded by alpha particles produced a positively charged hydrogen nucleus. He called it a proton.

THE PROTON AND THE NEUTRON

Rutherford suggested that protons were the basic stuff from which nuclei were built—hydrogen has one proton, helium two, lithium three, and so forth. But helium has considerably more than twice the atomic weight of hydrogen, and lithium is much more than three times as massive as hydrogen. Therefore Rutherford suggested that the nucleus also contained an uncharged particle of the same mass as a proton to make up the difference in weight. And the neutron was soon found, but the group that first found it did not know they were looking for it, and when they found it, they did not know what they had.

Frederick Joliot and Irene Curie

Irene Curie was the daughter of Marie and Pierre Curie, and she was 1 year old when radium was identified. She was 9 years old when her father died, and she was brought up in large part by her paternal grandfather. Her parents were politically liberal, with social reform leanings, and her grandfather, having taken part in the 1848 French Revolution against the monarchy, reinforced their teachings. Irene was given no religious education, but she attended a cooperative school, privately organized and staffed by Curie, Langevin, and others. Once educated she joined her mother at the Radium Institute, where she was, by all reports, a very sober, serious researcher.

Frederick Joliot was likewise from a nonreligious family. Educated as an engineer, during his studies he came in contact with Paul Langevin, who encouraged him to pursue a research career. It was Langevin who recommended Joliot to Marie Curie, who put him to work at the Radium Institute. There Joliot, a "Maurice Chevalier,"[1] met Irene Curie, a "block of ice,"[2] and the two married. They took Joliot-Curie as the family name, though Joliot continued to sign his scientific papers Joliot, and Curie continued to sign Curie.

Their work also remained separate until they decided to collaborate on an investigation of a new phenomenon: In Germany Walther Bothe and coworkers had seen radiation ten times stronger than expected when light elements were bombarded by alpha particles. To investigate this Bothe radiation the Joliot–Curies used a strong alpha source built from polonium accumulated by Marie Curie and a relatively new device: a cloud chamber. Invented by Charles Wilson just as World War I was beginning, a cloud chamber makes it possible to observe the path of charged particles. (When electrically charged particles pass through supersaturated water vapor, the water condenses in "clouds" in their wake.) The Joliot–Curies put paraffin, which is essentially carbon and hydrogen in about a 1 to 2 ratio, in the path of the Bothe radiation and found hydrogen atoms in their cloud chamber.

Because the Bothe radiation passed through everything, including several thicknesses of lead, they mistakenly assumed that they were dealing with high-energy gamma radiation (light waves). Accordingly in 1932 they reported that this "gamma radiation" knocked protons out of paraffin. Rutherford said, "I do not believe it."[3] A less kind competitor said, "What fools. They have discovered the neutral proton and they do not recognize it."[4]

The problem was one of billiard balls and bowling balls. Photons of gamma radiation can knock electrons from material, but then photons and electrons are both billiard balls. A proton (hydrogen nucleus) is almost 2000 times as heavy as an electron, making it a bowling ball. A billiard ball impinging on a bowling ball should not have much effect on it, let alone knock it out of the way.

James Chadwick in England found that the Bothe radiation was stronger in the forward direction than the backward direction. Because light should radiate in all directions, this indicated to him that the Bothe radiation was probably particles rather than light. The radiation made no trail in a cloud chamber, which indicated that it was an uncharged particle, and Chadwick suspected he had the neutron. The problem was detecting and measuring the mass of this uncharged particle. Rutherford advised, "How could you find the Invisible Man in [a crowded] Picadilly Circus? . . . by the people he collided with, by the reactions of those he pushed aside . . ."[5] Chadwick used the Bothe radiation to knock around nitrogen and helium atoms, as well as hydrogen, and by comparing the rebound, he calculated that the neutral particle had about the same mass as a proton. He declared the discovery of the neutron.

The Joliot–Curies had their day: They found that boron and aluminum atoms bombarded by alpha radiation continued to emit radiation after the alpha source had been removed. Their target atoms had absorbed the alpha particles and transmuted into artificially radioactive elements. For this work they received the Nobel Prize. Marie Curie wrote to her daughter, "We have returned to the glorious days of the old laboratory."[6] She inserted a paragraph on artificial radioactivity into the new edition of her treatise on radioactivity—published posthumously.

NUCLEAR FISSION

In Italy Enrico Fermi, a physicist studying radioactivity, watched these developments with interest. He reasoned that neutrons would be even more effective in penetrating and transmuting nuclei than alpha particles because they were uncharged and therefore would not be repelled by the charged nucleus (an idea said to have come to him while playing tennis). He began a program of radiating all known elements with neutrons to see what was produced, starting with the lightest elements first. He saw no new elements nor radioactivity with the first few, but with fluorine he observed induced radioactivity. This observation spurred a flurry of research in which Fermi and his group, Emilio Segré,[7] Edoardo Amaldi,[8] Franco Rasetti,[9] produced a wealth of heretofore unknown nuclei for known elements—nuclei with a different number of neutrons than the naturally occurring elements.

Finally around the spring of 1934 Fermi's group had worked its way down the Periodic Table to uranium, the heaviest element then known. In the process the group had found several radioactive nuclei that it could not identify as nuclei of currently known elements—at least down to atomic number 82 (lead). Not having reason to believe that anything lighter than atomic number 82 could be formed, the group declared that it had created elements heavier than uranium, or *transuranic* elements. Fermi was criticized for this conclusion, notably by Ida Noddack (the chemist who with her husband is credited with the discovery of rhenium). That Fermi himself was not entirely comfortable with his results is reflected in the title of his paper: "The Possible Production of Elements of Atomic Number Higher than 92."[10] But Fermi may have been subject to other influences. The newborn Italian fascism, looking for national heroes and national glory, may have pressured Fermi to claim remarkable results. Only in his Nobel Prize lecture did he offer names for the new elements, but then as Emilio Segré, a coworker on the project commented

> The moment was unfortunate; at that very time Hahn and Strassmann were discovering nuclear fission, thus proving that those elements [of Fermi's] were composed (to put it diplomatically) of poor chemistry.[11]

The work that Segré referred to was the team of Meitner, Hahn, and Strassmann.

Lise Meitner and Otto Hahn

The original team was Meitner and Hahn. Born in Germany in Frankfort on the Main, Otto Hahn planned on being an industrial chemist and went to work for Sir William Ramsay so that he could improve his English. Ramsay was working in radiochemistry, though without much success. Hahn, working with him on the decay of thorium, thought he had found a new element, which he called radiothorium. Bertram Boltwood, a friend of Rutherford and a leading radiochemist at Yale, knew that radiothorium was chemically inseparable from thorium and called it a compound of thorium and stupidity. This was too harsh because there were several such inseparable materials being identified at the time. It remained for the chemist Soddy to note that these chemically identical elements were slightly different in atomic weight.

We know now that chemically identical elements must have the same number of protons in the nucleus, but they can have varying numbers of neutrons. Different numbers of neutrons cause atomic weights—the sum of the weights of the protons and neutrons—to vary. Soddy called these chemically identical elements with slightly varying atomic weights *isotopes*, Greek for *same place*, because these elements occupied the same place in the Periodic Table. In fact atomic weights measured for normal samples are really averages over all the natural isotopes of the element being weighed. Weights for most elements appear constant because essentially all natural samples of an element have the same distribution of isotopes, called the *natural abundance*. (This explains some observed anomalies in Mendeleev's Periodic Table: Two elements next to each other could have an atomic weight inversion—for instance nickel, which follows cobalt, has a smaller atomic weight—because atomic weights are averages over isotopic weights, and a greater abundance of a heavier isotope could understandably tip the scales.) When this was shown, Hahn was chagrined, but youth will recover. On returning from Rutherford's laboratory Hahn obtained a position at the University of Berlin, and there he began to study radioactivity. Trained as a chemist (and freely admitting that he was more interested in beer halls than in physics as an undergraduate), he felt the need for a physicist's perspective on his data. He met Lise Meitner through Planck, and they began working together.

Lise Meitner, from a middle-class Austrian family and exactly the same age as Hahn, had gone to Germany to study physics, obtaining special permission (as a woman) to attend Planck's lectures, though she was not allowed in the laboratories. When Meither and Hahn first proposed working together, they were refused permission to use the laboratories (Meitner was again restricted), but with persistence they were granted the use of an abandoned carpenter's shop. Both Hahn and Meitner were skilled experimentalists, and the collaboration was productive. Soon Hahn was asked to head a radiochemistry program at the Kaiser Wilhelm Institute. In the new laboratory, free of the background radiation from contamination in their old facilities, Hahn and Meitner were able to make more sensitive measurements. They started looking for the naturally occurring "mother substance of actinium," the radioactive substance that would become actinium after giving off an alpha particle.

The work was interrupted by World War I. Hahn was conscripted and assigned to the chemical warfare unit, but Meitner continued the work on her own. Letters written by Meitner to Hahn from this period have been preserved and indicate a superficially formal relationship (Meitner always used a formal title when addressing Hahn), but under this was an easygoing exchange.[12] From the letters we can surmise that Hahn was anxious for results, and when the war drew to a close, Meitner had them ready. In 1918 they published the discovery of protoactinium (meaning *before actinium*), though the name was shortened to protactinium in 1949.

Until this time Meitner remained unsalaried, living off an allowance from her father. After the discovery she was give a position in physics at the Kaiser Wilhelm Institute and facilities to direct research. She worked productively with her own group, and Hahn with his, until Meitner decided to investigate Fermi's provocative results from the neutron bombardment of uranium and convinced Hahn to join her.[13] In 1933 the first anti-Semitic laws were passed in Germany, and Meitner considered leaving (she was of Jewish heritage, though she had been raised a Protestant), but Planck and Hahn urged her not to go. She was after all an Austrian citizen and baptized, so not in immediate danger. But then Hitler annexed Austria, and Meitner was classified as a German Jew, required to wear the Star of David, and subjected to harassment on the street.

By this time Meitner and Hahn were both entering their sixties and had been friends for 30 years. With the help of Hahn and Debye, a trip abroad was arranged for Meitner, with the tacit understanding that she would not return. Hahn gave her his grandmother's diamond ring so she would have something to use as a bribe, should it be necessary. She extracted a promise that he would send results to her as he achieved them and then took her tactical retreat. She went to Copenhagen, was housed by Neils Bohr, then accepted a position at the Nobel Institute in Stockholm.

The following Christmas, Meitner's nephew, Otto Frisch, came to visit her. His visit coincided with a letter from Hahn in which he reported that he and their associate, Fritz Strassmann, had achieved such bizarre results that he would for the time communicate them only to her. He wrote

> Our radium isotopes act like barium! Perhaps you can come up with some fantastic explanation. We ourselves know [uranium] can't actually burst into barium. You see, you will do a good deed if you can find a way out of this.[14]

Chemists by that time had reconciled themselves to small pieces like a proton or an alpha particle being chipped from a nucleus by bombardment, but these results indicated that the nucleus of uranium, with 92 protons, had split into fragments the size of barium. Barium, with 56 protons, was a pretty big chunk. Frisch and Meitner went on a walk to discuss the result. They discussed Bohr's model of the nucleus: a water drop held together by surface tension. Opposing the surface tension of the nuclear drop was the electrical repulsion of the positively charged protons squeezed into close contact. It could be envisioned then that an extra absorbed neutron might destabilize the water-drop nucleus to split it into two. They made a rough calculation of the energy that would be released from the mass difference of the two fragments and the original uranium nucleus, using Einstein's $E = mc^2$. They decided that the energy would be enough to blow the two fragments apart. They communicated their explanation to Hahn, who then published his results under the names of Hahn and Strassmann.

Lise Meitner wrote to her brother:

> Hahn has just published absolutely wonderful things based on our work together. As much as these results make me happy for Hahn personally and scientifically, many people here must think I contributed absolutely nothing to it.[15]

Why Hahn did not acknowledge Meitner's contribution to the project remains a mystery. It can be argued that to have done so would have been foolhardy, given the political situation. The work of a scientist like Meitner, classified as less than fifty percent Aryan might have been suppressed. (The National Socialists' classification of people as "Aryan" or "non-Aryan" is one of history's great misnomers: Aryan implies a language group, not an ethnic group, just as Jewish implies a religion, not a race.) Hahn, despite his anti-Nazi feelings, managed to retain his position through the war, and diplomacy may have helped his survival. (Strassmann's career was destroyed when he refused to join the Nazi-controlled Association of German Chemists.) But even after the war Hahn maintained his silence. He was awarded the Nobel Prize in 1946 for the work, and at that time Meitner wrote: "I find it quite painful that in his interviews Otto did not say one word about me or our thirty years together . . ."[16]

But rationalization is a way of surviving, too, and Meitner may have understood this. She and Hahn remained friends and correspondents until their deaths, months apart, when they were in their nineties. Recently though, Meitner's contribution has received some acknowledgment. The museum in Munich that displays the apparatus she designed for the neutron irradiation of uranium has changed the plaque from "Worktable of Otto Hahn" to "Worktable of Otto Hahn, Lise Meitner, and Fritz Strassmann."

Frisch and Meitner eventually published their theoretical explanation, but the idea was too amazing to keep the lid on for long. Returning to Copenhagen, Frisch told Bohr. In New York at a conference, Bohr told Fermi. Fermi had just left Italy with his Jewish wife, stopping only in Sweden to pick up his Nobel Prize. Accepting a position at Columbia University in New York, he immediately began work on this new idea, now called *nuclear fission*. He found, as did others, that another product of fission was more neutrons.

Soon it began to occur to several people that if neutrons caused fission and fission produced neutrons, then one fission reaction might cause another. If a fission reaction released more than one neutron, then one fission reaction might cause two more, then four more, and so on; that is, a chain reaction might occur. Under the right conditions this reaction might be self-sustaining, like combustion, and it might also, like combustion, provide a source of energy. Fermi began working on this possibility, as did the Joliot–Curies. At first feeling they were in a race for priority, the French laboratory did hurried measurements and rushed to publish its idea that a chain reaction was possible. But on the fall of France in World War II, the laboratory ceased publication. Its further results were hidden in the vaults of the Academy of Science, and Joliot became a member of the French Resistance.

Joliot found that neutrons had to be slowed down to be absorbed efficiently by fissioning nuclei (as a slow ball is easier to catch than a fast ball), and he found the best material for slowing them down was the heavy isotope of hydrogen called deuterium, recently discovered by a U.S. chemist, Harold Urey. Ordinary hydrogen has a tendency to absorb incident neutrons, but deuterium, which already has an extra neutron in the nucleus, works quite well. In addition *heavy water*, water in which the hydrogen is replaced by deuterium, can conveniently be poured around the neutron target. Joliot secured the only available stock of heavy water in France and kept it out of commission for the duration of the war. It occurred to him, as it occurred to others, that energy from a fission chain reaction might also make a very good bomb.

The United States however was still at peace, so ideas about superexplosives developed more slowly. Albert Einstein, prompted by three Hungarian-born American physicists, including Edward Teller (whom we meet later) wrote to President Roosevelt explaining that a fission bomb might be possible: the letter known as the Einstein Letter. Roosevelt assigned a group to study the problem in the spring of 1940, and the group found that a chain reaction might be possible, but it was unproven. The U.S. government made some funding available.

Work begun at Columbia University, where Fermi had his laboratory, established that a particular isotope of uranium, uranium-235, was responsible for the fission reaction observed in uranium and that graphite, the soft, easily worked form of carbon, was a neutron moderator. Across the continent at the University of California at Berkeley, the official discovery of the first transuranic element was made. As it turned out Fermi had been making transuranics—though products he originally claimed to be transuranics were not—and the first transuranic, neptunium (atomic number 93), was found in fission products from a Fermi-type reaction. Edwin McMillan found that light fission fragments flew off the bombarded uranium sample, which he collected on sheets of cigarette paper placed around the sample. He found one fission product that was so heavy that it did not fly off the uranium target, and this proved to be a product of neutron absorption by uranium-238, which decayed to an isotope of element number 93, neptunium.

Glenn Seaborg

Glenn T. Seaborg, a chemist, decided it would be a good idea to study the chemistry of this new transuranic, especially because nuclear theory predicted that a decay product of element number 93, an isotope of element number 94, would be *fissionable* (subject to fission when bombarded with low-energy neutrons), like uranium-235. In 1940 Seaborg and another Berkeley instructor, Joseph Kennedy, used the newly developed Lawrence cyclotron—a device that uses an alternating electric field to accelerate charged particles—to produce enough element 94 to study its chemistry. Seaborg suggested to a second-year student, Arthur Wahl, that a chemical separation scheme for neptunium and the new element would make a good Ph.D. dissertation and suggested a collaboration to McMillan.

Seaborg's group placed a uranium nitrate sample in the path of the neutron-seeded cyclotron particle beam and made a sample. In the sample they found an element chemically different from all other known elements: They had made plutonium, the name eventually given to the new element 94. (During the war they had to refer to both elements 93 and 94 in code, so neptunium became "silver" and plutonium became "copper"—as distinguished from "honest-to-god" copper.[17]) The sample was stored in a cigar box donated by G. N. Lewis.

When Seaborg's group accumulated enough plutonium to study, they found that a certain isotope of plutonium, plutonium-239, was even more potent than uranium-235 in the fission process. But plutonium-239 could be produced only in minute amounts—until the Japanese bombed Pearl Harbor. Then suddenly funds and resources for work on fissionable elements became available.

In spring 1942 the U.S. government decided to proceed with the development of all viable methods for producing plutonium-239 and separating uranium-235. The Army Corps of Engineers was brought in to help direct the project, and it set up an office in Manhattan in New York City. This became the Manhattan Engineer District Office, and the project became known as the Manhattan Project.

THE MANHATTAN PROJECT

The Manhattan Project was massive: It spent by 1944 more than a billion dollars per year and eventually built facilities scattered from South Carolina in the southeast to Washington in the northwest. It provided challenges for engineering and science and opportunities for engineers and scientists, including quite a few African-American chemists such as Moddie Daniel Taylor and Lloyd Albert Quarterman.

There were two choices for bomb material—uranium and plutonium—and both had associated problems. The problem with uranium was to show it could produce a sustained chain reaction, and the problem with plutonium was to produce enough to build a bomb. As it turned out these two goals were linked: To produce a lot of plutonium, a lot of free neutrons are needed, and the best place to find free neutrons is in a self-sustaining fission chain reaction. The reactor experiments, now moved to the University of Chicago, were under the direction of Fermi.

The uranium reactor was built on a squash court of the university (and it was here that the guards donned discarded raccoon coats to protect themselves from the cold). Dubbed CP-1 (Chicago pile number 1), it was called an atomic pile because it was just that: The reactor consisted of a pile of containers of uranium oxide interspersed with graphite bricks. There was a neutron source at the bottom to initiate the reaction, and neutron-absorbing cadmium rods could be inserted for control. Fermi's group demonstrated the feasibility of a uranium chain reaction by the end of 1942.

But Fermi's reactor just demonstrated the principle. To produce the amount of plutonium needed for a bomb, the Hanford plant was built in the state of Washington. The scale and speed of the project were such that the contract was signed for Hanford before the Fermi reactor actually ran. For the Hanford plant Seaborg and coworkers had to devise separation schemes for kilogram quantities of plutonium based on chemical properties deduced from microgram quantities. This might have been a reasonable challenge if plutonium were in a position in the Periodic Table where its properties could have been extrapolated from those of the other elements. But it was not yet entirely clear where plutonium belonged in the Periodic Table. Based on their chemical properties, Seaborg, around 1944, proposed that these elements formed another 14-element block analogous to the lanthanide series. Calling this the actinide series, he used these new positions for the new elements to predict their chemistry. Advised after the war not to publish his new version of the Periodic Table because it would ruin his scientific reputation, Seaborg remembered, "I didn't have any scientific reputation so I published it anyway."[18]

The properties of microgram samples do not always translate directly to kilogram samples, and an industrial scale-up of such a magnitude would normally proceed in stages. But the war effort had no time for stages, and at Hanford they did it all in one big leap. There were many technical challenges in addition to the scale-up. Once uranium targets had been irradiated with neutrons, plutonium had to be separated from the uranium matrix. It was present only in very small amounts, which would imply the need for sensitive techniques, but the separation schemes had to be rugged—insensitive to moderate error—because the whole separation process had to be handled remotely. Radiation from the material, the target, and the fission products was so intense that human operators could not approach the separation tanks. But amazingly a chemical separation process was found, developed, and it worked from the beginning.

Uranium processing proved to be as challenging. Here chemists were, in essence, separating uranium from itself because the desired uranium-235 is an isotope chemically identical to the common isotope, uranium-238, and comprises less than one percent of natural uranium. Physical processes that were devised included allowing corrosive gas-phase uranium hexafluoride to diffuse through porous nickel barriers (the heavier uranium-238 material diffused more slowly, so that enough stages eventually produced a fairly pure compound of uranium-235) and using a large-scale mass spectrometer. Materials needed for the uranium separation process included polymers resistant to uranium hexafluoride, a material nearly as reactive as fluorine itself.

Plunkett's Teflon found its first major application, and many other highly fluorinated polymers were developed for the project.

Even then, armed with materials and an idea, it remained to design and assemble a bomb.

THE BOMB

The branch of the project devoted to actual bomb design and assembly was called Project Y. The Project Y site, selected in fall 1942, was in the New Mexico desert in an abandoned boy's school named Los Alamos. Simultaneous with site selection came personnel recruitment. Bohr was smuggled with his family to Sweden, then made his harrowing flight to Britain, as previously described. His son, Aage, made the trip with him and fared better. Setting immediately to work on the Manhattan Project, they were called by the code names Nicholas Baker and his son James. Consequently at Los Alamos Bohr was known as old Nick. Otto Frisch was hurriedly made a British citizen, then just as hurriedly exempted from the military service that citizenship required. He was given a passport and a United States visa and shipped off to the Manhattan Project, as was another German exile named Klaus Füchs (of whom we hear more later). Research was directed by Julius Robert Oppenheimer, a U.S. physicist, described by Segré as "extraordinarily quick . . . [but] conscious of his abilities and somewhat prone to arrogance, a weakness that made him many enemies."[19] But with the perceived urgency of the project, personal differences were for the most part put aside. Oppenheimer was an effective head for the project.

Chemists on the Manhattan Project had plenty to do. The purity of the fissile material was important for several reasons. The purest material was the lightest, and bombs have to be carried by planes with limited payloads. In addition many impurities can absorb neutrons, so their presence could poison any chain reaction; therefore analytical techniques had to be developed to measure small amounts of impurities. The impression is sometimes given that a nuclear chain reaction is easy to achieve and it occurs spontaneously with any fissile material, but chain reactions actually require some minimum amount of material (the *critical mass*), or the chain reaction fizzles—as does any other reaction lacking fuel. The bomb would not have been possible if the amount of material needed were several tons (as some early calculations based on imperfect data indicated). As it turned out, the amount of uranium-235 or plutonium-239 needed for a bomb was only a few pounds. But the material has to be in the right shape. If it is in a long sheet, the reaction may fizzle because the majority of the neutrons escape harmlessly without encountering other nuclei to fission. Because the best shape is a sphere, chemists had to investigate the metallurgy of plutonium and uranium (which are metals) so that bomb parts could be properly machined to form spheres. Plutonium has five distinct allotropic forms between room temperature and its melting point, and the problem was finding a form that was malleable and stable at room temperature. Surfaces of the metals also had to be protected from corrosion, because plutonium is especially easily corroded by water and air. Plutonium surfaces were cleaned and then coated with an electroplated or evaporated metal coating, such as silver. With what little spare time they had, chemists also investigated the properties of this first transuranic to be collected in significant amounts. They found plutonium to be very electropositive and to have the highest electrical resistivity (lowest conductivity) of any metal found so far.

A big problem facing physicists and engineers was assembling the critical mass quickly enough so that it would explode rather than pre-ignite and fizzle. For this they used two designs: for uranium, a gun assembly that fired two pieces together to form a critical mass; for plutonium, an implosion arrangement in which conventional explosives compressed a barely subcritical sphere of material into a critical mass. Chemists on the Manhattan Project found themselves spending a good deal of time worrying about the chemistry of conventional explosives.

The speed and accuracy of the work (a rare combination in science) were amazing. But the motivation was there: By then it was obvious that the Axis powers had access to all the starting information the Allies had. There was a bomb effort in Germany as well as in Japan, but the Japanese effort was underfunded and had little uranium to work with. The German effort, headed by Heisenberg, was laboring under the assumption that it would require several tons of uranium for an explosion and that heavy water was the only workable neutron moderator. (There has been speculation, which has been vigorously challenged, that Heisenberg did not want to put a nuclear weapon at Hitler's disposal and was purposefully misdirecting the effort). But fruitful or not, the Axis efforts inspired the efforts of the Allies, who proceeded with remarkable efficiency, fortitude, and skill.

As with all new processes the work had hidden hazards. Otto Frisch had a close call while conducting experiments to determine the minimum amount necessary for a critical mass. One of these experiments, called the *Dragon series*, derived its name from the expression "tickling the tail of the Dragon." A subcritical mass fell down a 10-foot chute, passing another subcritical mass on the way. In the instant that the two masses were close together, they formed a critical mass, and when they did, indicator lights on radiation detectors showed the level of radiation achieved. In another experiment, called the *Godiva* series, Frisch put a small neutron source on a sphere of naked uranium-235 metal (Lady Godiva), then added layers of uranium, carefully watching a Geiger counter. In one of these experiments he bent over the assembly to add the twelfth layer, which should have still been less than a critical mass, when one of his assistants informed him that the counter had quit working. Actually the counter had received so much radiation that it was off scale. But Frisch did not realize this; he just knew that the assistant was about to move the counter, which would have changed the experimental configuration. He turned to stop the assistant, then saw that all the indicator lights were also on, signaling maximum radiation. He realized his torso was reflecting neutrons back into the assembly, making it approach criticality. He straightened up and swept the added layers of uranium off the source. He calculated the amount of radiation he had received and determined that he would live (which he did) and that he now also had the data he needed. He was able to complete his calculations on critical mass.

The complexity of the project for all the effort and support meant that atomic bombs (actually a misnomer, a more accurate name would be nuclear bomb) would not be available until 1945 and made the proposed target shift from Germany to Japan. But by July 1945 an implosion bomb was ready to be tested several days ahead of schedule.

Trinity

The test was code-named Trinity, though it is not clear whose choice this was or why. It was carried out on top of a steel tower at Alamagordo, New Mexico. Frisch reported

> Suddenly, the hills were bathed in a brilliant light; it was as though someone had turned the sun on with a switch. There was no sound.... It surprised me; I had expected a brief flash, but this stayed for some seconds, and only then did it begin to dim ... Then, when I thought it safe, I turned to see this pretty, perfect, red ball of fire about the size of the sun, connected to the ground by a short grey stem ...[20]

The whole area was seared by an intense light. Another observer, Kenneth Bainbridge, felt the heat on the back of his neck, but he did not mind. If the bomb fizzled, he would have been the one who had to find out why. Watching the explosion he turned to Oppenheimer and said, "Now we are all sons of bitches!"[21]

The first (and so far only) use of the uranium-235 gun-assembly weapon, nicknamed Little Boy, was the bombing by a B-29 named Enola Gay on Monday, August 6, 1945, at 8:15 A.M. (local time), in Hiroshima, Japan. An eyewitness recalls

> The people of Hiroshima had just begun their day's work. Suddenly, the sirens sounded, warning that a plane was approaching, but the sirens soon stopped and everyone went about their work.... I thought I heard the sound of the plane, but it seemed a long way off and very high up.
>
> I was hit by a thunderous flash and an explosion of sound. My eyes burnt—everything went black. I held my sister. Everything faded away—I thought I was dying.
>
> I woke up. I was alive. But my home was completely destroyed. When I crawled outside, I found that the whole of Hiroshima was destroyed. Everything was blown away, torn apart. Everything was burning.
>
> The banks of the river were crowded with people, everyone wanted to be near the water. There was a child screaming, trying to wake up her dead mother.

> I was very lucky, my family were all alive and we were together, sheltering in a cave.
>
> Father's face was badly burnt and swollen. My brother's back was full of pieces of glass from the window he was sitting beneath. My eldest sister had her teeth sticking through her lip, she had been using chopsticks.[22]

The plutonium-235 implosion assembly, nicknamed Fat Man, was used in the bombing of Nagasaki 3 days later. On the morning of the bombing, Japan's Supreme Council for the Direction of the War met in the prime minister's bomb shelter. The meeting was deadlocked because some still wanted to continue the war. After the second use of an atomic bomb, Japan surrendered.

There was then—and there still is—argument over whether or not the atomic bomb should have been used, but this seems an odd sort of argument. Civilian or military, fast or slow, death is death. It is strange to argue about who should die, how many, and in what manner.

Aside from the wartime bombing, concerns were raised about wartime and post-war testing of nuclear weapons and the effects of nuclear bomb residues, or *fallout*. The fact of this fallout was well-established. (One of the authors has in her possession a "Radiation Detection Kit"[23]—purchased for her when she was . . . well . . young—and the set of experiments that came with it.[23] One of these experiments suggests making a series of measurements several days after an atom bomb test has been announced. Experimenters are assured that they might detect substantial increases in the background count days or even weeks after the test.) But though the fallout was real, there was substantial debate over its effects. Linus Pauling argued that fallout would cause an increase in cancers, but Edward Teller argued that a little radiation might even be beneficial or at worst as dangerous as being a little overweight. Pauling led a campaign to stop atmospheric testing of nuclear weapons, which included nonviolent civil disobedience, and he was awarded the Nobel Peace Prize on October 10, 1962, the date the Nuclear Test Ban Treaty went into effect.

Another concern after World War II was whether the last bomb had been dropped. Soviet Russia built and tested its own bomb in

1949. It was learned that a British scientist on the Project Y team, Klaus Fuchs, a former member of the Communist party, had given information to Russian scientists. In the resulting investigation more spies turned up—Harry Gold, a chemist from Philadelphia; David Greenglass, a former army sergeant who had worked at Los Alamos; and Ethel and Julius Rosenberg. In the resulting concern about Communists working on nuclear projects, Joliot, who with his wife had joined the Communist party during the German occupation of France, was removed as head of the French Atomic Energy Commission. Curie was detained by the immigration authorities on a trip to the United States and denied membership in the American Chemical Society. Robert Oppenheimer was dismissed from atomic energy projects because of past political associations, and others were summoned for questioning by congressional committees or the FBI and subsequently lost their jobs and/or reputations.

In this atmosphere President Truman (Roosevelt had died a few months before the first atomic bomb was dropped on Japan) had no difficulty securing funding for research on an even more powerful weapon known as the Super, or hydrogen bomb.

THE SUPER

The principle behind the Super was atomic *fusion* rather than atomic *fission*. Atomic fusion is the joining together of two atomic nuclei, and it is a light-element rather than a heavy-element phenomenon: The Super had isotopes of hydrogen as its fuel. Fusion, like fission, releases vast amounts of energy; the fusion of hydrogen to helium in the sun accounts for the sun's energy. The idea of using fusion to power a bomb came up during a lunchtime conversation between Fermi and Teller in 1942. There was some work done on fusion as part of the Manhattan Project, but the main emphasis at that time remained on uranium and plutonium fission bombs. After the war, funding on all bomb projects was cut, but after the announcement of the Soviet atomic bomb test, Los Alamos went on a 6-day work week, and chemists at Hanford started coming up with ways of working with tritium, an isotope of hydrogen.

In the final design for the Super the heat to initiate the fusion reaction is provided by a fission reaction, and once ignited, the reaction spreads like a normal thermochemical reaction (such as combustion), hence *thermonuclear*. The first test of a thermonuclear device, codenamed Mike, was in 1952, and it completely leveled the small atoll in the Marshall Islands on which it was detonated. This weapon may have fulfilled the prediction of Alfred Nobel: "Perhaps my factories will end war sooner than your peace conferences."[21] War in the 1950s became "Cold" and "Limited." There was another bit of fallout from the Mike test, chemical instead of political: Sifting through debris from the Mike explosion, scientists found elements number 99 and 100: einsteinium and fermium.

NEW ELEMENTS

By the time the dust of World War II had begun to settle, gaps in the Periodic Table up to plutonium had been filled in. Marguerite Perey, a 30-year-old technician working for Marie Curie found element number 87 and named it francium. She accomplished this before earning the equivalent of a bachelors degree. Francium was the last naturally occurring element to be found in other than trace amounts. Using cyclotron bombardment, Segré and coworkers prepared element number 43, which they named technetium; and element number 61, promethium, which is also completely missing from the Earth. It is aptly named because it has been found in the spectrum of a star in the Andromeda galaxy and is in a sense stolen from the heavens, just as Prometheus stole fire from the gods. Astatine (meaning *unstable*), the missing halogen element, number 85, was also created by bombardment.

After their success with neptunium and plutonium, Seaborg and his collaborators[25] continued to look for more transuranics. (These collaborators included other scientists and graduate students who contributed many ideas and most of the work, and we regret that they must be consigned to a footnote.) They used cyclotron bombardment, a variety of targets, and microchemical techniques developed by Hahn. Americium and curium (elements number 95 and 96) were discovered in wartime at the Metallurgical Laboratory of the University of Chicago, but these were kept secret and not officially given names until after the war. As a reflection of the intense work that surrounded their discovery, they were jokingly referred to as "delirium" and "pandemonium." The first public announcement of the discovery of elements 95 and 96 was made in a national radio broadcast, the "Quiz Kids Program," where Seaborg appeared as a guest on Armistice Day, November 11, 1945. Americium was named in analogy to europium, which it falls under in the Periodic Table; and curium was named for the Curies, in analogy to gadolinium, thought to be named after Johan Gadolin. (Though these authors would never argue with Seaborg's chemistry, we note that gadolinium was actually named after the mineral gadolinite, not the man. So curium could have been called pitchblendium, but even if it had occurred to anyone, this has certain aesthetic limitations.)

Berkelium (number 97) was created after the war in the Berkeley cyclotron, as was californium (number 98). Commenting on the announcement of these discoveries in its column "Talk of the Town," the *New Yorker* magazine, said

> New atoms are turning up with spectacular, if not downright alarming frequency nowadays, and the University of California at Berkeley, whose scientists have discovered elements 97 and 98, has christened them berkelium and californium, respectively. While unarguably suited to their place of birth, these names strike us as indicating a surprising lack of public relations foresight on the part of the university, located, as it is, in a state where publicity has flourished to a degree matched perhaps only by evangelism. California's busy scientists will undoubtedly come up with another atom or two one of these days, and the university might well have anticipated that. Now it has lost forever the chance of immortalizing itself in the atomic tables with some such sequence as universitium (97), offium (98), californium (99), berkelium (100).[26]

The discovery team replied

> "Talk of the Town" has missed the point in their comments on naming of the elements 97 and 98. We may have shown lack of confidence but no lack of foresight in naming these elements "berkelium" and "californium." By using these names first, we have forestalled the appalling possibility that after naming 97 and 98 "universitium"

and "offium," some New Yorker might follow with the discovery of 99 and 100 and apply the names of "newium" and "yorkium."[26]

The *New Yorker* responded: "We are already at work in our office laboratories on 'newium' and 'yorkium'! So far we have just the names."[26]

As the number of new elements grew, so it seemed did the number of researchers involved with each discovery. But there were reasons for this: elements were becoming increasingly unstable and shorter lived as they became heavier, and millisecond as well as milligram procedures became important.[27] Mendelevium (number 101) was produced literally one atom at a time, and because of its short half-life, it had to be raced from bombardment to analytical laboratory. Seaborg recounted

> Since this was quite an event at the Rad Lab, [they] connected a fire bell in the hallway to the counters so that the alarm would go off after every time an atom of element 101 decayed. This was a most effective way of signaling the occurrence of a nuclear event, but quieter means of communication were soon substituted, following a suggestion put forward by the fire department.[28]

The groups became increasingly cosmopolitan, including men and women of Asian, Asian-American, African, African-American, European, South American, and other origins. And the cyclotron at Berkeley was not the only one in the world. Nobelium (102) was announced by researchers at the Nobel Institute in Sweden in 1957, but the work could not be reproduced. A group in Moscow announced their discovery of 102 in 1958, but it could not provide enough information to make the identification positive. More definitive work was done at Berkeley, and the Berkeley group is generally given credit for the discovery of nobelium.

Lawrencium (103) was sifted out of a bombardment mixture in 1961, but the priority for its discovery was again disputed by the Moscow group. Dubnium (104) the first of the transactinide series (coming after actinium), which is now believed to be another transition block, was shown to exist in Dubna, Russia, as well as Berkeley, California, and so was joliotium (105). The name neilsbohrium had been proposed for element 107 by the Russian group that produced it, but bohrium is now the accepted name. Hahnium and meitnerium are now the official names for elements 108 and 109, respectively. At one time the name seaborgium was proposed for element number 106, but as of August 31, 1994, the International Union of Pure and Applied Chemistry (the internationally recognized committee on nomenclature) has chosen the name rutherfordium, though this choice met with some controversy.[29] Both a Russian group and a German group claim to have produced element 110, but no name has been proposed pending confirmation of these claims.

Aside from basic scientific and philosophical interests, the longer lived isotopes of some of these exotic elements have found practical applications, including Space Nuclear Auxiliary Power units (SNAP units), transuranic thermoelectric units used to generate electricity on satellites, and as power sources on Earth in cardiac pacemakers. Though radioactive sources in pacemakers have largely been replaced by lithium batteries due to lower cost and easier construction (and probably a bit of nuclearphobia, too), lithium batteries only last about 10 years while the radioactive sources last 30. As prosthetic technology improves, perhaps these longer lived sources will come back into demand.

Radioactive isotopes of elements of the main body of the Periodic Table have also found biological applications in diagnosis and treatment of disease, an application that was pioneered in the mid-1900s by the physicist/chemist/biologist, George von Hevésy.

George von Hevésy

There is a story that may or may not be true about Otto Frisch searching for a word to describe the process he and Lise Meitner were about to propose when he ran into Hevésy and asked him what biologists call the process of cell division. Hevésy supposedly answered "fission."[30] A more verifiable story concerning Hevésy however is that when the Nazis overran Denmark, Hevésy saved Nobel medals entrusted to Bohr for safekeeping by dissolving them in acid and storing the solutions in old beer bottles. After the war the gold was recovered and the medals were recast.

By all accounts George Hevésy had an eventful life. In addition to being the codiscover of hafnium, he was the first to use the wealth of isotopes produced by Fermi as biological tracers; for this he received his own Nobel medal in 1943. Because isotopes of an element are chemically indistinguishable, the body takes radioactive isotopes of elements in and uses them as it would nonradioactive isotopes. The advantage is that gamma rays from appropriate isotopes penetrate body tissues and exit, so that radioactive elements inside the body can be detected from without. This allows the physician to see into the body without painful, invasive surgery—a complementary technique to X-ray imaging. In addition, because certain elements concentrate in certain parts of the body, their radioactive isotopes can be used to treat specific regions. For example iodine concentrates in the thyroid gland and radioactive iodine can be used for thyroid-specific treatments.

Despite these positive contributions radiochemistry still has to fight bad press. With the threat of nuclear holocaust, finding fallout in milk, and the disaster of Chernobyl,[31] public paranoia has increased to such an extent that the near-miracle technique of Nuclear Magnetic Resonance body imaging, which depends on the magnetic, not radioactive, properties of nuclei, has to be described as Magnetic Resonance Imaging (MRI). And a potential fountain of energy is hamstrung by public fear: Nuclear power is still arguably the cleanest form of electrical power production, and it has the potential for producing the least amount of waste. In the wake of the Chernobyl accident, it may be very difficult to maintain popular support for nuclear power plants, but body for body, and flower for flower, nuclear power still has a better safety and environmental record than coal-burning or oil-burning power plants. Of course the best thing would be to use less energy all around, and that is a challenging problem for policy makers—and sociologists.[32]

It can only be added that the most sensitive, reliable smoke detectors use americium-241—a fact to be remembered the next time a child is saved from burning.

Radiochemistry—born in the last years of the 1800s but matured in the 1900s—has become the signature technology of the twentieth century: the Nuclear Age. It owes its rapid development in large part to its application in weapons, but materials evolving from this effort have enhanced the quality of life and show promise for the future.

After all, the best is yet to come.

chapter
TWENTY

The Best Is Yet To Come

And so we conclude this history of chemistry, choosing as our cutoff the mid-1900s, (the second half of the century spans the careers of your authors, neither of whom cares to be considered history.) But chemical history continues to be created as we write, so we cannot resist a parting survey of some of the events of the last 50 years and a peek over the horizon at the future.

The world setting for chemistry has continued to change. The center of chemical activity has moved from Africa to India to Asia to Arabia to Europe to North America—if a center, that is, still exists. Chemistry has become increasingly delocalized and is now better described as a global enterprise. There have also been major technical advances affecting the course of chemistry—computers, transistors, lasers, space travel—and these have opened completely new areas of study. With such probes as Voyager we learned of planets with seas of hydrocarbons and bedrock of water and ammonia ice.[1] Spectroscopists have found evidence of molecules in the "empty" space between stars, including carbon monoxide, ammonia, formaldehyde, cyanoacetylene, acetaldehyde, methyl mercaptan, and even ethanol—more than a hundred different molecules have been detected so far.[2] The era of astrochemistry has begun.

Here on Earth chemistry in the late 1900s had political problems. It started the century as an apparent savior of society but then turned in many minds into a modern devilment. In modern society a distaste for the chemical arts, called by some *chemophobia*, has developed that has been partly deserved. Sometimes there has not been sufficient research and materials have been marketed before their effects were fully explored.

In some cases the urgency has had its justifications, as with the development of the nuclear arsenal before methods for storage and disposal of its accompanying wastes were available. The insecticide DDT (*d*ichloro*d*iphenyl*t*richloroethane) is another example: In 1939 the chemist Paul Müller showed that this compound was an effective insecticide, and his work with DDT and similar compounds earned him a well-deserved Nobel Prize in medicine in 1948. In 1943 DDT was used to stop a typhus epidemic in Italy that could have defeated the Allies as surely as any German army. The insecticide was similarly effective in controlling mosquito-borne malaria in the Far East. It and other insecticides (as well as artificial fertilizers, hybrid seed, and improved farming techniques) increased food production worldwide and reduced world hunger, leading to what was labeled the *Green Revolution*.[3] But then DDT began showing up in high concentrations in fish, and it was realized that it accumulated in biological systems. When birds ate the insects sprayed with DDT, DDT accumulated in the birds and weakened the egg shells of some species. These species began dying out, and the insect population began to grow again. The biologist and naturalist Rachael Carson brought this problem to worldwide attention in her book *Silent Spring*.[4] In the 1970s DDT was outlawed and the use of similar materials severely restricted.

Other incidents—the oily, black, carcinogen-laden liquids that heavy rains brought to the ground surface of Love Canal, a community built on a toxic chemical dump; mercury dumped from a chemical plant into the estuary at Minamata, Japan, which caused paralysis and mental disorders in thousands of people and death for several hundred; and in 1984 the methyl isocyanate leak in Bhopal, India, that killed some 2000 people and injured tens of thousands more—have not helped the negative image of chemistry.

Then there is *acid rain*. The combustion of coal and oil in power stations and the high temperatures of internal combustion engines lead to emissions of oxides of sulfur and nitrogen, and mixed with moist air, these oxides form sulfuric and nitric acids. In addition to eroding buildings and works of art, acid rain can severely damage plants, including trees. Rain as acidic as lemon juice has been observed. There is also the potential, though not proven, problem of global warming caused by human- and animal-generated atmospheric gases, particularly carbon dioxide and methane, which could reflect radiation that would otherwise escape the Earth—sometimes called the *greenhouse effect.*

Are chemical innovations the culprits in these cases, or the way they are used—or overused? Necessity may be the mother of invention, but invention appears to be the parent of need. Once an innovation is introduced, it rapidly seems to become a "necessity": no one wants to give up improvements in the quality of life and demand for a product can foster irresponsibility in the manufacture. Regulations for many kinds of materials and processes have been proposed, but these are met with protest when the regulations result in reductions in jobs, services, or production. It is gradually being realized that it is easier to modify technology than human nature, and the pendulum is swinging back: To borrow a phrase from the American Chemical Society, it will be chemists who have the solutions.

ENVIRONMENTAL CHEMISTRY

The study of environmental chemistry is complex because it involves an uncontrolled, many-variable ecosystem, but chemists have tackled complex systems before. Creative techniques for dealing with wastes are already being found, including bioremediation—the use of microorganisms (or *bugs* in bioremediation parlance) to digest waste.[5] One variety of bug eats the particularly nasty explosive used to implode plutonium in bombs, and others ingest TNT and nitroglycerine, though it may be imagined that they do so very carefully.

Handling nuclear waste (from power plants and weapons facilities) has also seen some creative approaches, including nuclear-fuel recycling, vitrification, and underground storage. Nuclear-fuel recycling uses the fact that plutonium-239 is produced by the neutron bombardment of uranium-238. This plutonium-239—along with unused uranium-235—can be extracted from spent fuel rods, collected, and made into more fuel elements. Unreacted uranium-238 can be separated and mixed with new fuels to be converted into more plutonium-239. A reactor capable of making more fuel than it uses is called a *breeder reactor*. This technology is used in several countries, though notably not the United States. In the vitrification process, fission waste is combined with molten glass. This glass hardens into logs that are very chemically inert and suitable for long-term storage. Plans have been made to store spent, unrecycled fuel and glass nuclear-waste logs in underground vaults. Although progress has been made toward building these vaults, scrutiny of their safety and environmental impact has been close. To date none of the vaults have been pronounced suitable to receive waste.[6]

Another approach to any waste problem is waste minimization, and chemists contribute to this effort, too. One recent development involves using supercritical carbon dioxide as a solvent for chemical reactions or cleaning—replacing environmentally objectionable organic solvents. Supercritical carbon dioxide is maintained at a temperature so high it has to be a gas, but under such high pressure that it still has liquid-like properties. It is a good solvent, and it can be recycled or allowed to dissipate as a nontoxic gas. Carbon dioxide can also be frozen into pellets to be used as a blasting agent. Because the pellets immediately turn into gas, this method leaves nothing to be disposed of but the material blasted. Carbon dioxide, a potential greenhouse gas, is a by-product of many manufacturing processes and its collection and reuse delays the release of carbon dioxide to the atmosphere. But recycling is not 100% efficient, and each step in the process requires energy. There is no free lunch.

Advancements in environmental chemistry are closely tied with developments in other areas, such as analytical chemistry. As particular compounds are found to be toxic at lower and lower levels, it becomes important to be able to detect them at these lower levels (though there are those who contend that being able to detect them at low levels is what causes the legislated tolerance level to drop). Analytical chemistry has pushed its detection limits for some contaminants from one particle in a million to one particle in a billion and even to one particle in a trillion.

Microanalytical techniques include new forms of microscopy—electron microscopy and scanning tunneling microscopy—which are able to generate images with atomic resolution, and microelectrodes, which act as sensors or initiate chemistry on the microscopic scale. Microelectrodes and other advances in electrochemistry have proved that this is a productive, vital, and indispensable area. Unfortunately in the later half of the 1900s, electrochemistry received a bit of bad press, too: cold fusion. Arguments can be made for excluding this episode from the history of chemistry because the only chemistry seems to have been the interaction of ambition and poor science. But it is also arguable that such episodes *must* be included in the history of chemistry because they sound a clarion call. They serve to remind us that valid chemical theory must be based on verified results obtained with sound scientific methods: the lessons of Lavoisier.

COLD FUSION: LOST LESSONS OF LAVOISIER

In 1989 the news hit a world hungry for positive results. The outlook for oil-burning and fission-power-generating plants was all gloom and doom, and fusion, a possible alternative source of power, looked about as far off as the stars that are still the only feasible fusion generators. Then Stanley Pons and Martin Fleischmann, electrochemists at the University of Utah, announced that they had created a fusion reaction in a beaker on a lab bench with common chemicals and a moderate current. It took an atomic explosion to ignite the fusion reaction of the hydrogen bomb, so the world was delighted to hear that fusion could now be achieved with a power supply plugged into an ordinary electrical outlet: So delighted that, at first, few cared that the announcement of the discovery came in a news conference rather than though the accepted route of peer-reviewed scientific publication, and so delighted that researchers all over the world dropped everything and rushed to duplicate the results. The experiment after all promised to be easy. Pons and Fleischmann said they had passed current with a palladium electrode and a platinum electrode through a cell containing deuterated water (heavy water). They had produced, they said, excess heat energy and gamma rays.

Fleischmann is a prominent electrochemist originally from Czechoslovakia, who arrived in England as a child when his family fled the Nazi occupation. By 1989 he had established a considerable reputation, having been elected to the Royal Society. Pons met Fleischmann when he joined Fleischmann's research group to complete his graduate studies. Pons, from the southern United States, is reported to be a pleasant, generous man with drawled speech and quiet manners. He took a 10-year detour in his career to work in his family business, but after finishing his Ph.D., he began a very productive career. He obtained a position at the University of Utah and started publishing at an enormous rate. He and Fleischmann continued their collaboration and finally got around to trying out an idea that Fleischmann had arrived at sometime in the 1970s but had never tried and had never been able to talk anyone else into trying—cold fusion.

Pons and Fleischmann reportedly financed the initial experiments themselves, believing the idea so fantastic that no one would fund it. They set up a cell and allowed it to electrolyze for several months while it was attended only by Joey Pons, the young son of Stanley Pons. Then one night with no one around, there was apparently a meltdown.[7] Something happened, because witnesses said later there was damage to the table where the cell had been sitting, but it was the explanation for the damage that was eventually questioned. Apparently afraid of being scooped in the research by another worker, Steven Jones at Brigham Young University (also in Utah), Pons, and Fleischmann convinced the University of Utah to hire security officers immediately to guard the laboratory, and they covered windows in the laboratory doors with black paper. They held a news conference; the press release read

> Two scientists have successfully created a sustained nuclear fusion reaction at room temperature in a chemistry laboratory at the University of Utah. The breakthrough means that world may someday rely on fusion for a clean, virtually inexhaustible source of energy.[8]

But details were slow in coming. Pons cautioned people against trying the reaction because of the possibility of explosions. When asked how to avoid explosions, he answered, "Avoid high currents and sharp edges"[9] (certainly good advice for life in general). Nonetheless there were published confirmations from other groups, followed gradually by retractions. Some groups were seeing increases in heat, but they were getting no other evidence of nuclear events; for instance they could find no ejected neutrons.

Pons explained the lack of neutrons by saying that cold fusion was a "nonclassical nuclear reaction."[10] Then someone pointed out that the gamma ray data given by Pons and Fleischmann did not contain signature features that should have been there and that data started to look like an experimental artifact. There was still the heat and the meltdown, but then that, too, had an alternative explanation. Palladium is a catalyst for the reaction of oxygen and hydrogen—easily glowing red hot in the process—and in an electrolysis cell containing water, there can be oxygen and hydrogen, too (as Nicholson and Carlisle showed in the early 1800s). A 1958 paper reported that the water solution in a palladium–platinum electrolysis cell develops hot spots if it is not stirred. Pons and Fleischmann had not stirred their cells.

Hopes began to die away, and researchers went back to their former pursuits. To date however Pons and Fleischmann have not withdrawn their claim. Pons now lives in Nice, France, and works for a Japanese company, doing undisclosed but well-funded research. Fleischmann, he says, may be joining him soon.

But electrochemistry in the late 1900s was much more than cold fusion: It developed storage batteries for electric cars, solar cells, and novel synthetic routes for inorganic and organic compounds. For even after all the years devoted to it, synthetic organic chemistry, with its huge potential for variety, is not yet cookbook chemistry, and much remains to be learned about synthetic possibilities.

BIOCATALYSIS

The big word in organic chemistry in the late 1900s is *biocatalysis*: The use of microorganisms to make chemicals. Although this is not new—yeast has been used to make ethanol since before written record, and Chaim Weizmann used fermentation to produce acetone for Britain in

World War I—now enzymes are being controlled and even engineered—designer bugs if you will. Sometimes biomolecules are found to work on biomass to produce industrially valuable compounds with more environmentally friendly byproducts than from conventional methods. In other words instead of metal catalysts and petroleum, future industries may use bugs and corn.[11] However the overall effectiveness of this approach must be critically evaluated—machines to harvest corn still use gasoline.

The synthetic route is important in inorganic chemistry, too, where the wealth of possible materials is hardly less than in organic chemistry. Inorganic synthesis has in the past often consisted of melting reactants together and cooling them under various conditions to obtain the hoped-for material (so-called shake-and-bake chemistry). This is being replaced by lower temperature, more controlled reaction schemes.[12] Such controlled synthesis is important in the production of molecular sieves and certain ceramic materials that have an unexpected property: They superconduct.

INORGANIC SUPERCONDUCTIVITY

In the early 1900s it was discovered that metallic mercury cooled to just 4 kelvins (4 degrees above absolute zero) superconducts; that is, the solid mercury offers zero resistance to the flow of electrons. At first this may seem like a perpetual motion machine, forbidden by thermodynamics, but the key is that the electrons do no work. The second law says nothing about perpetual motion being impossible so long as it is not in a machine that does no work.

In addition to conducting current over long distances without heat loss, superconductors completely repel magnetic fields, so that magnets placed above them remain suspended as if in defiance of gravity. This property could be used someday to elevate trains above conducting rails. With no train-to-rail contact, frictional energy loss is minimal, so "maglev" trains use less power than conventional trains. Demonstration prototypes have already been built.

Superconductors are already being used to create high magnetic fields for use in instrumentation. But conventional superconductors, all metals, need to be cooled to very low temperatures before they become superconducting. Liquid helium is used to cool them, but it is expensive and quickly boils away. Beginning in the late 1980s some new ceramics were found to superconduct at 35 kelvins, which is 35 degrees above absolute zero. The target temperature for superconductivity then became 77 kelvins, because at that temperature liquid nitrogen could be used to cool superconductors, and liquid nitrogen (nitrogen gas is the main component of air) is cheaper than a soft drink from a grocery store. Paul C.-W. Chu at the University of Houston broke the barrier with a ceramic with a formula of 1 part yttrium to 2 parts barium in a matrix of varying amounts of copper oxide. The material is an example of a *berthollide*, a material with a nonintegral ratio of elements. These compounds are named for Berthollet, the chemist in the 1800s who maintained for so long that compounds could have continuous composition. When Chu first published his results, ytterbium was somehow substituted for yttrium, and the ratios were reported incorrectly. Though we would never suggest that the mistakes were made purposefully (Chu blamed both mistakes on his secretary), the effect was to establish his priority while still buying him time alone with the true formulation.[13]

Making a ceramic into a wire that can carry current is still a formidable difficulty, though some progress is being made. One successful device that has been made from superconductor material is called a Superconducting Quantum Interference Device (SQUID).[14] It is extremely sensitive to magnetic fields and could someday be used as a noninvasive electromagnetic probe of the body's heart and brain activity. Recently a superconductor was fabricated that is claimed to superconduct at 153 kelvins—within the range of coolants used in air conditioners and refrigerators.[15]

Another piece of big news for the inorganic chemist in the late 1900s turned out to be big news for organic, physical, analytical, and biochemists, too. This was the discovery of a third solid form of carbon—the other two being diamond and carbon. This one forms a sphere shaped like the geodesic dome structure designed by the U.S. philosopher and engineer R. Buckminster Fuller; it is named appropriately buckminsterfullerene.

BUCKMINSTERFULLERENE

The naming of buckminsterfullerene follows a tradition in organic chemistry, a field where the number and variety of compounds at times defies systematic nomenclature. Instead of stringing together a list of prefixes and suffixes, it is sometimes simpler to give a new, nonsystematic name to a compound, then name its derivatives from there. Accordingly we find in the literature *soccerane* (shaped like a soccer ball) and *basketane* (shaped like a basket). The steroid researcher Russell E. Marker of birth control fame found so many compounds in his yams that he named one pennogenin, after Pennsylvania State University and another rockogenin, after a fellow chemist at Pennsylvania State nicknamed Rocky. And someone returned the compliment. There is a marcogenin.[16]

The original buckminsterfullerene has 60 carbons linked into a sphere and was nicknamed buckyball; buckybabies with 32 and 44 carbons have also been found, as well as larger spheres with up to 960 carbons in a single spherical molecule; and buckytubes, tough, atomic-scale, carbon cylinders. Although few applications have yet developed, these compounds display novel properties. They can trap atoms and molecular complexes in their spherical cages, and they have been envisioned as a way of transporting materials across membranes or to sequester compounds selectively. It has been suggested that the cage could protect water-sensitive materials, and it has been found that other atoms can be attached to the outside of the cage. In the early 1990s it was found that compounds of buckyballs with potassium or rubidium are superconducting.[17]

The investigation and explanation of the properties of these materials overlap with the interests of physical chemists, too. There are catalytic possibilities, which always perks up the ears of a good *kineticist* (a chemist studying reaction rates). In the late 1900s there are plenty of other things to keep the kineticist interested: Lasers now permit kinetic investigations of processes that occur in a quadrillionth of a second—femtosecond chemistry which has been used to probe the elusive transition state and fast intermolecular interactions.[18] Also supported by all the advances in techniques and instrumentation, the field of biochemistry has really taken off in the late 1900s, and it has had quite a few Nobel Prizes dedicated to it. The list of achievements includes the development of methods for recombining fragments of genetic material from different species: recombinant DNA.

BIOCHEMICAL ENGINEERING

The use of recombinant DNA has become known as *genetic engineering*, and in the late 1970s the Genentech Corporation programmed a bacterium to produce human insulin by splicing the insulin gene into the bacterial DNA, and it is now also possible to produce the human growth hormone in this way. This type of work has raised ethical and safety questions among the scientific community and the public, but researchers have gone a long way toward self-regulation.[19] In the future recombinant DNA techniques may be used to treat genetic diseases. The design of drugs, initiated by Paul Ehrlich, has also evolved and now includes structure-based drug design. In this process the three-dimensional structure of the target enzyme, virus, or other disease-causing entity is found, then a drug is designed that attaches to and alters this three-dimensional structure.[20] The hope of this research is eventually to develop treatments for many diseases, including cancer, psoriasis, glaucoma, and maybe even the common cold.

Many of the successful efforts we have been describing have been the efforts of large, well-funded laboratories and large groups of researchers (a recent scientific paper listed 22 coauthors),[21] which poses the question: Is the day of Dalton gone? Is there no more room for the maverick worker with a good idea? The story of the polymerase chain reaction provides an answer.

THE POLYMERASE CHAIN REACTION

In 1983 a method was discovered for making virtually unlimited copies of DNA from a single piece of material in a short time—one hundred billion copies in an afternoon. It requires only standard biochemical laboratory equipment and reagents. In the words of Kary B. Mullis, corecipient of the 1993 Nobel Prize, "The DNA may come from a drop of dried blood, the tissues of a mummified brain, or from a 40,000-year-old woolly mammoth frozen in a glacier."[22]

The enzyme DNA polymerase has a role in DNA replication, and Mullis was considering its use in a *sequencing* scheme (a method for deducing gene sequences in DNA). He thought of heating the DNA to break apart the strands, adding primers (small gene sequences that attach to their complementary bases), and then using DNA polymerase to cap the strands with an isotope-labeled cap to the primer. In the solution would be a selection of caps, and the one chosen to add to the primer, identified by its isotopic signature, would give information on how the primer fit the original DNA sequence. Then in Mullis' words:

> One Friday evening late in the spring I was driving to Mendocino County with a chemist friend. She was asleep. U.S. 101 was undemanding . . . That night the air was saturated with moisture and the scent of flowering buckeye. The reckless white stalks poked from the roadside into the glare of my headlights . . .
>
> some questions still nagged at me. . . . What if [the reaction] extended to create a sequence that [continued to the end of the DNA fragment]? Surely that would cause trouble . . .
>
> No, far from it! I was suddenly jolted by a realization: the strands of DNA in the target and the extended [pieces] would have the same . . . sequence. In effect, the . . . reaction would have doubled the number of DNA targets in the sample!
>
> Suddenly, for me, the fragrance of the flowering buckeye dropped off exponentially.
>
> Yet in the morning I was too tired not to believe that someone, somewhere, must have tried this idea already.[23]

It turned out no one had.

Mullis now works out of his home as a consultant and commands a comfortable fee. No one questions the value of his contribution, but there are some who argue that his trick is not Nobel-quality work. The answer has come back, "Oh, yeah? Then how come [you] didn't think of it?"[24]

The polymerase chain reaction is credited with making possible the development of a whole new field of study, the study of ancient DNA. Labeled by some *molecular paleontology*, these studies recently served as the basis for a popular movie, *Jurassic Park*, in which dinosaur DNA is retrieved from dinosaur blood in an ancient mosquito preserved in amber, repaired, cloned, and then made into dinosaurs. Though a fanciful idea at the beginning of the film's production, by the time the film was released, DNA from a 40-million-year-old bee trapped in amber and a 25-million-year-old termite had been replicated.[25] (But this is a long way—an enormously long way—from the production of a complete organism.)

In 1989 the Nobel Prize in chemistry went to Thomas R. Cech and Sidney Altman for independently discovering that a molecule does not have to be a protein to be an enzyme. The nonprotein enzymes that they found were RNA. Of the many questions on the origin of life, one recurrent one is how did the first DNA and RNA come together? They are both needed for self-replication. A reaction self-catalyzed by RNA suggests that there may have once been a 1-molecule machine that could live on its own.

The popular interpretation of these results in terms of the origin of life shows how much this question is still on the researcher's mind. There are now questions about the composition of Stanley Miller's primordial soup, but lipidlike organic molecules, amino acids, and nucleic acid bases have been found in meteors, which means they exist extraterrestrially, which means that the stuff of life is out there, somewhere else.[26]

But for all the amazing work done by chemists to date, there is still interesting, worthwhile work to do—from learning how sulfur-eating bacteria exist without light[27] to deciding how to deactivate and dispose of chemical weapon warheads.[28] A recent issue of *Accounts of Chemical Research* offers a list of current "Holy Grails in chemistry"[29]— goals for chemical research that are now at least partially realized and whose future prospects are good—including the isolation and manipulation of individual atoms and molecules, bond-selective chemistry (the making and breaking of selected individual bonds), room temperature superconductors, and solar splitting of water to hydrogen and oxygen for use in fuel cells. Certainly there are enough to keep all comers to the chemical field busy. And who knows what lies ahead? After all it has now been shown that mercury-201 bombarded with electrons undergoes a process known as electron capture, and in the process mercury—the element with 80 protons—is transmuted into element number 79, gold: The alchemists would have cheered.

But have all the basics now been covered? Is chemistry now just a matter of reworking the same old elements into new materials to serve biology, medicine, the environment, and industry? Read on!

SUPERHEAVY ELEMENTS

A better name for the Periodic Table might be the Periodic List because the list goes on. Elements immediately beyond fermium (atomic number 100) are so unstable that they must be made one atom at a time and their chemistries studied on a millisecond time scale—when they can be studied at all—but this may not be true for elements of higher atomic mass. Certain combinations of protons and neutrons have special stability; for instance uranium-235 is a very efficient fissioner and nuclear weapon material, but the much more stable uranium-238 is a fission dud and useless as a bomb. The numbers of protons and neutrons that seem to give nuclei special stability are called *magic numbers*, and a shell theory of the nucleus explaining these magic numbers was developed by the physicist Maria Gertrude Goeppert Mayer (who reportedly would get so caught up in her calculations that she would try to use her cigarette for chalk).

Beginning in the 1960s people began using this shell model of the nucleus to predict stabilities for elements beyond the known elements on the Periodic Table. Their calculations seemed to indicate that elements beyond lead should have diminishing stability (and indeed everything after bismuth is radioactive in all isotopic forms) but that an element with atomic number 114, with 184 neutrons—one of the so-called Superheavies—might show some special stability.[30] Graphs of nuclear stability resemble topographical maps, because of their three-dimensional form (number of protons versus number of neutrons versus nuclear stability), and the region of greatest stability became known as the "magic island," with other features known as the "stability peninsula," the "magic ridge," and the "magic mountain" (see Fig. 20.1).

Initial calculations predicted the stability of nuclei on the magic island to be as great as 100 million years, and because this is within

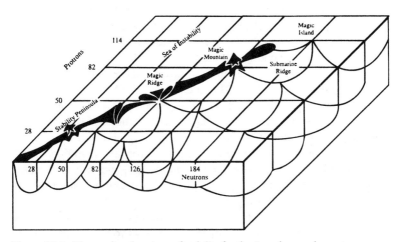

Figure 20.1. The predicted regions of stability for the Superheavy elements.

an order of magnitude of the lifetime of the Earth (about 4 billion years), scientists began speculating that these isotopes might be found on Earth. Efforts in this direction included investigating the chemical composition of meteorites with anomalous xenon-isotope distributions (which could have been caused by Superheavy fission), anomalous *halos* (tracks in mineral formations caused by the migration of fission products from decaying radioisotopes), and anomalous alpha activity or tracks in photographic film carefully shielded from cosmic radiation and other known natural radiation. The possibility of manufacturing the Superheavies was also immediately investigated because of the successful manufacture of plutonium and other artificial elements.

In conjunction with these efforts, calculations were made to predict the chemical properties of the Superheavies so that likely ores could be chosen for investigation and separation schemes devised. Separation techniques were developed to purify and identify elements with lifetimes as short as a thousandth of a second. Models were developed to predict such aggregate properties as entropies from samples as small as 500 atoms. Ground-state electron configurations, oxidation states, ionization energies, metallic radii, ionic radii, densities, melting points, and boiling points were predicted. In particular eka-mercury and eka-lead were predicted to be nonreactive (noble) volatile liquids or gases.

Dirac's relativistic treatment of quantum mechanics was also applied to the Superheavies, with the unexpected outcome of predicting that positrons, though of the same (positive) charge, could be attracted to a heavy enough nucleus. This meant that the Superheavies could have a positron as well as an electron spectrum, which would open the door for unusual chemical behavior. Calculations of the type used in quantum mechanics to predict electron energy levels were used to predict neutron and proton energy levels. From these energy levels gamma- and X-ray spectra—which would be the signatures of the new elements—were predicted. Fission product distributions, another signature property, were computed.

Although most of the anomalous natural behavior originally thought to be attributable to Superheavies was found actually to be the result of more mundane occurrences, such as fallout from aboveground nuclear weapons testing, the quest for their artificial production by bombarding targets with heavy ions continues.[31]

Calculations have shown that the gap between the last stable elements in the Periodic Table and the island of stability may not be so great as at first thought and there might be a sandbar of elements with measurable lifetimes leading toward the magic island. This would mean that leaping the gap might not be required and Superheavies might be formed by stepping neutron by neutron along the sandbar to the magic isle. The sandbar also holds the promise that even if Superheavy lifetimes turn out to be short compared to the lifetime of the Earth, their production in nature in such exotic environments as the gravitational fields of black holes in the vicinity of neutron-emitting stars or the interior of the Earth is not completely ruled out.

So perhaps someday some chemist will discover a peculiar component in an odd new ore or in a meteoric fragment from a dying star. Then fires will be stoked and vats will bubble. Chemists will bend over beakers with acids and oils to dissolve, digest, distill, and sublime . . .

Who knows what spirit will soar—or beast will crawl—from this cauldron's brew?

Endnotes

APOLOGIA

[1] Maurice P. Crosland, *Historical Studies in the Language of Chemistry* (Dover Publication, New York, 1978), p. 44.

CHAPTER 1

[1] Christopher B. Stringer, *Scientific American* (Dec. 1990), p. 98; Keith Stewart Thomson, *American Scientist* (Nov.–Dec. 1992), p. 519; Luigi Luca Cavalli-Sforza, *Scientific American* (Nov. 1991), p. 104; James Shreeve, *Discover Magazine* (Aug. 1990), p. 52; James Shreeve, *Discover Magazine* (Sept. 1992), p. 76; Allan C. Wilson and Rebecca L. Cann, *Scientific American* (Apr. 1992), p. 66; Christopher B. Stringer, *Scientific American* (Dec. 1990), p. 98.

[2] Jared Diamond, *Discover Magazine* (May 1989) p. 50; Michael Szpir, *American Scientist* (July–Aug. 1993) p. 36.

[3] Edward McNall Burns, *Western Civilizations: Their History and Their Culture*, 8th ed. (Norton, New York, 1973). Copyright W. W. Norton.

[4] Leonard John Goldwater, *Mercury: A History of Quicksilver* (York, Baltimore, 1972), p. 73.
[5] J. M. Roberts, *The Penguin History of the World* (Penguin, New York, 1990).
[6] Mary Elvira Weeks and Henry M. Leicester, *Discovery of the Elements*, 7th ed. (Journal of Chemical Education, Austin, TX, 1968).
[7] Ibid.
[8] Ibid.
[9] Tomas Bass, *Discover Magazine* (Dec. 1991), p. 63.
[10] Will Durant, *Our Oriental Heritage* (Simon and Schuster, New York, 1963), p. 104.
[11] H. W. F. Saggs, *The Greatness That Was Babylon* (Hawthorn, New York, 1962).
[12] Martin Levey, *Chemistry and Chemical Technology in Ancient Mesopotamia* (Elsevier, New York, 1959).
[13] Samuel Noah Kramer, *The Sumerians—Their History, Culture, and Character* (University of Chicago Press, Chicago, 1963), pp. 96–97. Copyright 1963, University of Chicago Press.
[14] A. Leo Oppenhein, *Ancient Mesopotamia—Portrait of a Dead Civilization* (University of Chicago Press, Chicago, 1964), p. 293.
[15] Martin Levey, *Chemistry and Chemical Technology in Ancient Mesopotamia* (Elsevier, New York, 1959), p. 108.
[16] William S. Ellis, *National Geographic* **184** (Dec. 1993), p. 37.
[17] J. M. Roberts, *The Penguin History of the World* (Penguin, New York, 1990), p. 109.
[18] Ruth Whitehouse and John Wilkins, *The Making of Civilization: History Discovered through Archaeology* (Knopf, New York, 1988), p. 23.
[19] Stephen W. Hawking, *Black Holes & Baby Universes & Other Essays* (Bantam, New York, 1993).
[20] *Dictionary of Scientific Biography*, vol. 1, Charles Coulston Gillispie, ed. (Scribner's, New York, 1976), p. 149.
[21] Henry M. Leicester, *The Historical Background of Chemistry* (Dover, New York, 1956), p. 25.
[22] This statement may or may not have originated with Aristotle; it is also accredited to Spinoza, and it is used as a proverb of unknown origin. It does however adequately summarize Aristotle's thinking on the matter.

CHAPTER 2

[1] Henry M. Leicester, *The Historical Background of Chemistry* (Dover, New York, 1956), p. 39.
[2] Mary Elvira Weeks and Henry M. Leicester, *Discovery of the Elements*, 7th ed. (Journal of Chemical Education, Austin, TX, 1968), pp. 23–24.
[3] Henry M. Leicester, *The Historical Background of Chemistry* (Dover, New York, 1956), p. 39.
[4] J. R. Partington, *A Short History of Chemistry*, 3rd ed. (Macmillan, New York, 1957), p. 16.

ENDNOTES

[5] John Maxson Stillman, *The Story of Alchemy and Early Chemistry* (Dover, New York, 1960), p. 22.
[6] Henry M. Leicester, *The Historical Background of Chemistry* (Dover, New York, 1956), pp. 44–45.
[7] Stephen F. Mason, *A History of the Sciences* (Collier, New York, 1962), pp. 66–67.
[8] Ibid.
[9] F. Sherwood Taylor, *The Alchemists* (Collier, New York, 1962), p. 31.
[10] John Maxson Stillman, *The Story of Alchemy and Early Chemistry* (Dover, New York, 1960), pp. 162–65.
[11] Ibid., pp. 163–65.
[12] Timothy Taylor, *Scientific American* (Mar. 1992), p. 84.
[13] Mary Elvira Weeks and Henry M. Leicester, *Discovery of the Elements*, 7th ed. (Journal of Chemical Education, Austin, TX, 1968), p. 10.
[14] Ibid.
[15] Joseph Needham, *Science and Civilization in China*, vol. 5, part 2 (Cambridge University Press, New York, 1983), pp. 294–304.
[16] Edwin O. Reishchauer and John K. Fairbank, *A History of East Asian Civilization*, vol. 1 (Houghton Mifflin, Boston, 1960), pp. 138–39. Copyright © 1960 by Houghton Mifflin Company.
[17] Joseph Needham, *Science and Civilization in China*, vol. 5, part 3, (Cambridge University Press, New York, 1983), pp. 66–67. Reprinted with the permission of Cambridge University Press.
[18] Ibid., pp. 38–42, 169–71.
[19] Ibid., p. 104.
[20] Praphulla Chandra Ray, *A History of Hindu Chemistry from Earliest Times to the Middle of the Sixteenth Century, A. D.* (Bengal Chemical and Pharmaceutical Works, Calcutta, 1903).
[21] Praphulla Chandra Ray, *History of Chemistry in Ancient and Medieval India* (Indian Chemical Society, Calcutta, 1956).
[22] Narendra Nath Kalia, *From Sexism to Equality* (New India Publications, New Delhi, 1986).
[23] Joseph Needham, *Science and Civilisation in China*, vol. 5, part 3 (Cambridge University Press, New York, 1983), p. 104.
[24] Ibid., p. 162.

CHAPTER 3

[1] Kenneth Bailey, *The Elder Pliny's Chapters on Chemical Subjects*, part 2 (Edward Arnold, London, 1932), p. 101.
[2] Ibid.
[3] Richard Olson, *Science Deified and Science Defied: The Historical Significance of Science in Western Culture* (University of California Press, Berkeley, 1982), p. 160. Copyright 1982, The Regents of the University of California.

4. Ibid., p. 162.
5. Ibid., p. 160.
6. Saul Bellow, *Discover Magazine* (Dec. 1987), p. 76.
7. Henry M. Leicester, *The Historical Background of Chemistry* (Dover, New York, 1956), p. 66.
8. Ibid., pp. 68–69.

CHAPTER 4

1. Theophilus, *On Divers Arts,* John G. Hawthorne and Cyril Stanley Smith, translators (Dover, New York, 1979), pp. 40–41.
2. Eric John Holmyard, *Makers of Chemistry* (Oxford at the Clarendon Press, 1931), p. 86.
3. Morris Kline, *Mathematics for the Nonmathematician* (Dover, New York, 1985).
4. Edward McNall Burns, *Western Civilizations, Their History and Their Culture*, 8th ed. (Norton, New York, 1973), p. 332. Copyright W. W. Norton.
5. Leonard John Goldwater, *Mercury: A History of Quicksilver* (York, Baltimore, 1972), p. 92.
6. J. R. Partington, *A History of Greek Fire and Gunpowder*, (W. Heffer, Cambridge, U.K., 1960), p. 82.
7. Ibid., p. 83.
8. *Dictionary of Scientific Biography*, vol. 1, Charles Coulston Gillispie, ed. (Scribner's, New York, 1976), p. 100.
9. Eric John Holmyard, *Makers of Chemistry* (Oxford at the Clarendon Press, New York, 1962), p. 91.
10. J. R. Partington, *A History of Greek Fire and Gunpowder* (W. Heffer, Cambridge, 1960), p. 65.
11. Stephen F. Mason, *A History of the Sciences* (Collier, New York, 1962), p. 115.
12. Eric John Holmyard, *Makers of Chemistry* (Oxford at the Clarendon Press, 1931), p. 96.
13. Ibid., p. 97.
14. F. Sherwood Taylor, *The Alchemists* (Collier, New York, 1962), p. 84.
15. Henry M. Leicester, *The Historical Background of Chemistry* (Dover, New York, 1956) p. 78.
16. J. R. Partington, *A History of Greek Fire and Gunpowder* (W. Heffer, Cambridge, U.K., 1960), p. 74.
17. Ibid., p. 73.
18. Ibid., p. 66.
19. Ibid., p. 69.
20. Assuming, we imagine, an alphabet that does not yet contain *w* or *j*. The Latin language was in a state of evolution—*j* was just a fancy way of writing *i* and not a separate letter until the 1600s—and the author of the cryptogram may not have used *w*.

ENDNOTES

[21] Eric John Holmyard, *Makers of Chemistry* (Oxford at the Clarendon Press, 1931), pp. 103–4.

CHAPTER 5

[1] Praphulla Chandra Ray, *A History of Hindu Chemistry from Earliest Times to the Middle of the Sixteenth Century, A. D.* (Bengal Chemical and Pharmaceutical Works, Calcutta, 1903), pp. 115–117.
[2] *A History of Women in the West, III Renaissance and Enlightenment Paradoxes*, Natalie Zemon Davis and Arlette Farge, eds. (Harvard University Press, Cambridge, MA, 1993), p. 449; John Mann, *Magic, Murder and Medicine*, (Oxford University Press, New York, 1994); Evelyn Reed, *Woman's Evolution from Matriarchal Clan to Patriarchal Family* (Pathfinder, New York, 1975); Rosalind Miles, *The Women's History of the World*, (Harper & Row, New York, 1989), p. 110.
[3] Leonard John Goldwater, *Mercury: A History of Quicksilver* (York, Baltimore, 1972), p. 21.
[4] Richard Olson, *Science Deified and Science Defied: The Historical Significance of Science in Western Culture* (University of California Press, Berkeley, 1982), p. .210. Copyright (c) 1982 The Regents of the University of California.
[5] John Maxson Stillman, *Theophrastus Bombastus Von Hohenheim Called Paracelsus. His Personality and Influence As Physician, Chemist, and Reformer,* (Open Court, Chicago, 1920).
[6] Eric John Holmyard, *Makers of Chemistry* (Oxford at the Clarendon Press, 1931), pp. 111–12.
[7] Jared Diamond, *Discover Magazine* (Oct. 1992), p. 64. On the other hand syphilis may have existed in Europe as far back as the early Greek civilization, but it may have been misdiagnosed as leprosy, see Rick Gore, *National Geographic* (Nov. 1994), p. 2.
[8] Will Durant, *Our Oriental Heritage* (Simon and Schuster, New York, 1963), p. 530–31.
[9] John Maxson Stillman, *Theophrastus Bombastus Von Hohenheim Called Paracelsus: His Personality and Influence As Physician, Chemist, and Reformer* (Open Court, Chicago, 1920).

CHAPTER 6

[1] Henry Hobhouse, *Seeds of Change, Five Plants That Transformed Mankind* (HarperCollins, New York 1987), p. 14. Copyright 1985, 1986 by Henry Hobhouse.
[2] Described as a man with "red hair and all together more children than a bachelor should have." See Joseph Bronowski, *The Ascent of Man* (Little, Brown, Boston, 1973), p. 200. It may be that Galileo should be identified more strongly than Descartes with

the transition from impressionistic thinking to true scientific thought due to his objective observations of nature, scrutiny of conclusions, and insistence on experimental verification. However, Descartes in France was a more effective spokesperson; Galileo was inhibited by his proximity to the Vatican.

[3] *Dictionary of Scientific Biography*, vol. 6 Charles Coulston Gillispie, ed. (Scribner's, New York, 1976), p. 254.

[4] J. R. Partington, *A Short History of Chemistry*, 3d ed. (Macmillan, New York, 1957), p. 46.

[5] Ibid., pp. 44–45.

[6] *Dictionary of Scientific Biography*, vol. 6, Charles Coulston Gillispie, ed. (Scribner's, New York, 1976), p. 254.

[7] J. R. Partington, *A Short History of Chemistry*, 3d ed. (Macmillan, New York, 1957), pp. 72–73.

[8] Ibid., p. 70.

[9] E. N. da C. Andrade, *Nature* **136** (1935), p. 358.

[10] Lloyd Motz and Jefferson Hane Weaver, *The Story of Physics* (Plenum, New York, 1989), p. 76.

[11] Eric John Holmyard, *Makers of Chemistry* (Oxford at the Clarendon Press, New York, 1962), p. 152.

[12] J. R. Partington, *A Short History of Chemistry*, 3d ed. (Macmillan, New York, 1957), p. 84.

[13] Eric John Holmyard, *Makers of Chemistry* (Oxford at the Clarendon Press, New York, 1962), p. 146.

[14] Bernard Jaffe, *Crucibles: The Story of Chemistry from Ancient Alchemy to Nuclear Fission*, 4th ed. (Dover, New York, 1976), p. 29.

CHAPTER 7

[1] R. R. Palmer and Joel Colton, *A History of the Modern World*, 6th ed. (Knopf, New York, 1984), p. 322.

[2] Ibid., p. 330.

[3] *Dictionary of Scientific Biography*, vol. 12, Charles Coulston Gillispie, ed. (Scribner's, New York, 1976), p. 145.

[4] Bernard Jaffe, *Crucibles: The Story of Chemistry from Ancient Alchemy to Nuclear Fission*, 4th ed. (Dover, New York, 1976), p. 40.

[5] Aaron J. Ihde, *The Development of Modern Chemistry* (Harper & Row, New York, 1964), pp. 49–50.

[6] J. R. Partington, *A Short History of Chemistry*, 3d ed. (Macmillan, New York, 1957), p. 118.

[7] *A Century of Chemistry, the Role of Chemists and the American Chemical Society*, Kenneth M. Reese, ed. (American Chemical Society, Washington, D.C. 1976).

[8] Bernard Jaffe, *Crucibles: The Story of Chemistry from Ancient Alchemy to Nuclear Fission*, 4th ed. (Dover, New York, 1976), p. 58. This same person, Jean-Francois

Pilatre de Rozier, was the first to attempt flight in a hydrogen balloon using hot air to control height; he was killed when the balloon caught fire. See Henry Monmouth Smith, *Torchbearers of Chemistry* (Academic, New York, 1949), p. 221.
9. *Dictionary of Scientific Biography*, vol. 8, Charles Coulston Gillispie, ed. (Scribner's, New York, 1976), p. 469.

CHAPTER 8

1. Aaron J. Ihde, *The Development of Modern Chemistry* (Harper & Row, New York, 1964), p. 61.
2. *Dictionary of Scientific Biography*, vol. 8, Charles Coulston Gillispie, ed. (Scribner's, New York, 1976), p. 75.
3. Aaron J. Ihde, *The Development of Modern Chemistry* (Harper & Row, New York, 1964), p. 64
4. Ibid., p. 68.
5. Ibid., p. 61.
6. *Dictionary of Scientific Biography*, vol. 8, Charles Coulston Gillispie, ed. (Scribner's, New York, 1976), p. 82.
7. Eric John Holmyard, *Makers of Chemistry* (Oxford at the Clarendon Press, 1931), p. 212.
8. W. A. Smeaton, *Fourcory, Chemist and Revolutionary (1755–1809)* (University College, London, 1962), p. 58.

CHAPTER 9

1. Aaron J. Ihde, *The Development of Modern Chemistry* (Harper & Row, New York, 1964), p. 100.
2. J. R. Partington, *A Short History of Chemistry*, 3d ed. (Macmillan, New York, 1957), p. 171.
3. Stephen F. Mason, *A History of the Sciences* (Collier, New York, 1962), p. 475.
4. *Dictionary of Scientific Biography*, vol. 14, Charles Coulston Gillispie, ed. (Scribner's, New York, 1976), p. 69.
5. Ibid., p. 77.
6. Ibid., p. 68.
7. Ibid.
8. J. R. Partington, *A Short History of Chemistry*, 3d ed. (Macmillan, New York, 1957), p. 198.
9. Mary Elvira Weeks and Henry M. Leicester, *Discovery of the Elements*, 7th ed. (Journal of Chemical Education, Austin, TX, 1968), p. 315.

CHAPTER 10

1. Lennard Bickel, *The Deadly Element, the Story of Uranium* (Stein and Day, New York, 1979), p. 21.
2. All these stories and more can be found in Mary Elvira Weeks and Henry M. Leicester, *Discovery of the Elements*, 7th ed. (Journal of Chemical Education, Austin, TX, 1968). Also see A. M. White and H. B. Friedman, J. Chem. Ed., 9, (1932), p. 238.
3. Sir Harold Hartley, *Humphry Davy* (Nelson, London, 1966), p. 31.
4. Anne Terneer, *Mercurial Chemist: A Life of Sir Humphry Davy*, (Methuen, London, 1963), p. 37.
5. Ibid.
6. Ibid., p. 41.
7. Stephen F. Mason, *A History of the Sciences*, (Collier, New York, 1962), p. 441.
8. Colin A. Russell, *Sir Humphry Davy* (Open University Press, Buckingham, England 1972), p. 70.
9. *Dictionary of Scientific Biography*, vol. 14, Charles Coulston Gillispie, ed. (Scribner's, New York, 1976), p. 79.
10. Joan Solomon, *Structure of Matter: The Growth of Man's Ideas on the Nature of Matter* (David and Charles, Newton Abbot, England, 1973), p. 71.
11. Anne Terneer, *Mercurial Chemist: A Life of Sir Humphry Davy* (Methuen, London, 1963), p. 98
12. K. M. Reese, *Chem. Eng. News* (Apr. 19, 1993), p. 56.
13. Ibid.
14. K. M. Reese, *Chem. Eng. News* (May 3, 1993), p. 64.
15. *Women in Chemistry and Physics: A Biobibliographic Sourcebook*, Louise S. Grinstein, Rose K. Rose, and Miriam H. Rafailovich, eds. (Greenwood Press, Westport, CT, 1993), p. 371.
16. L. Pearce Williams, *Michael Faraday* (Da Capo, New York, 1965), p. 28.
17. Ibid., p. 44.
18. Ibid.
19. Sidney Ross, *Nineteenth-Century Attitudes: Men of Science* (Kluwer Academic Publishers, Dordrecht, Netherlands, 1991), pp. 150–58. Reprinted by permission of Kluwer Academic Publishers.
20. Ibid.
21. Colin A. Russell, *Sir Humphry Davy* (Open University Press, Buckingham, England, 1972), 70.
22. Anne Treneer, *Mercurial Chemist: A Life of Sir Humphry Davy* (Methuen, London, 1963), p. 219.
23. *Dictionary of Scientific Biography*, vol. 5 Charles Coulston Gillispie, ed. (Scribner's, New York, 1976), p. 322.
24. F. A. Miller, *J. Chem. Ed.* **63** (1986), p. 685.

CHAPTER 11

[1] Walter J. Moore, *Physical Chemistry*, 4th ed. (Prentice-Hall, Englewood Cliffs, 1972), p. 119.
[2] Bernard Jaffe, *Crucibles: The Story of Chemistry from Ancient Alchemy to Nuclear Fission*, 4th ed. (Dover, New York, 1976), p. 76.
[3] D. S. L. Cardwell, *From Watt to Clausius, the Rise of Thermodynamics in the Early Industrial Age* (Heinemann, London, 1971), p. 80.
[4] Walter J. Moore, *Physical Chemistry*, 4th ed. (Prentice-Hall, Englewood Cliffs, NJ, 1972), p. 40.
[5] *Dictionary of Scientific Biography*, vol. 7, Charles Coulston Gillispie, ed. (Scribner's, New York, 1976), p. 382.
[6] Joseph H. Noggle, *Physical Chemistry*, 2d ed. (Scott, Foresman, Boston, 1989), p. 129.
[7] *Dictionary of Scientific Biography*, vol. 14, Charles Coulston Gillispie, ed. (Scribner's, New York, 1976), p. 185.
[8] Ibid., vol. 16, p. 186.
[9] Ibid., vol. 2, p. 267.
[10] Vivian Ovelton Sammons, *Blacks in Science and Medicine* (Hemisphere, New York, 1990), p. 32.
[11] Lynde Phelps Wheeler, *Josiah Willard Gibbs, the History of a Great Mind*, rev. ed., (Yale University Press, New Haven, CT, 1952), p. 181.
[12] G. N. Lewis and Merle Randall, *Thermodynamics and the Free Energy of Chemical Substances* (McGraw-Hill, New York, 1923), p. 26.
[13] *Dictionary of Scientific Biography*, vol. 15, Charles Coulston Gillispie, ed. (Scribner's, New York, 1976), supplement 1, p. 463.

CHAPTER 12

[1] Norman L. Allinger, Michael P. Cava, Don C. De Jongh, Carl R. Johnson, Norman A. Lebel, and Calvin L. Stevens, *Organic Chemistry*, 2d ed. (Worth, New York), 1976.
[2] Bernard Jaffe, *Crucibles: The Story of Chemistry from Ancient Alchemy to Nuclear Fission*, 4th ed. (Dover, New York, 1976), p. 138.
[3] *Dictionary of Scientific Biography*, vol. 14, Charles Coulston Gillispie, ed. (Scribner's, New York, 1976), p. 474.
[4] Alan J. Rocke, *Quiet Revolution, Hermann Kolbe and the Science of Organic Chemistry* (University of California Press, Berkeley, 1993).
[5] Reprinted with permission from O. Theodor Benfey, *From Vital Force to Structural Formulas* (American Chemical Society, Washington, D.C., 1975), p. 46. Copyright 1975 American Chemical Society.
[6] Ibid.
[7] Ibid., p. 40.

8. Aaron J. Ihde, *Development of Modern Chemistry* (Harper & Row, New York, 1964), p. 196.
9. C. A. Russell, *The History of Valency* (Leicester University Press, Leicester, UK 1971), pp. 39–40.
10. Leonard Dobbin, *J. Chem. Ed.* **11** (1934), p. 335.
11. Reprinted with permission from O. Theodor Benfey, *From Vital Force to Structural Formulas* (American Chemical Society, Washington, D.C., 1975), pp. 76–77. Copyright 1975 American Chemical Society.
12. Ibid., p. 102.
13. Ibid., p. 107.
14. Ibid., p. 107.
15. Emmett Reid, *My First Hundred Years* (Chemical Publishing Co., New York, 1972), p. 60.
16. Frederick H. Getman, *The Life of Ira Remsen* (Journal of Chemical Education, Easton, PA, 1940), p. 9.
17. Dean Stanley Tarbell and Ann Tracy Tarbell, *Essays on the History of Organic Chemistry in the United States, 1875–1955* (Folio Publishers, Nashville, TN, 1986), p. 43.

CHAPTER 13

1. Mary Elvira Weeks and Henry M. Leicester, *Discovery of the Elements*, 7th ed. (Journal of Chemical Education, Austin, TX, 1968), p. 662.
2. *Dictionary of Scientific Biography*, vol. 11, Charles Coulston Gillispie, ed. (Scribner's, New York, 1976), p. 279.
3. Bernard Jaffe, *Crucibles: The Story of Chemistry from Ancient Alchemy to Nuclear Fission*, 4th ed. (Dover, New York, 1976), pp. 159–60.
4. Radon has received some publicity because of relatively high concentrations found in some homes built on soils or near rocks rich in uranium. This is not considered a major source of radiation exposure however because homes can be tested and the problem controlled by proper ventilation because radon is a gas.
5. Emmett Reid, *My First Hundred Years* (Chemical Publishing Co., New York, 1972), p. 56.
6. Eve Curie, *Madame Curie*, translated by Vincent Seean, (Doubleday Doran, New York, 1937), p. 154.
7. Ibid., p. 499.
8. Ibid., p. 172.
9. Mary Elvira Weeks and Henry M. Leicester, *Discovery of the Elements*, 7th ed. (Journal of Chemical Education, Austin, TX, 1968), p. 810.
10. Eve Curie, *Madame Curie*, translated by Vincent Seean (Doubleday Doran, New York, 1937), p. 278.
11. Emilio Segré, *From X-Rays to Quarks—Modern Physicists and Their Discoveries* (Freeman, San Francisco, 1980), p. 44.

ENDNOTES

[12] Lennard Bickel, *The Deadly Element, The Story of Uranium* (Stein and Day, New York, 1979), p. 30.
[13] Bernard Jaffe, *Crucibles: The Story of Chemistry from Ancient Alchemy to Nuclear Fission*, 4th ed. (Dover, New York, 1976), p. 211.
[14] Ibid., p. 169.
[15] George B. Kauffman, *Alfred Werner, Founder of Coordination Chemistry* (Springer-Verlag, New York, 1966).
[16] Ibid., p. 52.
[17] Ibid., p. 47.

CHAPTER 14

[1] Henry Hobhouse, *Seeds of Change, Five Plants That Transformed Mankind* (HarperCollins, 1987), pp. 12–13.
[2] George B. Kauffman and Paul M. Priebe, *J. Chem. Ed.* **66** (1989), p. 397.
[3] Fred Aftalion, *A History of the International Chemical Industry* (University of Pennsylvania Press, Philadelphia, 1991), p. 95.
[4] Gerald Weissmann, *Scientific American* (Jan. 1991), p. 84.
[5] Eldon J. Gardner, *History of Biology*, 3d ed. (Burgess, Minneapolis, 1972), pp. 169–76.
[6] Joseph Needham, ed., *The Chemistry of Life: Eight Lectures on the History of Biochemistry* (Cambridge at the University Press, Cambridge, England, 1970), p. 39.
[7] Ibid., p. 48
[8] Gerald Weissmann, *Scientific American* (Jan. 1991), p. 84.
[9] Colin A. Russell with Noel G. Coley and Gerrylynn K. Roberts, *Chemists by Profession, The Origins and Rise of the Royal Institute of Chemistry* (Open University Press, Milton Keynes, England, 1977), p. 103.
[10] G. Kass-Simon and Patricia Farnes, eds., *Women of Science—Righting the Record* (Indiana University Press, Bloomington, IN, 1990), p. 152.
[11] Nicholas Halasz, *Nobel, A Biography* (Orion Press, New York, 1959), p. 111.
[12] Ibid., p. 114
[13] George W. Gray, *Scientific American* (Dec. 1949), p. 11.
[14] Ibid., p. 12.
[15] Ibid., p. 13.
[16] Ibid.
[17] Max F. Perutz, *Is Science Necessary? Essays on Science and Scientists* (Barrie and Jenkins, London, 1989), p. 184.
[18] Sterling Seagrave, *Yellow Rain* (Evans, New York, 1981), p. 42. From Yellow Rain © 1981 by Sterling Seagrave. Reprinted by permission of the publisher, M. Evans and Company, Inc.
[19] Ibid., p. 43.

CHAPTER 15

[1] Personal communication, Dr. David Karraker.
[2] Emilio Segré, *From X-Rays to Quarks—Modern Physicists and Their Discoveries* (Freeman, San Francisco, 1980), p. 76.
[3] Armin Hermann, *The Genesis of Quantum Theory*, translated by Claude W. Nash (Massachusetts Institute of Technology, Cambridge, 1971), p. 15.
[4] Max Jammer, *The Conceptual Development of Quantum Mechanics* (McGraw-Hill, New York, 1966), p. 77.
[5] Ibid., p. 136.
[6] *Dictionary of Scientific Biography*, vol. 10 Charles Coulston Gillispie, ed. (Scribner's, New York, 1976), p. 423.
[7] Emilio Segré, *From X-Rays to Quarks—Modern Physicists and Their Discoveries* (Freeman, San Francisco, 1980), p. 152.
[8] Max Jammer, *The Conceptual Development of Quantum Mechanics* (McGraw-Hill, New York, 1966) p. 258.
[9] David C. Cassidy, *Scientific American* (May 1992), p. 106.
[10] Ibid., p. 109.
[11] David Z. Albert, *Scientific American* (May 1994), p. 58.
[12] Reprinted with permission from *Nobel Laureates in Chemistry, 1901–1992*, Laylin K. James, ed. (American Chemical Society, Chemical Heritage Foundation, Washington, D.C., 1993), p. 472. Copyright 1993 American Chemical Society.
[13] Linus Pauling, *The Nature of the Chemical Bond and the Structure of Molecules and Crystals: An Introduction to Modern Structural Chemistry*, 3d ed. (Cornell Universtiy Press, Ithaca, New York,1960), p. 387.
[14] Adapted from an analogy suggested by Professor Marshal Cronyn, Reed College, Portland, Oregon.
[15] Reprinted with permission from *Nobel Laureates in Chemistry, 1901–1992*, Laylin K. James, ed. (American Chemical Society, Chemical Heritage Foundation, Washington, D.C., 1993), p. 368. Copyright 1993 American Chemical Society.
[16] Ibid., p. 471.
[17] Ibid., p. 474.
[18] Walter J. Moore, *Physical Chemistry*, 4th ed. (Prentice-Hall, Englewood Cliffs, NJ, 1972), p. 688.
[19] Paul G. Hewitt, *Conceptual Physics*, 7th ed. (HarperCollins, New York, 1993), p. 591.
[20] James L. Stokesbury, *A Short History of WW II* (William Morrow, New York, 1980), p. 377.

CHAPTER 16

[1] Peter J. T. Morris, *Polymer Pioneers, A Popular History of the Science and Technology of Large Molecules* (Beckman Center for the History of Chemistry, publication no. 5 Philadephia, Pennsylvania, 1986).

ENDNOTES

[2] Royston M. Roberts, *Serendipity, Accidental Discoveries in Science*, (Wiley, New York, 1989), p. 175.
[3] Reprinted with permission from *Nobel Laureates in Chemistry*, 1901–92, Laylin K. James, ed. (American Chemical Society, Chemical Heritage Foundation, Washington, D.C., 1993). Copyright 1993, American Chemical Society.
[4] P. A. Lehmann, A. Bolivar, and R. Quintero, *J. Chem. Ed.* **50** (1973), p. 195.
[5] J. D. Watson and F. H. C. Crick, *Nature* **171** (1953), p. 737. Reprinted with permission from *Nature* copyright 1953, Macmillan Magazine Limited.
[6] Wilkins and Franklin/Gosling have their papers immediately following: M. H. F. Wilkins, *Nature* **171** (1953), p. 738 and Rosalind E. Franklin and R. G. Gosling, *Nature* **171** (1953), p. 740.
[7] Anne Sayre, *Rosalind Franklin and DNA* (Norton, New York, 1975), p. 151. Copyright 1975 by Anne Sayre.
[8] Ibid., p. 104
[9] Ibid., p. 105
[10] Ibid., p. 128.
[11] James D. Watson, *The Double Helix—A Personal Acount of the Discovery of the Structure of DNA* (Atheneum, New York, 1968), pp. 75–76. Copyright 1968, James D. Watson.
[12] Ibid., pp. 167–69.
[13] James D. Watson, *The Double Helix—A Personal Acount of the Discovery of the Structure of DNA* (Atheneum, New York, 1968), p. 20. Copyright 1968, James D. Watson.
[14] Joseph S. Fruton, *A Skeptical Biochemist* (Harvard University Press, Cambridge, MA, 1992), p. 224.
[15] André Lwoff, *Scientific American* (July 1968), p. 133.
[16] Personal communication from Dr. Clarence M. Cobb.
[17] Robert Olby, *The Path to the Double Helix* (University of Washington Press, Seattle, 1974), p. 362.
[18] Merriley Borell, *Album of Science: The Biological Sciences in the Twentieth Century* (Scribner's, New York, 1989), p. 181.

CHAPTER 17

[1] For an interesting account of the development of Beckman Instruments from a storefront enterprise to a multimillion-dollar industry, see Beckman Instruments, Inc. "There is no satisfactory substitute for excellence," an address by Dr. Arnold O. Beckman to the Newcomen Society (Newcomen Society in North America, New York, 1976).
[2] Peter Hayes, *Industry and Ideology, I.G. Farben in the Nazi Era* (Cambridge University Press, Cambridge, 1987).
[3] *Great Chemists*, Eduard Farber, ed. (Interscience, New York, 1961), pp. 1439–40.
[4] Ron Dagani, *Chem. Eng. News* (May 23, 1994), p. 39.
[5] For an interesting discussion of the war effort that includes an investigation into incendiaries and the feasibility of using bats to carry bombs, see Louis F. Fieser, *The*

Scientific Method: A Personal Account of Unusual Projects in War and in Peace (Reinhold, New York, 1964).
6. Ron Dagani, *Chem. Eng. News* (May 23, 1994), p. 40.
7. Anne Sayre, *Rosalind Franklin and DNA* (Norton, New York, 1975), p. 138. Copyright 1975 by Anne Sayre.
8. *Women in Chemistry and Physics: a Biobibliographic Sourcebook*, Louise S. Grinstein, Rose K. Rose, and Miriam H. Rafailovich, eds. (Greenwood Press, Westport, CT, 1993), p. 333.
9. Reprinted with permission from *Nobel Laureates in Chemistry, 1901–1992*, Laylin K. James, ed. (American Chemical Society, Chemical Heritage Foundation, Washington, D.C., 1993), p. 306. Copyright 1993, American Chemical Society.
10. Ibid., p. 307.
11. William L. Jorgensen and Lionel Salem, *The Organic Chemists Book of Orbitals* (Academic, New York, 1973). This book is dedicated to the people of Vietman, and it opens with the statement, "Science sans conscience n'est que ruine de l'âme."
12. Reprinted with permission from *Nobel Laureates in Chemistry, 1901–1992*, Laylin K. James, ed. (American Chemical Society, Chemical Heritage Foundation, Washington, D.C., 1993), p. 466. Copyright 1993, American Chemical Society.
13. Shekhar Hattangadi, *Industrial Chemist* (February, 1988) pp. 22.
14. Reprinted with permission from *Nobel Laureates in Chemistry, 1901–1992*, Laylin K. James,ed.(American Chemical Society, Chemical Heritage Foundation, Washington, D.C., 1993), p. 561. Copyright 1993, American Chemical Society.
15. James E. Huheey, Ellen A. Keiter, and Richard L. Deiter, *Inorganic Chemistry: Principles of Structure and Reactivity*, 4th ed. (Harper Collins, New York, 1993),p. 813

CHAPTER 18

1. Walter J. Moore, *Physical Chemistry* (Prentice-Hall, Englewood Cliffs, NJ, 1972), p. 324.
2. *Chem. Eng. News* (Feb. 26, 1951), p. 763.
3. *Women in Chemistry and Physics: A Biobibliographic Sourcebook*, Louise S. Grinstein, Rose K. Rose, and Miriam H. Rafailovich, eds. (Greenwood Press, Westport, CN, 1993), p. 502.
4. Ibid., p. 503.
5. Ibid.
6. Ibid.
7. Ibid., p. 504.
8. Keith J. Laidler, *Chemical Kinetics*, 3d ed. (Harper Collins, New York, 1987), p. 496.
9. G. K. Vemulapalli, *Physical Chemistry* (Prentice-Hall, Englewood Cliffs, NJ, 1993).
10. See W. M. Kaushik, Zhi Yuan, and Richard M. Noyes, *J. Chem. Ed.* **63** (1986), p.76; and Lee R. Summerlin and James L. Ealy, Jr., *Chemical Demonstrations: A Sourcebook for Teachers* (American Chemical Society, Washington, D.C., 1985), pp. 81–83.

ENDNOTES

[11] Reprinted with permission from *Nobel Laureates in Chemistry, 1901–1992*, Laylin K. James, ed. (American Chemical Society, Chemical Heritage Foundation, Washington, D.C., 1993), p. 504. Copyright 1993, American Chemical Society.

[12] Ibid.

[13] Ilya Prigogine, *Order out of Chaos, Man's New Dialogue with Nature* (Bantam, New York, 1984).

CHAPTER 19

[1] Emilio Segré, *From X-Rays to Quarks—Modern Physicists and Their Discoveries* (Freeman, San Francisco, 1980), p. 182. Copyright 1980 by Emilio Segre. Used with permission of W.H. Freeman and Company.

[2] Lennard Bickel, *The Deadly Element, The Story of Uranium* (Stein and Day, New York, 1979), p. 60.

[3] Emilio Segré, *From X-Rays to Quarks—Modern Physicists and Their Discoveries* (Freeman, San Francisco, 1980), p. 183.

[4] Ibid., pp. 183–184.

[5] Lennard Bickel, *The Deadly Element, The Story of Uranium* (Stein and Day, New York, 1979), p. 66.

[6] Emilio Segré, *From X-Rays to Quarks—Modern Physicists and Their Discoveries* (Freeman, San Francisco, 1980), p. 199.

[7] Emilio Segré later worked in United States. He was codiscoverer of technetium, and he worked with Seaborg on plutonium. He won the Nobel Prize in 1959 for creating antiproton.

[8] Edoardo Amaldi was a chemist who opted to stay in Italy when others left.

[9] There were many other workers involved in the project, and we regret that we can not include all the names here. See Emilio Segré, *From X-Rays to Quarks—Modern Physicists and Their Discoveries* (Freeman, San Francisco, 1980).

[10] E. Fermi, *Nature* **133**, 1934, p. 898.

[11] Emilio Segré, *From X-Rays to Quarks—Modern Physicists and Their Discoveries* (Freeman, San Francisco, 1980), p. 205.

[12] Ruth Lewin Slime, *J. Chem. Ed.* **63** (1986), p. 653.

[13] Ruth Lewin Slime, *J. Chem. Ed.* **66** (1989), p. 373.

[14] *Who Found the Missing Link*, a video from the series The *Periodic Table and the Human Element* (Films for the Humanities & Sciences, Princeton, NJ, 1994).

[15] Ibid.

[16] Ibid.

[17] Glen T. Seaborg, *J. Chem. Ed.* **66** (1989), p. 379.

[18] For a discussion of this and other interesting historical developments in actinide chemistry, see George Kauffman, Chem. Eng. News, (Nov., 19, 1990), pp. 18–29.

[19] Emilio Segré, *From X-Rays to Quarks—Modern Physicists and Their Discoveries* (Freeman, San Francisco, 1980), p. 215.

[20] Lennard Bickel, The Deadly Element, the Story of Uranium (Stein and Day, New York, 1979), p.257
[21] Ibid.
[22] from MY HIROSHIMA by Junko Morimoto. Copyright © 1989 by Junko Morimoto. Used by permission of Viking Penguin, a division of Penguin Books USA, Inc. This is a well-written and well-illustrated children's book.
[23] Produced by the Library of Science, 59 Fourth Ave., New York.
[24] George W. Gray, *Scientific American* (Dec. 1949), p. 11.
[25] For a full list of collaborators (and there were many), see Glenn T. Seaborg and Walter D. Loveland, *The Elements Beyond Uranium* (Wiley, New York, 1990).
[26] Ibid., p. 27. Reprinted by permission; © 1950, 1978, The New Yorker Magazine, Inc.
[27] Darlene Hoffmann, *Chem. Eng. News* (May 2, 1994), p. 24.
[28] Glenn T. Seaborg and Walter D. Loveland, *The Elements Beyond Uranium* (Wiley, New York, 1990), p. 44.
[29] Ron Dagani, "Shuffling of Heavy-Element Names by IUPAC Panel Provokes Outcries," *Chem. and Eng. News*, December 5, 1994, p. 25; *Science News*, October 22, p. 271.
[30] Lennard Bickel, *Deadly Element, the Story of Uranium* (Stein and Day, New York, 1979), p. 97.
[31] David Holloway, *Chemtech* (Feb. 1991), p. 80.
[32] See Bernard L. Cohen, *Before It's Too Late: A Scientist's Case for Nuclear Energy* (Plenum, New York, 1983); and Max F. Perutz, *Is Science Necessary: Essays on Science and Scientists* (Barrie and Jenkins, London, 1989).

CHAPTER 20

[1] Jonathan I. Lunine, *American Scientist* **82** (1993), p. 134.
[2] G. K. Vemulapalli, *Physical Chemistry* (Prentice-Hall, Englewood Cliffs, NJ, 1993), p. 425.
[3] There is some controversy as to whether the Green Revolution was all that good a thing. Some societies became dependent on the new technology, and they were unable to return to their previous methods when problems with the new technology were found. Just as Davy's invention of the safety lamp caused coal miners to dig deeper, more dangerous mines, the Green Revolution caused some countries to expand in a manner that again stressed their food supply.
[4] Rachel L. Carson, *Silent Spring* (Houghton Mifflin, New York, 1987).
[5] Susan Chollar, *Discover Magazine* (April 1990), p. 76; and *Discover Magazine* (Aug. 1993).
[6] Darleane C. Hoffman and Gregory R. Choppin, *J. Chem. Ed.* **63** (1986), p. 1059; and Bernard L. Cohen, *Scientific American* (June 1977), p. 21.
[7] Gary Taubes, *Bad Science: The Short Life and Weird Times of Cold Fusion* (Random House, New York, 1993), p. 4.
[8] Ibid., p. xviii.
[9] Ibid., p.124.

ENDNOTES

[10] Ibid., p. 123.
[11] Gary Stix, *Scientific American* (Nov. 1993), p. 104; and John Rennie, *Scientific American* (March 1994), p. 107.
[12] Andreas Stein, Steven W. Keller, and Thomas E. Mallouk, *Science* **259** (1993), p. 1558.
[13] Gary Taubes, *Bad Science: The Short Life and Weird Times of Cold Fusion* (Random House, New York, 1993), p. 165.
[14] John Clarke, *Scientific American* (August 1994), p. 46.
[15] Philip Yam, *Scientific American* (Dec. 1993), p. 118; and Philip Yam, *Scientific American* (Jan. 1994), p. 18.
[16] Alex Nickon and Ernest F. Silversmith, *The Name Game, Modern Coined Terms and Their Origins* (Pergamon, New York, 1987).
[17] H. W. Kroto, J. R. Heath, S. C. O'Brien, R. F. Curl, and R. E. Smalley, *Nature* **318** (1985), p. 162; W. Krätschmer, Lowell D. Lamb, K. Fostiropoulos, and Donald R. Huffman, *Nature* **354** (1990) p. 283; and *Science News* **140** (1993), p. 84.
[18] Ian W. M. Smith, *Nature* **362** (1993), p. 498; and I. Peterson, *Science News* (July 31, 1993), p. 71.
[19] Michael D. Jones and Jeffrey T. Feyerman, *J. Chem. Ed.* **64** (1987), p. 337.
[20] Charles E. Bugg, William M. Carson, and John A. Montgomery, *Scientific American* (Dec. 1993), p. 92.
[21] F. N. Spiess, Ken C. Macdonald, T. Atwater, R. Ballard, A. Carranza, D. Cordoba, C. Cox, V. M., Diaz Garcia, J. Francheteau, J. Guerrero, J. Hawkins, R. Haymon, R. Hessler, T. Juteau, M. Kastner, R. Larson, B. Luyendyk, J. D. Macdougall, S. Miller, W. Normark, J. Orcutt, and C. Rangin, *Science* **207** (1980), p. 1421.
[22] Kary B. Mullis, *Scientific American* (Apr. 1990), p. 56.
[23] Ibid.
[24] Ivan Amato, *Science* **262** (1993), p. 506.
[25] Sharon Begley, *Newsweek* (June 14, 1993), p. 57; Svante Pääbo, *Scientific American* (Nov. 1993), p. 86; and Philip E. Ross, *Scientific American* (May 1992), p. 115.
[26] Raloff, *Science News* (Aug. 6, 1986), p. 71.
[27] Cindy Lee Van Dover, *Discover Magazine* (Sept. 1993), p. 37.
[28] Lois R. Ember, *Chem. Eng. News* (Mar. 21, 1994), p. 16.
[29] *Accounts of Chemical Research*, Mar. 1995.
[30] Peter Armbruster and Gottfried Münzenberg, *Scientific American* **260**, 1989 p. 66; Darleane C. Hoffman, *Chem. Eng. News* (May 2, 1994), p. 24; S. G. Thompson and C. F. Tsang, *Science* **178**, 1972 p. 1047; and *The Chemistry of the Actinide Elements*, J. J. Katz, G. Seaborg, and L. R. Morss, eds. (Chapman and Hall, New York, 1986). Figure 20.1 is adapted from Charles H. Atwood and R. K. Sheline, *J. Chem. Ed.* **66** (1989), p. 391.
[31] Michael Freemantle, *Chem. Eng. News* (Mar. 13, 1995) p. 35.

Annotated Bibliography

For readers who want to read further in the history of chemistry we offer the following bibliography. Only sources in English and sources likely to be available (even if out-of-print) in good public libraries or university or college libraries are listed.

REFERENCE WORKS

C. C. Gillispie, ed., *Dictionary of Scientific Biography* (Scribner's, New York, 1970–85); over 20 well-indexed volumes cover most major figures in the development of chemistry (and other sciences) as long as they are dead. Articles range in length from less than a page to tens of pages, and they include references to sources.

J. R. Partington, *A History of Chemistry* (St. Martin's, London, 1962–70), 4 vol; cannot be read as a connected history. Excellent articles on individuals and on some themes. Partington never completed or published a projected volume on alchemy, but the first volume on early science is well done. Lists of publications of important chemists are thorough.

REPRINTS OF CLASSIC WORK IN CHEMISTRY

O. T. Benfey, *Classics in the Theory of Chemical Combination* (Dover, New York, 1963), paperback; articles by founders of nineteenth-century organic structural theory.

D. Knight, ed., *Classical Scientific Papers: Chemistry, Series 1 and 2* (American Elsevier, New York, 1968, 1970); facsimile reprints of papers tracing themes in chemical history.

H. M. Leicester, and H. S. Klickstein, *A Source Book in Chemistry, 1400–1900* (Harvard University Press, Cambridge, 1965); excerpts from classic papers.

A. Lavoisier, *Elements of Chemistry*, (Dover, New York, 1965); paperback; facsimile of the 1790 Robert Kerr translation of Lavoisier's influential textbook.

OTHER RECOMMENDED WORKS

O. T. Benfey, *From Vital Force to Structural Formulas* (Houghton-Mifflin, Boston, 1964), reprinted by the American Chemical Society in paperback; excellent history of the early development of organic chemistry.

M. P. Crosland, *Historical Studies in the Language of Chemistry* (Harvard University Press, Cambridge, MA 1962); interesting survey of why chemical compounds and concepts are called what they are.

E. Farber, ed., *Great Chemists* (Interscience, New York, 1962); useful compilation of short biographies.

L. S. Grinstein, R. K. Rose, and M. H. Rafailovich, ed., *Women in Chemistry and Physics: A Biobibliographic Sourcebook* (Greenwood Press, Westport, CT, 1993).

E. Holmyard, *Alchemy* (Penguin, London, 1957) paperback.

H. M. Leicester, and M. E. Weeks, *Discovery of the Elements* 7th. ed., (American Chemical Society, Easton PA, 1968); Still the best popular account of these fascinating stories. (Someone out there should update this classic.)

R. P. Multhauf, *The Origins of Chemistry* (Oldbourne, London, 1966); scholarly investigation of the threads that were woven together into chemistry.

J. M. Stillman, *The Story of Alchemy and Early Chemistry* (Dover, New York, 1960), paperback.

COMPETITION (OTHER CURRENT SHORT HISTORIES)

W. H. Brock, *The Norton History of Chemistry* (W. W. Norton, New York, 1993).

J. Hudson, *The History of Chemistry* (Chapman and Hall, New York, 1992), paperback.

D. Knight, *Ideas in Chemistry: A History of the Science* (Rutgers University Press, New Brunswick, NJ, 1992), paperback.

J. R. Partington, *A Short History of Chemistry* (Dover, New York, 1990), paperback.

H. W. Salzberg, *From Caveman to Chemist* (American Chemical Society, Washington D.C., 1991).

Index

Abel, Frederick, 298
Abortificant, 56
Abraham, 59
Absolute zero, 222
Abu Bakr Mohammad ibn Zakariyya al-Razi, 64
Academy
　Athens, 24
　Plato's, 32
Academy of Science, 162
　French, 109, 154, 156, 164, 166, 171, 247, 271, 297, 400
　Royal Prussian, 187
　Russian, 13, 260
Accum, Fredrick, 293
Acetaldehyde, 415
Acetic acid: *See* Acid, acetic
Acetone, 302, 303, 421,
Acetyl-choline, 383
Acetylene, 236, 249
Acetylsalicylic acid: *See* acid, acetylsalicylic
Acid, 36, 112, 135, 159, 164, 187, 194, 195, 203, 207, 313, 413, 430

Acid (*cont.*)
　acetic 52, 87, 236
　acetylsalicylic 290
　amino, 289, 343, 356, 427
　aqua regia, 15, 87, 104
　barbituric, 287
　benzoic, 241
　carbolic: *See* phenol
　carbonic, 158
　citric, 87, 236
　cyanic, 239
　formic, 236
　fulminic, 239
　glutamic, 289
　hydrochloric, 15, 87, 111, 204
　hydrofluoric, 260
　isocyanic, 239
　lactic, 236
　mineral, 49, 86–89
　nitric, 15, 87, 96, 113, 144–146, 158, 192, 255, 303, 298, 417
　nucleic: *See* nucleic acid
　organic, 158, 236

Acid (*cont.*)
  racemic, 252
  ribonucleic: *See* ribonucleic acid
  salicylic, 290
  sulfuric, 87, 104, 113, 125, 126, 140, 158, 165, 188, 205, 289, 298, 327, 417
  sulfurous, 165
  tartaric, 250, 252, 291
  uric, 289
Acid meters, 360
Acid rain, 417
Aconite, 42
Actinide series, 403
Actinium, 397
Adelard of Bath, 73
Adenine, 356
Adrenaline, 302, 349
Affinity, 210
Affinity, dualistic theory, 209, 241, 277
Africa, 9, 17, 41, 52, 127, 302, 304, 415
Age of Oil, 358
Agricola, Georgius, 103
Air, 134, 156, 164
Air, 296
Air pump, 180
Al-Khwarizmi, 73
Al-Razi, 64, 73
Albert the Great, 79, 83
Albertus Magnus: *See* Albert the Great
Alchemist, 188, 236
  European, 39
  female, 37, 46, 49
Alchemy
  Alexandrian, 29
  Arab, 62
  Chinese, 41–50
  European, 92–97
  Indian, 41–50
Alchymia, 104
Alcohol, 12, 47, 86, 89, 164, 222, 236, 245, 246, 285, 286, 291, 346
Alexander the Great, 25–29
Alexandria, 28, 29
Algebra, 73
Ali al-Husayn ibn Sina, 66
Alkali, 12, 195
Alkali metals, 165

Alkaline, 65
Alkaline earths, 182
Alkalinity, 210
Alkalis, 164
Alkaloid, 285, 289
Alkanes, 236
Alkenes, 236
Alkynes, 236
Allotropes, 184
Allotropic forms, 405
Alloys, 103
Alpha particle, 273, 398
Alpha radiation: *See* Radiation, alpha
Altman, Sidney, 427
Alum, 10, 104
Aluminum, 374, 394
Amaldi, Edoardo, 394
Amalgamation, 43
Amber, 427
Americium, 410, 411, 414
American Chemical Journal, 255
American Chemical Society, 142, 362, 385, 409, 417
American Journal of Science, 255
Amine, 246
Amino acid: *See* Acid, amino
Ammonia, 104, 141, 246, 277, 303, 305, 356, 360, 415
Ammonium chloride, 10, 36, 65, 113, 141
Ammonium cyanate, 240
Analogy, 22
Analytical chemist: *See* Chemist, analytical
Analytical instrumentation, 360
Anaxagoras, 18, 19
Anaximander, 18, 19
Anaximenes, 18, 19
Andromeda galaxy, 410
Androsterone, 349
Anesthetics, 45
Angular momentum, 320, 322, 329
Animal electricity: *See* Electricity, animal
Anion, 199
Annales de Chimie, 165
Anode, 199
Anthracene, 284
Anthrax, 292

INDEX

Antibiotics, 350
Antimony, 24, 188, 374
Antiseptics, 48, 350
Apeiron, 18
Apreece, Mrs., 195
Aquamarine, 250
Aqua regia: *See* Acid, aqua regia
Archaeus, 99
Arcueil, 173, 202, 207
Argon, 134, 144, 265, 301, 310
Aristotle, 24, 27, 29, 57, 63, 73, 76, 79, 80, 109, 112, 118, 121, 124, 126, 133, 152, 158, 161, 168, 177
Aristotleism, 33
Army Corps of Engineers, 401
Arndy, F. G., 328
Aromatic compounds, 249, 284
Arrhenius, Svante, 276, 301, 380
Arsenic, 8, 36, 242
Arsenic sulfides, 35
Ars Probandi, 104
Ash, 55
Asphalt, 62
Aspirin, 290, 306, 350
Association of German Chemists, 399
Assyrians, 8, 15
Astatine, 410
Astrochemistry, 416
Atom, 21, 177, 210, 240, 259, 272, 317, 318, 326
 cubic, 312
 planetary model, 273
 plum-pudding model, 319
Atomic bomb, 373, 404
Atomic number, 274
Atomic theory, 177–179
Atomic weight, 177, 179, 183, 258, 259, 274, 278, 392, 396
Aufbau principle, 322
Automobile, 335
Avicenna, 66, 73, 90, 97, 100, 101
Avogadro, Amedeo, 208
Avogadro's law, 208, 210, 259

Bacon, Francis, 108, 214
Bacon, Roger, 81–85
Baden Dye and Soda Company, 287
Badische Anilin und Soda-Fabrik, 287
Baeyer, Adolf, 287, 301, 347, 362, 363
Bain-marie, 37
Bainbridge, Kenneth, 407
Baker, James, 404
Baker, Nicholas, 404
Balances, 104
Ballooning, 202, 206, 207
Balmer, Johann, 319, 320
Banks, John, 181
Banks, Joseph, 197, 201
Banting, Frederick, 349
Barbituates, 287
Barium, 182, 398, 423
Barium chloride, 270
Barium platinum cyanide, 267
Bartholomew the Englishman, 77
Bartlett, Neil, 374
Barton, Derek, 364–366
Base, 203, 313
Base, nitrogen-containing, 352
BASF, 287, 303
Basketane, 424
Batteries, 421
Bayen, Pierre, 157
Bayer, 290
Beakers, 184
Becher, John Joachim, 124, 165
Becquerel, Henri, 267, 301, 390
Beddoes, Thomas, 190
Beer, 11, 86, 292
Bell Telephone laboratories, 323
Belousov, B. P., 387
Benzaldehyde, 241, 243, 339
Benzene, 198, 249, 284, 328, 366, 373
Benzoic acid, 241
Benzoin, 241
Benzoyl amide, 241
Benzoyl bromide, 241
Benzoyl chloride, 241, 243
Benzoyl cyanide, 241
Benzoyl ethyl ester, 242
Benzoyl iodide, 241
Benzoyl radical, 242
Bergius, Friedrich, 360
Berkelium, 411
Berlin Institute of Technology, 287
Bernard, Sarah, 198

Bernoulli, Daniel, 177
Bernoulli, Jean, 114
Berthelot, Marcellin 87, 265
Berthollet, Claude, 162, 165, 172, 202, 204, 231, 298, 379, 423,
Berthollide, 176, 423
Berzelius, Jöns, 182, 186, 200, 209, 237, 239, 244, 261, 277, 291, 306, 327
Best, Charles, 349
Beta radiation: *See* Radiation, beta
Bhopal, 416
Biocatalysis, 421
Biochemical Engineering, 425
Biochemistry: *See* Chemistry, biochemistry
Biological tracers, 413
Biomass, 375, 422
Biomolecules, 422
Bioremediation, 417
Biot, Jean, 252
Biringuccio, Vannoccio, 103
Birkbeck College, 356, 365
Birth-control pill: *See* Pill, the
Bismuth, 428
Bitumen, 55
Blackbody, 314, 317
Black, Joseph, 134, 154, 157, 216
Blagden, Charles, 159
Blanc, Le, 323
Blodgett, Katharine, 387
Blomstrand, Christian, 278
Bloom, 9
Bloomington shuffle, 373
Bodenstein, Max, 380, 383
Bohr, Aage, 404
Bohr, Henrik, 274
Bohr, Niels, 310, 319, 320, 322–324, 326, 327, 329, 334, 397–399, 404, 413
Bohrium, 412
Boiling point, 430
Boiling point elevation, 275
Boisbaudran, Paul, 263
Boltwood, Bertram, 395
Boltzmann, Ludwig, 225, 233, 325
Bomb, 400
Bomb test, 408

Bonaventura, 82
Bond
 angles, 332
 chemical, 248, 306, 326–328, 345
 covalent, 312, 327, 345
 double, 249
 hydrogen, 345, 353
 length, 327, 332
 nonpolar, 312
 peptide, 343
 polar, 310
 resonance, 328
 shared-electron-pair: *See* Bond, two-electron
 triple, 249
 two-electron, 312, 313, 326, 327
Bond-selective chemistry, 427
Book of Fires for Burning Enemies, 84
Book of the Preservation of the Solemn Seeming Philosopher, 47
Book of the Properties of Things, 78
Boraces, 65
Borlase, John, 190
Born, Max, 325, 326, 329
Boron, 374, 394
Bosch, Carl, 303, 360
Bothe, Walther, 393
Bouchet, Edward, 228
Boullay, Pierre, 241
Boulton, Matthew, 217
Bourdelain, 153
Boyle's law, 117, 224
Boyle, Robert, 114–119, 121, 122, 177, 214
Bragg, William Henry, 342
Bragg, William Lawrence, 327, 342
Brahe, Tycho, 109
Brand, Hennig, 109
Brass, 43, 47
Breeder reactor, 418
Brethren of Purity, 64
Berzelius, 197
British Alkali Act, 298
Broglie, Louis de, 322
Bromine, 257, 381
Bronze, 8, 43, 47, 55
Brooks, Harriet, 273

INDEX

Brownian motion, 318, 381
Büchner, Eduard, 301
Buckminsterfullerene, 423, 424
Buckybabies, 424
Buckyball, 424
Buckytubes, 424
Bucquet, Jean, 163
Bugs, 417, 422
Bunsen burner, 242
Bunsen, Robert, 242, 262, 287
Burning glass, 114
Butenandt, Adolf, 349
Butter of antimony, 163

Cacodyl cyanide, 242
Cacodyl radical, 242
Cacodylic compounds, 287
Cadmium, 186, 188
Caffeine, 289
Calcination, 35, 38, 117, 122, 124, 133, 161
Calcium, 182
   carbonate, 14, 15, 43, 55, 136, 172, 236
   hydroxide, 53, 136
   oxide, 14, 61, 65, 135
   sulfate, 10, 43, 298
Californium, 411
Calips, 59
Callinicus of Heliopolis, 61
Calomel, 43
Caloric, 160, 191, 211, 212, 215, 217
Calx, 123, 124, 126
Calxes: See Metal oxides
Camphor, 43
Cancer, 425
Cannabis, 43
Cannizzaro, Stanislao, 210, 258, 259
Canon, 73, 101
   of medicine, 67
Carbohydrates, 288, 343
Carbolic acid: See phenol
Carbon, 11, 62, 165, 179, 184, 198, 236, 237, 245, 247–249, 253, 258, 280, 284, 288, 289, 310, 371, 373, 393, 400, 423
   chains, 247–250
   tetrahedral, 328
   three-dimensional, 301

Carbonate salts, 135
Carbon dioxide, 54, 113, 133, 135, 137, 139, 140, 156, 160, 179, 190, 204, 245, 265, 290, 291
   supercritical, 418
Carbon monoxide, 62, 113, 179, 192, 284, 361, 415
Carboxylic acid group, 236
Carburizing, 9
Cardanus, 122
Carlisle, Anthony, 181, 421
Carneades, 118
Carnot, Lazare, 167, 217
Carnot, Sadi, 217, 218, 222, 223
Carothers, Wallace, 339
Carson, Rachael, 416
Carver, George Washington, 375
Castor oil, 42
Catalysis, 184, 302, 303, 361, 384, 421, 422
Cathode, 199
Cathode rays, 263, 266
Cathode ray tube, 267
Cation, 199
Caustic alkalis, 47, 48
Caustic potash: See potassium hydroxide
Caustic soda: See sodium hydroxide
Cavendish, Henry, 142–145, 159, 165
Cells, 345
Cellulose, 298, 339
Celsus, 54, 97
Cephalosorin, 369
Ceramics, 423
Ceria, 261
Cesium, 263
Chadwick, James, 393, 394
Chain, Ernst, 351
Chalk: See Calcium carbonate
Challenger, 376
Champagne, 114
Charaka, 47
Charcoal, 55, 84, 125, 126, 156, 160, 204
Charles' law, 207
Charles, Jacques, 144, 207, 220
Charles–Gay-Lussac law, 207
Chaulmoogra oil, 43
Chemical affinity, 306
   dualistic theory of, 182, 184

Chemical bond: *See* Bond, chemical
Chemical Revolution: *See* Revolution, chemical
Chemical warfare, 304
Chemical Warfare Service, 312
Chemical weapon, 427
Chemist
  analytical, 332, 187
  biochemist, 332
  inorganic, 332, 359
  organic, 233, 332, 359
  physical, 233, 376
Chemist, professional: *See* Professional chemist
Chemistry, 109
  analytical, 211, 281, 293–297, 404, 418
  biochemistry, 211, 281, 288, 345, 424
  definition, 4
  environmental, 417–419
  industrial, 281
  inorganic, 211, 241, 253, 257–283, 372–375, 391, 422
  organic, 203, 211, 235–256, 286, 301, 362–372, 421
  physical, 211, 286, 301, 382
  polymer, 341
  quantum, 345
  shake-and-bake, 422
  surface, 385
  synthetic, 256
  synthetic inorganic, 281
  word origins, 29
Chemophobia, 416
Chemotherapeutic agents, 374
Chernobyl, 414
Chibnall, Charles, 344
Chicago pile number 1 (CP1), 402
Chiral compound, 250
Chirality, 250, 289
Chlorides, 47
Chlorine, 113, 197, 198, 201, 204, 242–244, 257, 265, 278, 304, 311, 347, 381
Chlorophyll, 277, 347, 370,
Cholesterol, 349, 365, 370
Chromatography, 360
Chu, Paul, 423

Chymia, 104
Cinnabar: *See* mercury sulfide
Cisplatin, 374
City Philosophical Society, 198
Civetone, 363
Clapeyron, Emile, 218
Clausius, Rudolf, 223, 224, 228, 230, 276
Cloud chamber, 393, 394
Coal, 254, 284, 302, 354, 360, 361
  gas, 262, 284–285
  tar, 292, 302
Cobalt, 396
Cocaine, 348
Codeine, 289
Coenzymes, 348, 374
Coffee, 107
Cold, common, 425
Cold fusion, 419
Cole, Sidney, 344
Combustion, 121, 122, 133, 156, 160, 212, 399
Compendium, 53
Compendium studii philosophiae, 84
Compound, organic, 236
Computer, 351, 415, 451
Comte, August, 267
Concerning the Nature of Things, 351
Concerning Pyrotechnics, 103
Conformation, 362, 363
  boat, 365
  chair, 365
Conservation of mass, 154, 156, 161, 168
Conservation by Sanitation, 296
Conservatory of Arts and Trades, 193
Constantine, 40
Contraceptive, 350
Contribution to the Constitution of Inorganic Compounds, 279
Contributions to Physical and Medical Knowledge, Principally from the West of England, 191
Conversations on Chemistry, 196
Cook, Florence, 263
Cooking, 4, 212
Coordination complexes, 277, 302
Coordination number, 279
Copernicus, 109

Copley Medal, 139
Copper, 6–8, 23, 31, 33, 47, 103, 104, 113, 164, 255, 257, 296, 322
Copper acetate, 31
Copper arsenite, 293
Copper carbonate, 43
Copper oxide, 53, 423
Copper sulfate, 36, 47, 55, 113
Copula theory, 243
Cork, 114
Cornforth, John, 346
Cornforth, Rita, 347
Corrosive sublimate, 43
Cortisone, 349, 350, 370
Cosmetics, 14, 375
Cosmic radiation, 429
Cost of Living, 296
Couper, Archibald, 247
Cours de chymie, 110
Courtois, Bernard, 205
Covalent bond: See Bond, covalent
Cowpox, 292
CP-1, 402
Cretinism, 100
Crick, Francis, 353–357
Critical mass, 405
Crookes, 266
Crookes, William, 263
Crown ethers, 371
Crucibles, 104
Crystallization, 35, 38
Crystals, 327
Cullen, William, 135
Curie, Irene, 390, 392–394, 399
Curie, Marie, 267–272, 301, 383, 392, 394, 409, 410
Curie, Pierre, 269, 272, 390, 392, 399
Curium, 410, 411
Cyanic acid, 237
Cyanoacetylene, 415
Cyclotron, 401, 412
Cystine, 289

Dacron, 340
Dalton, John, 177, 188, 208, 209, 214, 221, 425
   law, 207
   unit, 342

Dalton and Gay-Lussac's law, 207
Darwin, Erasmus, 142
Davy, Humphry, 182, 189, 203, 204, 214, 384
DDT, 416
Death in the Pot, 293
Debye, Peter, 397, 398
De re metallica, 103
De Mirabilibus Mundi, 80
Democritus, 21, 177
Denatured proteins, 346
Density, 429
Deoxyribonucleic acid, 352–357, 425, 426, 356
Dephlogisticated air, 138, 141, 143, 159
Dephlogisticated nitrous air, 140
De la pirotechnia, 103
Descartes, René, 108, 117, 132, 177, 325
Deuterium, 400
Dewar, Katherine, 230
Dexio, 199
Diamagnetism, 201
Diamond, 184, 376, 423
Dichlorodiethyl sulfide, 305
Dichlorodifluoromethane, 372
Dichlorodiphenyltrichloroethane, 416
Dichloroethane, 361
Dickson, James, 340
Didymia, 261
Digestion, 91, 430
Dioscorides, 53
Dirac, Paul, 333, 430
Displacement, 112
Dissenting academy, 139, 193
Dissolution, 430
Distillation, 86, 91, 93, 104, 430
Djerassi, Carl, 350
DNA: See Deoxyribonucleic acid
Doctor Mirabilis, 82
Doctrine of Phlogiston Established, The, 192
Domagk, Gerhard, 350
Dome (distilling apparatus), 38
Dorn, Friedrich, 266
Double bonds: See Chemical bond, double
Double Helix, The, 354
Dragon series, 406
Drown, Thomas, 296

Dubnium, 412
Dumas, Jean, 240, 242–244, 247
Du Pont, 339, 340, 371
Dye, 13, 31, 34, 36, 55, 72, 236, 238, 277, 285, 290, 292, 347, 375
  German industry, 286
  synthetic, 287
Dynamite, 299
Dysprosia, 261

Eacheria, 104
Earths, 7, 55
École Polytechnique, 173, 202– 204, 217, 269
Ecosystem, 417
Edgeworth, Richard, 142
Ehrlich, Paul, 292, 425
Einstein, Albert, 318, 323, 326, 333, 334, 400
  letter (to President Roosevelt), 400
Einsteinium, 410
Eka-aluminum, 260
Ekaboron, 260
Ekalead, 430
Ekamercury, 430
Ekasilicon, 260
Electric force, 200
Electricity, 170, 194, 198, 199, 221
Electricity, animal, 180
Electrodes, 199
Electrolysis, 182, 198, 199
Electromagnetic field, 201
Electromagnetic spectrum, 331
Electromagnetic wave, 314
Electromagnetism, 200
Electron, 266, 273, 274, 311, 321, 323, 325, 326, 332, 393
Electron configuration, 322, 429
Electroscope, 181
Electrum, 7
Elektron, 179
Element, 20, 43, 48, 63, 188, 258, 274, 281, 306, 309
  Aristotle and, 132
  artificial, 429
  Becher's theory, 124, 125, 132
  four-element theory, 20, 24, 78, 112, 118, 122, 124, 168

Element (*cont.*)
  Lavoisier's 165
  radioactive, 414
  sulfur-mercury theory, 64, 78, 100, 118, 132
  superheavy, 428
  transuranic, 395, 400, 410
  tria prima, 100
Elements of Natural History and Chemistry, 164
Eleutherius, 118
Elixir, 44, 49, 87, 99
Embalming, 172
Empedocles, 20, 21
Empiricists, 70, 81
Encyclopedia, 53, 77
Encyclopedia Britannica, 196
Encyclopedists, 70, 77–78
Energy, 399
  rotational, 331
  translational, 216
  vibrational, 331
Energy density, 315
Energy/mass equivalence, 323
Engine knock, 372
English Royal Society, *See* Royal Society
Enola Gay, 407
Ensembles, 227
Entropy, 317, 223, 275, 429
Environmental chemistry, 417–419
Enzyme, 289, 384, 416, 422, 425, 427, 346, 348
Epicureanism, 33
Epilepsy, 102
Epsom salt: *See* Salt, Epsom
Equilibrium, 173, 231, 388
Equivalent weight, 199
Erbia, 261
Ergastia, 104
Essay on Combustion, 165
Esson, William, 379, 380
Estradiol, 350
Ethane, 249
Ethanol, 11, 86, 99, 416, 421
Ether, 25, 371, 372
Ethyl alcohol, 361
Ethylene, 236, 241, 249, 339
Ethylene glycol, 361

INDEX

Ethylene oxide, 361
Ethyl iodide, 246
Eudiometer, 140
Euler, 196
Europe, 128, 239, 329, 415
Europium, 263, 411
Exclusion Principle, 321
Exothermic process, 232
Experiments and Observations of Different Kinds of Air, 142
Explosives, 103, 405
Extraction, 91
Eyring, Henry, 382

Factitious airs, 143
Faithful Brethren, 64
Fallout, 408, 430
Faraday, Michael, 196–198, 243
  unit (faraday), 198
Fats, 343
Fatty acids, 245
Fatty earth: *See* Elements, Becher's theory
Femtosecond, 424
Fermentation, 11, 291, 297, 303, 421
Fermentation, cell-free, 301
Fermi, Enrico, 394, 397, 399, 400, 409, 413
Fermium, 410, 428
Ferric oxide, 43
Ferric sulfate: *See* Iron sulfate, 104
Ferrocene, 372, 373
Field theory, 201
Field, 200
Fieser, Louis, 365
Fieser, Mary, 365
Filter paper, 184
Filtration, 35, 38
Fire, 6
Fire-air, 145,146, 157, 168
Fireworks, 103
First law of thermodynamics, 223
Fischer, Emil, 288, 301, 344, 346, 372
Fischer, Franz, 361
Fischer–Tropsch process, 361
Fissile material, 401, 404
Fission, 409, 413, 419, 428, 430
Fission products, 400

Fission reaction, 400
Fixed air, 136, 139, 156, 157
Fixed sulfuric acid, 126
Fleischmann, Martin, 419–421
Fleming, Alexander, 351
Florey, Howard, 351
Fluorescence, 332
Fluorine, 260, 301, 345
Flux, 9, 14, 104
Fluxional motion, 373
Foolish Wisdom and Wise Folly, 124
Formaldehyde, 415
Formulas, chemical, 176, 183
Frontier orbital theory of reactions, 369
Fourcroy, Antoine, 162, 165–167, 202, 203, 235
Fox, Sidney, 356
Francium, 410
Frankland, Edward, 246, 253, 277
Franklin, Benjamin, 139, 180
Franklin, Rosalind, 353–357
Freezing point depression, 275
French Atomic Energy Commission, 409
French Revolution: *See* Revolution, French
Freon, 372
Fresenius, Karl, 293
Frisch, Otto, 398, 399, 404, 406, 407, 413
Froben, Johannes, 101, 102
Fructose, 288
Fruton, Joseph, 356
Füchs, Klaus, 404, 409
Fukui, Kenichi, 369
Fulhame, Elizabeth, 165
Fuller, R. Buckminster, 423
Functional groups, 246
Funnels, 184
Furnaces, 104
Fusion, 35, 38, 409, 419

Gadolinia, 262
Gadolinium, 262, 411
Gadolin, Johan, 262, 411
Galactose, 288
Galen, 56, 57, 63, 67, 73, 90, 97, 98, 100, 101, 319
Galileo, 108, 109, 116

Gallium, 260
Galvani, Luigi, 180
Gamma radiation, 393
Gamma ray, 392, 414, 420, 421, 430
Gangrene, 291
Gas, 62, 113, 122, 206
  ideal, 222
  inert, 384
  kinetic theory of, 230
  noble, 263–266, 312, 374
  rare: *See* Gas, noble
  war, 329
Gas gangrene, 291
Gas masks, 347
Gasoline: *See* Petroleum
Gay-Lussac, Joseph, 202, 203, 239, 241
Gay-Lussac's law, 224
Geber, 64
Geiger counter, 406
Geiger, Hans, 273
  counter, 406
Gems, 31, 55
Genentech Corporation, 425
General Electric Company, 384
Genetic engineering, 425
Genetic information, 352
Gerald of Cremona, 73
Gerhardt, Charles, 245
Germanium, 260, 374
German Physical Society, 317
Gibbs function, 228, 231
Gibbs, Josiah, 226, 232, 295, 325
Gilbert, Davies, 190
Gilding, 72
Glass, 14, 55, 72, 86, 103, 114, 297
  organic, 339
Glauber, Johann, 110
Glauber's salt: *See* Salt, Glauber's
Glaucoma, 425
Global warming, 417
Glucose, 288
Glutamic acid: *See* Acid, glutamic
Glycerol, 298
Glyceryl trinitrate, 298
Godiva series, 406
Goiters, 100
Gok, 285

Gold, 6–8, 15, 23, 29, 30, 33, 34, 36, 41–44, 47, 55, 94, 103, 257, 322, 413, 428
  false, 32, 36, 47, 55, 66, 80, 109
  Inca, 42
  potable, 46
Gold, Harry, 409
Gomberg, Moses, 255
Goodyear, Charles, 338
Gosling, Raymond, 354
Goudsmit, Sam, 321
Graphite, 184, 376, 400
Great Mirror, 78
Great Work, 83
Greek fire, 61, 84
Greenglass, David, 409
Greenhouse effect, 417
Greenhouse gas, 418
Green Revolution, 416
Grignard, Victor, 302
Guericke, Otto von, 114, 180
Guettard, 153
Gunpowder, 11, 62, 83, 89, 96, 114, 158, 164, 166, 175, 205, 298, 305, 384
Gypsum, 10

Haber, Fritz, 303, 305, 334, 360, 384
Hagar, 59
Hahnium, 412
Hahn, Otto, 395–401, 410
Halogen, 184, 374
Halos, 429
Handedness, 250
Hanford, 402, 409
Harcourt, Augustus, 379, 380
Harpoon mechanisms, 383
Hassel, Odd, 363–365
Hata, Sahachiro, 293
Haüy, Abbé, 186
Healers, 92
Heat engine, 217
Heat, 159, 160, 164, 231, 233, 317
  mechanical equivalent, 221
Heavy metal pigments, 293
Heavy water, 400, 406, 420
Heisenberg, Werner, 324, 326, 334, 405
  uncertainty principle, 325

INDEX

Heitler–London model, 328
Heitler, Walter, 326
Helium, 263, 265, 272, 301, 310, 321, 322, 392, 394, 409, 423
Helix, 352
Helmholtz, Hermann von, 223
Helmont, Franciscus Mercurius van, 110, 111
Helmont, Johannes van, 110, 122, 154
Hemoglobin, 277, 344, 374
Henry, Joseph, 200
Hermetic art, 39
Herschbach, Dudley, 383
Hertzberg, 334
Hertz, Heinrich, 314, 318
Hess, Sophie, 299
Hevésy, George, 413
Hexamethylbenzene, 366
Hill, Julian, 340
Hindu, 15, 49, 60, 62
  physicians, 9
Hippocrates, 97, 100
Hiroshima, 407
History of Electricity, 139
Hodgkin, Dorothy, 351
Hoff, Jacobus van't, 252–254, 276, 301, 362, 379
Hoffmann, Roald, 286, 298, 369–371
Hofmann, August von, 285
Hofmann, Felix, 290
Hofmann, Fritz, 338
Holmia, 261
Home Sanitation, 296
Homologous series, 245
Honest-to-god copper, 401
Hooke, Robert, 116, 119–121, 177, 291
Hooke's law, 120
Hoover, H. C. and L. H. (President and Mrs.), 103
Hopkins, Frederick, 344
Hopkins, Johns, 231, 254, 255
Hormones, 349–351
  sex, 350, 365
Humboldt, Alexander, 207
Hume, David, 137
Hund, Frederick, 329–331, 369
Hydrocarbons, 236, 284, 361, 415
  rings, 373, 362

Hydrogen, 11, 113, 117, 133, 141, 158, 159, 165, 179, 183, 192, 195, 207, 236, 237, 243, 244, 249, 259, 266, 280, 284, 302, 303, 310, 321, 322, 326, 333, 346, 356, 360, 361, 376, 381, 382, 384, 392, 393, 400, 409, 421, 427
Hydrogen bomb, 409, 419
Hydrogen chloride, 140, 242, 298
Hydrogen cyanide, 356
Hydrogen peroxide, 204
Hydrogen spectrum 319
Hydrogen sulfide, 298
Hydroxide, 36, 270
Hypatia, Lady, 40

Iatrochemistry, 90, 91, 109
Iatrochemists, 83, 100
Ibn Hayyan, 64
Ice, 250
ide (suffix) 165
Ideal gas: *See* Gas, ideal
Identification, 391
I. G. Farben, 287, 290, 304, 340, 361
Imperial Chemical Industries, 339, 340
Improvement of the Mind, 196
Incendiaries, 305
Indiana University, 353
Indicators, 119
Indigo, 13, 55, 287, 301, 362
Indium 263
Industrial Water Analysis, 296
Industry, chemical, 103
Inflammable air, 143, 159
Inflammable earth: *See* Elements, Becher's theory
Ingold, Christopher, 366–369
Insecticide, 416
Inspiration Copper Company, 382
Institute of France, 171, 205
Institute for Physical Chemistry, 380
Instrument Revolution, 360
Insulin, 344, 349, 351
Interference pattern, 314
Intermolecular forces, 277, 345
International Union of Pure and Applied Chemistry, 413

Interpolation, 317
Introduction to the Study of Chemisty, 255
Invisible College, 116
Iode, 205
Iodine, 201, 204, 205, 257, 381, 414
Ion, 199, 200, 266, 275, 276, 311, 312, 372
  heavy, 430
Ionic radii, 429
Ionization, 281
Ionization energy, 429
Iowa State College of Agriculture and Mechanical Arts, 375
Irenaeus, 57
Iron, 33, 41, 42, 47, 55, 103, 207, 284, 303, 373
  false, 80
  oxide, 5, 159, 361
  sulfate, 36, 47, 102
Irone, 363
Isomerism, 237, 239
Isomers, 184
Isotopes, 266, 396
  radioactive, 414

Jabir ibn Hayyan, 64
  Corpus, 64
Jardin du Roi, 153, 163
Jasmone, 363
Jenner, Edward, 292
John of Rupescissa, 87, 99
Joliot, 399, 400, 409
Joliot, Frederick, 392–394
Jones, Steven, 420
Jørgensen, Sophus, 278
Joule, James, 220, 221
  unit (Joule), 221
Journal of the American Chemical Society, 255
Journal of Analytical Chemistry, The, 294
Journal of Organic Chemistry, 362
Journal of Physical Chemistry, 232
Julian, Percy, 349
Jurassic Park, 427
Justinian, 58

Kaiser Wilhelm Institute, 304, 361, 396, 397
  for Chemistry, 286

Kaiser Wilhelm Institute (*cont.*)
  for Coal Research, 286
  for Electrochemistry, 286
  for Physical Chemistry, 286
Kaolin, 42
Karlsruhe Conference, 210, 259
Kekulé, Friedrich, 247, 253, 277, 279, 287, 301, 306, 366
  sausages, 248
Kelvin, Lord, 201, 221, 266, 271
  unit (Kelvin), 222
Kepler, 109
Kermes, 13
Kerosene, 222
Kinetics, 376, 377–390, 424
Kinetic theory of gas: *See* Gas, kinetic theory of
King, Victor, 280
Kinship of Three, 45
Kinsky, Bertha, 299
Kirchhoff, Gustav, 262
Klaproth, Martin, 187
Klug, Aaron, 356
Ko Hung, 46, 47
Kolbe, Hermann 246, 253
Kossel, Walther, 312
Krypton, 265, 301, 374

Laboratory, 90, 104
Lake Natron, 172, 173
Lamp, miners: *See* Miner's lamp
Langevin, Paul, 271, 323, 392, 393
Langmuir–Blodgett films, 387
Langmuir, Irving, 312, 329, 384, 385, 387
  isotherms, 385
  theory, 312
Lanthana, 261
Lanthanides, 261
Lanthanide series, 403
Lanthanum, 261
Laplace, Pierre-Simon, 160
Lapworth, Arthur, 367
Laser, 415
Latent heat, 216
Laudanum, 99
Laue, Max, 342
Laughing gas, 192, 194

INDEX

Laurent, Auguste, 243–245, 250
Lavoisier, Antoine, 141, 153–169, 173, 175, 177, 184, 187, 190, 195, 204, 220, 237, 241, 298, 419
Lavoisier, Marie, 156, 158
Law of definite proportions, 123, 175–177, 203
Law of mass action, 172
Law of multiple proportions, 177–179
Lawrencium, 412
Lead, 33, 47, 103, 393, 395
  white, 32, 245
Lead acetate, 32, 53
Lead carbonate, 43
Lead chromate, 293
Lead oxide, 43, 46
Lead poisoning, 58
Lead sulfate, 55, 102
Leather tanning, 15
Le Bel, Joseph, 252–254, 362
Le Blanc, Nicolas, 297
Le Châtelier, Henri, 269
Le Châtelier's principle, 231
Lee, Yuan, 383
Leeuwenhoek, Antony, 291
Leibig, Justus, 220, 237–240, 243, 244, 247, 285, 291
Leiden jar, 180
Lemery, Nicholas, 110
Lesser Work, 83
Letters to a German Princess, 196
Leucine, 289
Leucippus, 21
Lewis, Gilbert, 310–313, 329, 332, 367, 401
  structures, Lewis dot , 367
Lewis–Langmuir theory, 312
Lewis, William, 381
Libau, Andreas, 103
Liber de proprietatibus rerum, 77
Liber ignium ad comburendos hostes, 84
Lichtentels, Canon Cornelius von, 102
Ligand, 277, 373, 374
Light, 159, 161, 313, 323
  bulb, 384
  infrared, 315
  plane-polarized, 362
  polarized, 250, 378

Light (*cont.*)
  ultraviolet, 314, 318, 325
  visible, 315, 331
  wave–particle duality, 319
Lime, 14, 55
  unslaked, 65
Limestone, 15, 61
Limewater, *See* Calcium hydroxide solution
Lindemann, Frederick, 382
Lippmann, Gabriel, 269
Lister, Joseph, 292
Lithium, 310, 392, 413
Little Boy, 407
Liver of sulfur, 126
Lock and key model, 346
Lomonosov, Mikhail, 146
London, Fritz, 326, 334
Lonsdale, Kathleen, 366
Los Alamos, 404, 409
Love Canal, 416
Lucite, 339
Lunar Society, 142
Lutetia, 261
Lwoff, André, 356
Lyceum, 32
Lye, 48
Lysergic acid, 370

Macintosh, Charles, 338
Macromolecules, 341
Madder, 13
Madhyamika logic, 49
Magic, 5, 12, 29, 39, 44, 49, 62, 78
Magic island, 428, 430
Magic mountain, 428
Magic numbers, 428
Magic ridge, 428
Maglev trains, 422
Magnesia alba: *See* Magnesium carbonate
Magnesium, 311
Magnesium carbonate, 135
Magnesium silicate, 43
Magnetic field, 320, 422
Magnetic moment, 329
Magnetic Resonance Imaging, 414
Magnetism, 199
Main group elements, 374
Malachite, 43

Malaria, 51, 285
Mandragora, 53
Manganese oxide, 5
Manganous acetate, 243
Manhattan Engineer District Office, 402
Manhattan Project, 402, 404, 409
Mansfield, Charles, 285
Marat, Jean, 166
Marcet, Jane, 196
Marcogenin, 424
Marcus Graecus, 84
Mariotte, Edmé, 117
Mariotte's law, 117
Mark, Hermann, 342
Marker, Russell, 349, 424
Mary the Jew, 37, 93
Mass spectrometer, 403
Mass spectroscopy, 360
Matissen, Sophie, 280
Mauve, 286, 290
Maxwell, James, 201, 229, 314
  demon, 230
Mayer, Julius, 219–221
Mayer, Maria, 428
Mayow, John, 122
McMillan, Edwin, 400
Mechanical equivalent of heat: *See* Heat, mechanical equivalent
Mechanism, electron theory of organic, 366–369
  reaction, 378
Meister, Joseph, 292
Meitnerium, 412
Meitner, Lise, 395–401, 413
Meltdown, 420, 421
Melting point, 429
Mendeleev, Dmitri, 210, 258–261, 265, 322, 396
Mendelevium, 412
Mephitic air, 137
Mephitis, 137
Mercury, 29, 36, 42–45, 47, 53, 55, 64, 93, 100, 103, 111, 112, 140, 188, 201, 222, 247, 422, 427
  gold amalgam, 55
  lead amalgam, 46
  poisoning, 102, 200
Mercury chloride (ic), 43, 104

Mercury chloride (ous), 43
Mercury fulminate, 299
Mercury oxide, 102, 141, 157
Mercury sulfide, 5, 29, 46, 72, 102
Metal, 55, 186, 203, 372, 374
  noble, 264
Metallic radii, 429
Metallurgical Laboratory, 410
Metallurgy, 6, 47, 103, 232, 384
Metal oxides, 125, 156
Metargon, 266
Meteor, 427, 429, 430
Methane, 113, 236, 356
Methyl mercaptan, 415
Methyl radical, 242
Metric system, 166, 173
Meyer, Lothar, 210, 258, 276
Microanalytical techniques, 419
Microelectrodes, 419
Micrographia, 121
Microorganism, 291, 292, 297, 302, 350, 421
Microscope, 291
Microscopic reversibility, 389
Microscopy, 419
Microspheres, 356
Midgley, Thomas, 372
Mild alkalis: *See* Carbonate salts
Miller, Stanley, 427
Minamata, 416
Miner's lamp, 201
Minerals, 104
Mineral waters, 104
Mista, Bhava, 100
Mitouard, Pierre, 156
Moissan, Henri, 260, 301
Molecular beam, 383
Molecular dynamics, 383
Molecular films, 385
Molecular-orbital model, 328–331
Molecular orbital theory, 330, 369
Molecular paleontology, 426
Molecule, 183, 210
Moles, 208
Momentum, 325
Monolayer films, 385
Monomers, 338
Monomolecular films, 385

INDEX

Monophysites, 59
Montesson, Marquise de, 162
Moore, Walter, 377
Mordant, 13, 31
Morphine, 289, 290
Mortar, 55, 104
Morveau, Louis, 162, 165–167, 202
Mosander, Carl, 261
Moseley, Henry, 274, 310
Mosquito, 427
Müller, Paul, 416
Mulliken, Robert, 329, 369
Mullis, Kary, 425
Mummification, 15
Muscone, 363
Musschenbroek, Pieter, 180
Mussel, 13
Mustard gas, 305
My First Hundred Years, 254

Nagasaki, 408
Nāgārjuna, 49
Naphtha, 62, 195
Naphthalene, 284
Natron, 15, 172
Natural abundance, 396
Natural History, 53
Natural and Mystical Things, 36
Natural philosophy, 18, 32, 43
Natural product, 240, 288, 301, 338, 368
Nature, 386
Nature of the Chemical Bond and the Structure of Molecules and Crystals, The, 328
Needham, Joseph, 45
Neon, 301, 310
Neptunium, 401, 410
Nernst, Walter, 226, 380
Neutron, 392, 394, 430, 428
 moderator, 400
Newcomen, Thomas, 215
New Experiments Physico-Mechanicall, Touching the Spring of the Air and Its Effects, 116
Newlands, John, 258
New quantum theory, 322–326
New System of Chemical Philosophy, 179

Newton, Issac, 108, 109, 120, 132, 146, 214, 317, 325
Nicholson, William, 181, 421
Nickel, 296, 396, 403
Nicotine, 289
Niter, 11, 43, 62, 65, 83, 84, 104, 164, 205
 potassium nitrate: *See* Potassium nitrate
 sodium nitrate: *See* Sodium nitrate
Nitric oxide, 140, 141, 157, 192, 209
Nitro-aerial, 123
Nitrocellulose, 298
Nitrogen, 62, 134, 138, 141, 144, 159, 165, 197, 209, 236, 264, 265, 280, 289, 289, 303, 313, 345, 374, 394, 417, 423
Nitrogen dioxide, 140, 179
Nitrogen monoxide, 179
Nitrogen oxides, 146, 159
Nitroglycerin, 298, 299, 417
Nitro groups, 298
Nitrosyl chloride, 113
Nitrous air, 121, 140
Nitrous oxide, 192, 194
Nobel, Alfred, 298–301, 410
Nobel, Immanuel, 299
Nobel Institute, 397, 412
Nobelium, 412
Nobel Prize, 298–302, 305, 349, 360, 363, 365, 368–372, 374, 383, 385, 389, 391, 394, 399, 408, 413, 414, 416, 425, 427
Noble gases: *See* Gas, noble
Noble metals: *See* Metals, noble
Noddack, Ida, 395
Nomenclature, 165, 259
Nonequilibrium systems, 390
Nonequilibrium thermodynamics, 388, 389
Normal saline solution, 158
19–Norprogesterone, 350
Nuclear Age, 414
Nuclear chain reaction, 404
Nuclear fission, 399
Nuclear-fuel recycling, 417
Nuclear Magnetic Resonance, 414
Nuclear power, 414

Nuclear Test Ban Treaty, 408
Nuclear weapon, 428
Nucleic acid, 343, 352, 356, 427
Nucleotides, 343, 352
Nucleus, 273, 274
Nucleus theory, 244
Nylon, 340

Octet, 313
Octet rule, 310
Oersted, Hans, 199
Oil, 99, 254, 343, 361, 430
    bitter almonds, of, 241
    Vitriol, of, 163
Old quantum theory, 313–322, 329
On Divers Arts, 72
On the Marvelous Things in the World, 80
On Metals, 103
Onsager, Lars, 388, 389
Oparin, Aleksandr, 356
Opium, 16, 53, 99, 128, 287
Opium Wars, 287
Oppenheimer, Robert, 404, 407, 409
Optical activity, 250
Opticks, 121
Opus Majus, 83
Opus Minus, 83
Opus Tertium, 83
Orbitals, 320, 373
    antibonding, 329
    bonding, 329
    d, 320, 373
    f, 320
    molecular, 329
    p, 320
    s, 320
Ore, 6, 7, 103, 186, 261, 284, 429, 430
    copper, 296
    iron, 295
Organic chemistry: *See* Chemistry Organic
Organic compounds: *See* Compounds, organic
Organic feedstock, 375
Organometallic chemistry, 373
Organometallics, 247, 302
Orgel, Leslie, 356

Oscillations, 387, 390
Osmosis, 275
Ossietzky, Otto, 349
Ostwald, Friedrich, 232, 276, 301, 303, 360, 379, 384, 387
Ourobouros, 35
Ovulation, 350
Oxford Chemists, 114
Oxidation, 35, 384, 429
Oxygen, 11, 14, 113, 117, 133, 141, 143, 146, 157–159, 161, 164, 165, 168, 179, 183, 207, 209, 220, 236, 237, 246, 258, 265, 280, 284, 289, 301, 322, 339, 345, 346, 371, 376, 384, 421, 427
Oxymuriatic acid: *See* Acid, hydrochloric

Palladium, 188, 419, 421
Pandemonium, 411
Pao Ku, 46
Pao Phy Tzu, 47
Paracelsus, 97–103, 111, 118, 121, 126, 293, 319
Paraffin, 393
Paramagnetism, 201
Paris Arsenal, 157
Paris, Marianna, 180
Parke-Davis, 350
Parson, Alfred, 312
Partington, J. R., 85
Pasteurization, 291
Pasteur, Louis, 250, 280, 301, 346
Patrick, Joseph, 339
Pauli, Wolfgang, 321
    exclusion principle, 321
Pauling, Linus, 327, 353, 369, 408
Paulze, Jacques, 155
Paulze, Marie: *See* Lavoisier, Marie
Peanuts, 375
Pedersen, Charles, 371, 372
Peizoelectricity, 269
Peligot, Eugène-Melchior, 188
Penicillin, 351
Penicillium notatum, 351
Pennogenin, 424
Pentose, 352
Peptide bond, 343

INDEX

Peptides, 289
Percolators, 10
Perey, Marguerite, 410
Perfumes, 14
Perignon, Dom, 114
Periodic Table, 236, 256, 257, 263, 274, 321, 395, 396, 403, 410, 411, 428, 430
Perkin, William, 285, 290, 303
Perpetual motion machine, 218, 422
Perrin, Jean, 381
Petroleum, 61, 62, 358, 360, 366, 372, 374–376, 422
  cracking, 361
Pharmacopoeia, 5, 12, 42, 45, 47, 53
Phase diagram, 229
Phase rule, 228, 232
Phenol, 284, 290, 292
Phenylhydrazine, 288
Philosopher of fire, 111
Philosophical Transactions, 120, 141
Philosophus per ignem, 111
Phlogisticated air, 138, 144
Phlogiston, 123, 124, 126, 127, 132, 133, 140, 146, 147, 157, 159, 162, 164, 165, 212
Phosgene, 305, 348
Phosphate group, 352
Phosphorescence, 267, 332
Phosphorescent substances, 84, 273
  ink, 80
Phosphoric salts, 356
Phosphorus, 109, 119, 156, 165, 366, 374
  pentachloride, 278
Photochemistry, 332
Photoelectric effect, 318
Photoluminescence, 332
Photon, 332
Photosynthesis, 347
Physica et mystica, 36
Physical Chemistry, 377
Physical properties, 362
Picardet, Claudine, 163
Pigment, 5, 13, 72
Pill, the, 349, 365
Pitch, 62
Pitchblende, 187, 270

Planck, Max, 316, 335, 396
Plaster of Paris, 33
Platinum, 260, 384, 419, 421
  hexafluoride, 374
Plexiglas, 339
Pliny, 37, 53–56
Plunkett, Roy, 340
Plutonium, 235, 239, 401–403, 405, 408, 410, 417, 418, 429
Pneumatic chemists, 134
Pneumatic Medical Institution, 190
Pockels, Agnes, 385
Pockles Point, 386
Polanyi, Michael, 382
Polonium, 270, 271, 301, 374, 393
Polyamide, 340
Polyethylene, 339
Polymerase chain reaction, 425
Polymers, 338, 343, 353, 356, 361, 404
  inorganic, 374
Polystyrene, 339
Polysulfides, 35
Polyvinylchloride, 339
Pons, Stanley, 419–421
Positrons, 430
Potash, 65, 304
  alum, 43
Potassium, 188, 194, 195, 203, 258, 265, 424
Potassium aluminum sulfate, 10, 43
Potassium carbonate, 297
Potassium chlorate, 175, 298
Potassium hydroxide, 126, 146, 194
Potassium nitrate, 11, 62, 121, 146, 158, 298
Potassium polysulfide, 126
Potential energy surface, 382
Predator–prey model, 387
Priestley, Joseph, 138–145, 157–159, 162, 165, 192, 338
Prigogine, Ilya, 389
Primal element, 19
Primal substance, 18
Principle of Le Châtelier, 231
Privatdozent, 279
Proactinium, 397
Probable Relief and Possible Cure for Pulmonary Consumption, 190

Professional chemist, 185
Progesterone, 350
Project Y, 404
Prometheus, 410
Prontosil, 350
Propylene, 361
Prosthetic technology, 413
Protein, 289, 343, 348, 352, 427
Protoactinium, 397
Proton, 392–394, 398, 428
Proust, Joseph, 175–177
Pseudo-Democritus, 36, 63
Psoriasis, 425
Purification, 391
Purines, 289
Putrid fever, 190
PVC: *See* Polyvinylchloride
Pythagoras, 20, 21

Qualities, 24
Quanta, 317–319
Quantum, 320, 323
Quantum chemistry, 309–335, 360, 390
Quantum mechanics, 313, 376, 381, 430
Quantum numbers, 320
Quarterman, Lloyd, 402
Quartz, 250, 252
Quicklime: *See* Calcium oxide
Quinine, 285, 289, 290, 370
Quiz Kids, 411

Radiation, 270, 272, 273, 406
  alpha, 272, 273, 392
  beta, 272, 392
  hypothesis, 381, 385
Radical, 256, 381
Radicals, organic theory, 241
Radioactivity, 266, 281
  artificial, 394
Radiochemistry, 266, 390–414
Radioisotopes, 429
Radiothorium, 395
Radium, 270–272, 301, 392
Radium Institute, 392
Radon, 266, 374
Raman, Chandrasekhara, 332
Raman scattering, 332

Raman spectroscopy, 332
Ramon Lull, 81, 85–86
Ramsay, William, 144, 264, 301, 397
Rare earths, 261
Rare gases: *See* Gas, rare
Rasaratnasamuchchaya, 90
Rasetti, Franco, 394
Rausolfia, 43
Rayleigh, Lord, 144, 264, 326, 329, 386
Reaction kinetics: *See* Kinetics
Reaction mechanism, 313
Rebound reactions, 383
Reducing agent, 284
Reduction, 8, 91, 161
Refining, 103
Reflections on the Motive Power of Fire, 218
Reflections on Phlogiston, 161
Reich, Ferdinand, 263
Reid, E. Emmet, 254, 266
Relativity, theory of, 318
Remsen, Ira, 254
Researches, Chemical and Philosophical;
  Chiefly concerning Nitrous Oxide, or
  Dephlogisticated Nitrous Air, and Its
  Respiration, 192
Reserpine, 370
Resonance bond, 328
Respiration, 160
Retort, 146
Reversible process, 173, 223
Revolution
  chemical, 152, 281
  French, 151, 171, 193, 202, 297
  French (1848), 392
  Industrial, 186, 212, 215, 217, 223, 293
Rey, Jean, 122
Rhazes, 64
Ribonucleic acid, 341, 352, 356, 427
Richards, Ellen Swallow, 294, 302
Richards, Robert, 295
Richter, Hieronymus, 263
Richter, Jeremias, 177, 178
Ring whizzing, 373
Río, Andrés Manuel del, 188
RNA: *See* Ribonucleic acid
Robert of Chester, 73
Robinson, Robert, 367–369

INDEX

Rockogenin, 424
Roger of Helmarshausen, 72
Roget, Peter, 192
Rohdewald, Margarete, 348
Roland, Marie, 164
Röntgen, Wilhelm, 267
Roots, 21
Rose, Valentin, 187
Rosenkranz, George, 350
Rothe radiation, 393
Rouelle, Guillaume-François, 153
Royal College of Chemistry, 285, 298
Royal Gunpowder Administration, 157
Royal Institution, 193, 198, 214, 293
Royal Prussian Academy of Science: See Academy of Science, Royal Prussian
Royal Society, 109, 120, 122, 139, 141, 143, 154, 181, 194, 197, 200, 201, 221, 224, 420
    of Edinburgh, 137
    of Medicine, 163
Rubber, 139, 335, 338, 358, 360, 361
    synthetic, 303, 338
    Thiokol, 339
    tubing, 184
Rubidium, 263, 424
Ruggiero, Count, 237
Rumford, Count, 168, 193, 214
Rust, 23
Rutherford, Daniel, 137, 144
Rutherford, Ernest 272, 301, 319, 392, 394
Rutherfordium, 413
Ruzicka, Leopold, 349, 363

Sabatier, Paul, 302, 384
Sachse, Herman, 363
Saffron, 13
Sala, Angelus, 113, 122
Sal ammoniac: See ammonium chloride
Salicin, 290
Salicylic acid: See Acid, salicylic
Salix alba, 290
Salt, 12, 55, 65, 86, 154, 182, 186, 195, 203, 258
    common: See Sodium chloride
    saltwater, 158

Salt *(cont.)*
    Epsom, 163
    Glauber's, 163
    nitrate, 303
Saltpeter: *See* niter
Salvarsan, 293, 262
Samarium, 263
Samarski, Colonel, 262
Sand, 14
Sanger, Frederick, 344, 349
Sanitation in Daily Life, 296
Sarrett, Lewis, 349
Saturated compounds, 361
Saturated hydrocarbon, 249
Scaio, 199
Scandium, 260, 261
Sceptical Chymist, The, 118
Scevasia, 104
Scheele, Carl, 134, 138, 145, 146, 157, 162, 165, 168, 204
    green, 293
School of Mines in Mexico City, 188
Schrödinger, Arnold 327, 323–325
Schwindler, 243
Seaborg, Glenn, 401, 402, 410, 411
Seaborgium, 413
Searle Pharmaceuticals, 350
Second law of thermodynamics, 224
Secret of Secrets, 65
Segré, Emilio, 394, 395, 404, 410
Séguin, Armand, 168
Selenium, 188, 257, 322
Self-organizing systems, 390
Self-replication, 352, 427
Semon, Waldo, 339
Sennert, Daniel, 377
Separation, 391
Seton, Alexander, 109
Sharp waters, 65
Shelburne, Lord, 139, 141, 157
Shell, 320
Shell theory of the nucleus, 428
Sickle-cell anemia, 344
Side chain, 343
Silent Spring, 416
Silicates, 14
Silicon, 14, 236, 258, 374
Silicon dioxide, 14, 252

Silk, 292, 350
  synthetic, 340
Silver, 6–8, 33, 43, 47, 103, 188, 258, 322
  false, 36
Silver chloride, 55
Silver cyanate, 237
Silver fulminate, 238
Silver oxide, 237
Sklodowska, Marie, *See* Curie, Marie
Sleeping sickness, 293
Smallpox, 292
Smelting, 8, 103
Smith, Adam, 137
Smog, 140
Smoke detectors, 414
Smokeless powder, 298, 302
SNAP, 413
Soap, 12, 164, 236, 297, 375
Soapstone, 43
Sobrero, Ascanio, 298
Soccerane, 424
Society for Encouraging Industry and Promoting the Welfare of the Poor, 193
Society for the Encouragement of National Industry, 269
Soda ash, *See* Sodium carbonate
Soda water, 139
Soddy, Frederick, 272, 396
Sodium, 195, 203, 258
Sodium ammonium tartarate, 252
Sodium carbonate, 10, 14, 65, 172, 297
Sodium chloride, 7, 10, 55, 65, 140, 158, 172, 173, 297
Sodium hydroxide, 175, 194, 195, 289, 290
Sodium nitrate, 11
Solar cells, 421, 427
Solubility, 346
Solute, 275
Solution, 38
Solvation, 35
Solvay conference, 322
Solvent, 164, 275, 346
  organic, 372, 418
Solvent cage, 346
Sommerfeld, Arnold, 322, 327, 329
Sommerville, Mary, 195

Soot, 55
Sorbose, 288
Southern, John, 219
Space Nuclear Auxiliary Power, 413
Space travel, 415
Spagyric physicians, 100
Spectrometer, 263, 360
Spectroscopy, 262–264, 331, 332
Speculus majus, 78
Speedy and Certain Cure for Pulmonary Consumption, 190
Speedy Relief and Probable Cure for Pulmonary Consumption, 190
Spin, 329, 332
Spinoza, 108
Spirits, 38, 65
Spiritus sylvestris, 113
Spontaneous process, 223
SQUID, 423
Stability peninsula, 428
Stahl, Georg Ernest, 124–126, 165, 187
Stained glass, 72
Standing wave, 315, 316
Stanley, Wendell, 351, 352
Stannic chloride, 104
Stark, Johannes, 334
Stationary states, 320
Statistical mechanics, 225, 226, 381
Staudinger, Hermann, 341, 363
Staudinger, Magda, 342
Steam engine, 217
Steel, 47, 284
Steroid, 349, 365
Stoichiometry, 178
Stokes, Jonathan, 142
Stone, Edmund, 290
Strain theory, 363
Strassmann, Fritz, 395–401
Stripping reactions, 383
Strohmeyer, Fredrich, 186
Strychnine, 370
Styrofoam, 339
Sublimation, 35, 38, 93, 430
Sublimatories, 103
Subshell, 322
Substitution theory, 242
Sugar, 107, 288, 291, 343, 352
  grape, 177

INDEX

Sukraniti, 77
Sulfa drugs, 351
Sulfate, 165, 270
Sulfite, 165
Sulfonamide, 350
   drugs, 351
Sulfur, 38, 43, 55, 62, 64, 100, 113, 121, 125, 156, 165, 258, 305, 322, 417, 427
Sulfur dioxide, 62, 113
Sulfuric acid: See Acid, sulfuric
Sumner, James, 347
Super, the, 409
Superconducting Quantum Interference Device, 423
Superconductivity, 422, 423
   room-temperature, 427
Superheavy elements: See Elements, superheavy
Surface films, 384
   balance, 385
Suspicions about the Hidden Realities of the Air, 121
Swallow, Ellen: See Richards, Ellen
Sweet potato, 375
Swindler (alchemical), 94
Syntex, 350
Synthesis, organic, 368
Syphilis, 100, 111, 293, 350

Table salt: See Sodium chloride
Tait, Peter, 207, 220
Takamine, Jokichi, 349
Talc, 43
Tartaric acid, 346
   salts, 250
Taube, Henry, 374
Taylor, Moddie, 402
Tea, 287
Tear gas, 305
Technetium, 410
Teflon, 340, 341, 404
Teller, Edward, 400, 408, 409
Tellurium, 187, 257
Terbia, 261
Terpenes, 289, 301
Terylene, 340
Testosterone, 350

Tetracyclines, 370
Tetraethyllead, 372, 374
Tetrahedron, 251, 363
Thales, 18, 19
The Book of Composition of Alchemy, 73
The Book of the Lover and the Beloved, 86
Themistus, 118
Thenard, Louis, 203–205, 241, 378
Theophilus, 71
Theophrastus, 32
Theory of residues, 245
Theory of Sound, The, 264
Theory of types, 244
Theosebeia, 37
Thermodynamics, 213–233, 309, 332, 376, 422
Thiokol rubber: See Rubber, Thiokol
Third law of thermodynamics, 226
Third Work, 83
Thomas, M. Carey, 256
Thompson, Benjamin: See Rumford, Count
Thomson, Joseph (J. J.), 266, 272, 319
Thomson, William: See Kelvin, Lord
Thorium, 270
Three-dimensional compounds, 250
Three-dimensional structure, 351, 360, 362
Thrice Great Hermes, 38
Thulia, 261
Thyroid gland, 414
Timaeus, 23
Tin, 33, 47, 103, 104, 164
Tin oxide, 165
Titanium, 187
TNT, 417
Tobacco, 107
Tolman, Richard, 381
Toluene, 284
Traité élémentaire de chimie, 165
Transactions of the Connecticut Academy, 229
Transistor, 415
Transition state, 424
Transmutation, 24, 29, 35, 42, 67, 73, 78, 88, 104, 109, 112, 119, 154
   water transmutation experiment, 154
Transuranic elements, 395, 410
Travers, Morris, 265

Tria prima, 100
Trichloroacetic acid, 244
Trinity, 407
Triple bonds: See Chemical bond, triple
Trithemius, Hans, 98
Tritium, 409
Tropsch, Hans, 361
Tshan Thung Chhi, 45
Tuberculosis, 66, 373
Tungsten, 384
Tuskegee Institute, 375
Tycho, Sophia, 109
Types, theory of, 244
Typhus, 416
Tyrosine, 289

Uhlenbeck, George, 321
Ultraviolet Catastrophe, 316
Unimolecular reactions, 381
United Alkali Company, 297
Universities, 69, 75, 104
Unsaturated compounds, 302, 361
Unsaturated hydrocarbon, 249
Uranium, 267, 270, 395, 398, 400–403, 405, 406, 418, 428
Uranium hexafluoride, 341
Uranium nitrate, 401
Uranium oxide, 270
Urea, 240
Urey, Harold, 356, 400
Urine, 61, 240

Valence, 246, 247, 256, 258, 277, 279, 281, 360
Valence-bond model, 328, 330
Vanadium, 188
Vapors, 38
Världarnas Utveckling, 277
Vauquelin, Nicholas, 164, 167, 186, 202, 203
Verdigris: See copper acetate
Vesicant, 305
Vials, 104
Vincent of Beauvais, 77
Vinegar, 32, 52, 55, 56, 58, 61, 375
Virus, 351, 352, 425
  polio, 356
  tobacco mosaic, 352

Viscosity, 229
Vital force, 240
Vitalism, 240
Vitamins, 348, 374
  B_{12}, 370, 351
Vitrification, 417, 418
Vitriol, 36, 65, 104
Volta, Assandro, 180
Voltaic pile, 181, 203, 208
Voltaire, 130
Voyager, 415

Wahl, Walter, 401
Wallach, Otto, 301
Walsh, Gertrude, 368
Wash bottles, 184
Waste minimization, 418
Water and Food, 296
Water gas, 361
Water spots, 154
Water, 23, 133, 154, 164, 179, 182, 183, 195, 237, 245, 246, 265, 277, 288, 346, 356, 376, 415, 427
Water-drop model of nucleus, 398
Waterston, John, 224
Watson, James, 353–357
Watt, Gregory, 190, 192
Watt, James, 142, 159, 190, 215, 219, 228
Wavelength, 315
Wave mechanics, 323
Wax, 343
Weapon, 61
Wedgwood, Josiah, 142, 181
Wedgwood, Thomas, 192
Wei Po-Yang, 45
Weizmann, Chaim, 303, 368, 421
Wends, 71
Werner, Alfred, 278–281, 302, 306, 372
Wheland, George, 369
Whewell, William, 199
Whinfield, Rex, 340
Whiting, Mark, 373
Wilcke, Johan 217
Wilhelmy, Ludwig, 378, 379, 383
Wilkins, Maurice, 354
Wilkinson, George, 372

Wilkinson, Mary, 139
William, Frederick, 130
Willstätter, Richard, 347
Wilson, Charles, 393
Windler, S. C. H., 243
Wine, 11, 55, 86, 99, 236, 292
Winkler, Clemens, 260
Winthrop, John the Younger, 188
Witches, 92, 227
Wöhler, Fredrich, 183, 237–240, 243, 254, 291
Wollaston, William, 188, 199–201
Women's Laboratory, 295
Wonderful Teacher, 82
Woodward–Hoffmann rules, 370
Woodward, Robert, 369–371, 373
World War I, 272, 274, 301, 302, 309, 322, 329, 333, 335, 338, 384, 392, 393, 397, 422
World War II, 333, 338, 339, 348, 351, 353, 358, 361, 363, 364, 365, 373, 374, 383, 400, 410

Worlds in the Making, 277
Wurtz, Charles, 247, 253

X ray, 430
X ray crystallography, 342, 360, 366
X ray diffraction, 342, 351, 354
X ray imaging, 414
Xenon, 265, 301, 374, 429
Xylenes, 284

Yeast, 291, 421
Ytterbium, 423
Yttria, 261, 262
Yttrium, 261, 423

Zeeman effect, 321
Zhabotinskii, A. M., 387
Ziglerin, Marie, 96
Zinc, 43, 246
Zinc oxide, 186
Zinc sulfate, 102
Zosimos, 29, 37, 63, 93